Cooperative Management

Series editors

For further volumes:
http://www.springer.com/series/11891

Constantin Zopounidis
Nikos Kalogeras · Konstadinos Mattas
Gert van Dijk · George Baourakis
Editors

Agricultural Cooperative Management and Policy

New Robust, Reliable and Coherent Modelling Tools

Springer

Editors
Constantin Zopounidis
Department of Production Engineering and Management
Technical University of Crete
Chania
Greece

Nikos Kalogeras
Department of Marketing, School of Business and Economics
Maastricht University
Maastricht
The Netherlands

Konstadinos Mattas
School of Agriculture
Aristotle University of Thessaloniki
Thessaloniki
Greece

Gert van Dijk
Center for Entrepreneurship
Nyenrode Business University
Breukelen
The Netherlands

George Baourakis
Business Economics and Management
Mediterranean Agronomic Institute of Chania
Chania
Greece

ISBN 978-3-319-06634-9 ISBN 978-3-319-06635-6 (eBook)
DOI 10.1007/978-3-319-06635-6
Springer Cham Heidelberg New York Dordrecht London

Library of Congress Control Number: 2014941201

Printed on acid-free paper

Springer is part of Springer Science+Business Media (www.springer.com)

Preface

The Book Series Cooperative Management provides an invaluable forum for creative and scholarly work on cooperative management, policy, economics, organizational, financial and marketing aspects of cooperative communities throughout the Mediterranean region and worldwide. The main objectives of this series are to advance knowledge related to collective management processes and cooperative entrepreneurship as well as to generate theoretical knowledge with the aim of promoting research within various sectors wherein market communities operate (agriculture, banking, real estate, insurance and other forms). Scholarly papers appearing in this series should relate to one of these areas, should have a theoretical and/or empirical problem orientation and should demonstrate innovation in theoretical and empirical analyses, methodologies and applications. Analyses of market communities' problems and phenomena pertinent to managerial research, extension and teaching (e.g., case studies) regarding cooperative entrepreneurship are equally encouraged. Further, this series encourages interdisciplinary and cross-disciplinary research from a broad spectrum of business economics disciplines.

With the increasing globalization and liberalization of commodity markets, it becomes more and more important to account for various factors that influence and shape the economic, market, institutional and thus policy conditions. A crucial issue is to develop robust methods, analytical methodologies, techniques and methods for analysing the information regarding the behaviour of various stakeholders (e.g. consumers, producers, managers, policy-makers) that operate in cooperative management networks and market communities engaged in the primary sector. Therefore, this very first issue of the book in the series of cooperative management focuses exclusively on the utilization of micro- and macro-level agricultural data to the development of new coherent, reliable and comprehensive modelling tools for conducting policy analysis of the rapidly changing agricultural and environmental conditions and complex decision problems of various stakeholders.

Among these changing conditions, one can mention the drastically increased price volatility of farm outputs, macroeconomic instability, climate change and the new post-2013 Common Agricultural Policy (CAP) in the European Union (EU). Throughout this unstable period, the impact of these changes on various stakeholders' income, profitability, efficient (sustainable) management of resources and operations and safety are key issues to be addressed by policy-makers in the EU. Moreover, these new agricultural market and policy conditions certainly have an effect on the use of natural resources and hence on the environmental concerns of farming activities. That is, the establishment of 'greening' processes and criteria utilized in many agricultural and environmental operational systems is another practical research direction in policy studies that has been rapidly emerging.

The material presented in this book describes models, methodologies and techniques in diverse areas of agricultural and environmental management and policy, including risk management and pricing approaches, economic efficiency analysis, classification techniques and behavioural aspects of various stakeholders' decisions, among others. We believe that this book will be of interest to both scholars and practitioners working in the fields of business economics, cooperative management, risk management and agricultural policy. EU policy-makers may gain useful insights on how to effectively monitor the changing environment and account for the rapidly changing patterns and their effects on farmers' behaviour and performance, so as to fittingly adopt the policy measures.

We would like to thank the assistant editor Georgios Manthoulis and English Prof. Maria Verivaki for the English proofreading. We extend appreciation to the authors and referees of these chapters, and Springer Academic Publications, for their assistance in producing this book.

March 2014

Constantin Zopounidis
Nikos Kalogeras
Konstantinos Mattas
Gert van Dijk
George Baourakis

Contents

Contributors

Charisios Achillas School of Economics, Business Administration and Legal Studies, International Hellenic University, Thessaloniki-N. Moudania, Greece

Marco Acutis Department of Agricultural and Environmental Sciences, University of Milan, Milan, Italy

Shaikh Mostak Ahammad Department of Business Administration of Food and Agricultural Enterprises, University of Patras, Patras, Greece

Foivos Anastasiadis School of Economics, Business Administration and Legal Studies, International Hellenic University, Thessaloniki-N. Moudania, Greece

Dimitrios Andreopoulos Faculty of Science and Technology, Free University of Bozen/Bolzano, Bolzano, Italy

Fragiskos Archontakis School of Economics, Business Administration and Legal Studies, International Hellenic University, Thessaloniki-N. Moudania, Greece

F. Arfini Department of Economics, University of Parma, Parma, Italy

George S. Atsalakis Department of Production and Management Engineering, Technical University of Crete, Crete, Greece

Georgios Banias School of Economics, Business Administration and Legal Studies, International Hellenic University, Thessaloniki-N. Moudania, Greece

G. Benedetto Department of Science for Nature and Environmental Resources, University of Sassari, Sassari, Italy

A. Bezat-Jarzębowska Institute of Agricultural and Food Economics-National Research Institute (IAFE-NRI), Warsaw, Poland

Julian Binfield FAPRI, University of Missouri, Columbia, MO, USA

A. Bouaziz Hassan II Agronomy and Veterinary Medicine Institute, Rabat, Morocco

James P. Breen School of Agriculture, University College Dublin, Dublin, Ireland

D. Carboni Department of Social and Human Science, University of Sassari, Sassari, Italy

Francesco Comiti Faculty of Science and Technology, Free University of Bozen/ Bolzano, Bolzano, Italy

G. L. Corinto Department of Education, Cultural Heritage and Tourism, University of Macerata, Macerata, Italy

Stefano Corsi Department of Economics, Management and Quantitative Methods, University of Milan, Milan, Italy

D. D'Agostino Mediterranean Agronomic Institute of Bari, CIHEAM-IAMB, Valenzano, Bari, Italy

Dimitrios Damigos School of Mining and Metallurgical Engineering, National Technical University of Athens, Athens, Greece

John Davis Agri-Food and Biosciences Institute, Belfast, UK; Queen's University Belfast, Belfast, UK

M. Donati Department of Economics, University of Parma, Parma, Italy

Trevor Donnellan Rural Economy Development Programme, Teagasc, Athenry, Co. Galway, Ireland

Nadja El Benni University of Applied Sciences HTW Chur, Chur, Switzerland; ETH Zurich, Zurich, Switzerland

D. El Chami School of Applied Sciences, Cranfield Water Science Institute (CWSI), Cranfield University, Bedfordshire, UK

Siyi Feng Agri-Food and Biosciences Institute, Belfast, UK

Robert Finger Wageningen University, Wageningen, The Netherlands; Bonn University, Bonn, Germany

Christian Fischer Faculty of Science and Technology, Free University of Bozen/ Bolzano, Bolzano, Italy

Lena Galata Mediterranean Agronomic Institute of Chania (MAICh), Crete, Greece

S. Ghinoi Department of Statistical Sciences "P. Fortunati", University of Bologna, Bologna, Italy

R. Gigante Department of Economics, University of Parma, Parma, Italy

Andrea Giussani Department of Agricultural and Environmental Sciences, University of Milan, Milan, Italy

Kevin Hanrahan Rural Economy Development Programme, Teagasc, Athenry, Galway, Ireland

R. Harbouze Montpellier 1 University, Montpellier, France; CIHEAM-IAM, Mediterranean Agronomic Institute of Montpellier, Montpellier, France; Hassan II Agronomy and Veterinary Medicine Institute, Rabat, Morocco

Sebastian Hess Swedish University of Agricultural Sciences (SLU), Uppsala, Sweden

Jørgen D. Jensen Department of Food and Resource Economics, University of Copenhagen, Copenhagen, Denmark

Kostas Karantininis Swedish University of Agricultural Sciences (SLU), Uppsala, Sweden

Paweł Kobus Faculty of Economic Sciences, Warsaw University of Life Sciences, Warsaw, Poland

N. Lamaddalena Mediterranean Agronomic Institute of Bari, CIHEAM-IAMB, Valenzano, Bari, Italy

Ph. Le Grusse CIHEAM-IAM, Mediterranean Agronomic Institute of Montpellier, Montpellier, France

A. C. Leucci Department of Statistical Sciences "P. Fortunati", University of Bologna, Bologna, Italy

J. C. Mailhol Joint Research Unit G-Eau, Cemagref, Montpellier, France

Claudio Marcianò Agraria Department, Mediterranean University of Reggio Calabria, Reggio Calabria, Italy

Anne Margarian Thünen-Institute, Braunschweig, Germany

Alan Matthews Department of Economics, Trinity College Dublin, Dublin, Ireland; Institute for International Integration Studies, Trinity College Dublin, Dublin, Ireland

Ana Corina Miller Agri-Food and Bioscience Institute, Belfast, UK; Institute for International Integration Studies, Trinity College Dublin, Dublin, Ireland

Federica Monaco Department of Agricultural and Environmental Sciences, University of Milan, Milan, Italy

S. Niavis Department of Planning and Regional Development, University of Thessaly, Volos, Greece

Cathal O'Donoghue Rural Economy Development Programme, Teagasc, Athenry, Co. Galway, Ireland

Frank Offermann Thünen-Institute, Braunschweig, Germany

Jens E. Ørum Department of Food and Resource Economics, University of Copenhagen, Copenhagen, Denmark

Stefano Pareglio Department of Mathematics and Physics, Catholic University of Sacred Heart, Milan, Italy

Myles Patton Agri-Food and Biosciences Institute, Belfast, UK

Alessia Perego Department of Agricultural and Environmental Sciences, University of Milan, Milan, Italy

M. Raki Hassan II Agronomy and Veterinary Medicine Institute, Rabat, Morocco

W. Rembisz Institute of Agricultural and Food Economics-National Research Institute, Warsaw, Poland

Anthony N. Rezitis Department of Business Administration of Food and Agricultural Enterprises, University of Patras, Patras, Greece

Giuseppa Romeo Agraria Department, Mediterranean University of Reggio Calabria, Reggio Calabria, Italy

P. Ruelle Joint Research Unit G-Eau, Cemagref, Montpellier, France

Guido Sali Department of Agricultural and Environmental Sciences, University of Milan, Milan, Italy

A. Scardigno Mediterranean Agronomic Institute of Bari, CIHEAM-IAMB, Valenzano, Bari, Italy

D. Sgargi Department of Statistical Sciences "P. Fortunati", University of Bologna, Bologna, Italy

A. Sielska Institute of Agricultural and Food Economics-National Research Institute, Warsaw, Poland

Andrea Tosini Department of Economics, Management and Quantitative Methods, University of Milan, Milan, Italy

G. Vlontzos Department of Agriculture Crop Production and Rural Environment, University of Thessaly, Volos, Greece

Adam Was Faculty of Economic Sciences, Warsaw University of Life Sciences, Warsaw, Poland

V. J. Wesz Junior Department of Development, Agriculture and Society (DDAS), Rural Federal University of Rio de Janeiro (UFFRJ), Rio de Janeiro, Brazil

Part I
Risk Management and Pricing Issues in Agricultural Policy Analysis

Chapter 1
VAR Models for Dynamic Analysis of Prices in the Agri-food System

A. C. Leucci, S. Ghinoi, D. Sgargi and V. J. Wesz Junior

Abstract An adequate understanding of the dynamics that characterize the agri-food market is fundamental for the development of really efficient economic policies, especially after the two recent hikes in the prices of food commodities. The econometric literature provides today advanced analysis tools such as VAR models: these models are based on a system of equations in which each variable is regressed on a set of deterministic variables, on a number of lags related to each covariate in the model. To test the effectiveness of this analytical tool at dealing with the issues related to agri-food economy, we applied a VAR analysis on prices of major food and energy commodities (oil and biodiesel) referring to the period January 2000–December 2012. Our results identified statistically significant intertemporal relationships between the price of corn, soybean oil, rapeseed and oil, and suggested the direction of these relationships; we could conclude that the price of corn and soybeans are influenced mainly in the energy market. Moreover, we focused on the United States market and we set as variables the share of commodities used for the production of biofuels: we could observe that important alterations on the food market are due to the convenience in producing ethanol and biodiesel, since the portion of the crops used for energy is in direct competition with that devoted to the feeding. This kind of model, therefore, deals adequately with data and issues of the agri-food system and provides an analytical basis to develop economic policies that can take into account the complexity of the global food system.

A. C. Leucci (✉) · S. Ghinoi · D. Sgargi
Department of Statistical Sciences "P. Fortunati", University of Bologna, Bologna, Italy
e-mail: annacaterina.leucci3@unibo.it

V. J. Wesz Junior
Department of Development, Agriculture and Society (DDAS),
Rural Federal University of Rio de Janeiro (UFFRJ), Rio de Janeiro, Brazil

C. Zopounidis et al. (eds.), *Agricultural Cooperative Management and Policy*,
Cooperative Management, DOI: 10.1007/978-3-319-06635-6_1,

1.1 Introduction

The use of biofuels derived from agricultural products (bioethanol, biodiesel, vegetal oils, etc.) has grown rapidly over the past 15 years, although in Brazil and the United States bioethanol was used for commercial purposes already from the 1970s, driven by the energy crisis of 1973. In the last few years, the international debate focused on agricultural products that can also be exploited for the production of biofuels (corn, sugar cane, soybean oil, canola oil): the question concerns the opportunity to allocate larger and larger shares of agricultural production for energetic purposes rather than for food. On the one hand, it is argued that the increasing use of biofuels would reduce emissions of carbon dioxide in the atmosphere; on the other, there is some evidence that the impact on gas emissions would be modest (Esposti 2008) while it may be considerable on food supply, given the competition for the use of land and resources.

This article has two main objectives: on the one hand, it aims to investigate a central issue in this debate, that is, the interaction between the price of oil, the production of bio-ethanol and the price of food commodities used in the production of biofuels; on the other, it seeks to identify a statistical and an econometric method that allows to analyze adequately the dynamics and interactions of food prices.

We decided to proceed using vector autoregressive models (VAR), seeking first to understand whether these models were suitable to treat the data, and then analyzing the results from an economic perspective, in order to catch the intertemporal relationship between these variables, and the extent to which they act on each other. The aim is to understand if the price of food commodities involved in the production of biofuels is mainly generated on the energetic market or on agrifood market.

Our work has the following structure:

In the beginning, an introductory chapter gives a brief overview on prices of agricultural commodities and energy, their variability and possible interactions; we present a review of the recent history and current situation of the agri-food and energy markets and propose recognition of the scientific production currently available on the topics in question.

The following chapter is dedicated to our analysis: the first and second paragraphs contain a methodological recognition of VAR models, while Sect. 1.3 is devoted to the analysis we carried out. In particular, we provide first of all a brief presentation of the data selected, then a description of the models built, and of the results obtained, all followed by comments, possible justifications and empirical, theoretical and legislative validation, and finally an interpretation.

The research ends with a conclusion, where we try to recompose the evidence and to draw an overall picture of the results obtained, and their possible implications for the present and the future of the global food industry.

1.2 Agricultural Commodities and Biofuels: An Overview

1.2.1 The Biofuels Market

The biofuels market is rapidly expanding, and recent forecasts suggest that this growth is likely to continue in the future (Rosegrant 2008; Frondel and Peters 2007). This is mainly due to the fact that, being alternative energy sources to fossil fuels, the increase in oil prices and the introduction of environmental policies for the reduction of CO_2 emissions have contributed to their growth (Zezza 2007).

The European Commission, with the recent strategy "20-20-20" (20 % cut in greenhouse gas emissions, reduction of energy consumption by 20 and 20 % of energy produced from renewable sources), is working to replace 10 % of the demand for fossil fuels from the transport sector; in order to succeed, it is focusing on tax breaks and subsidies that have as an incentive the production of biofuels. The United States and Brazil, the world's major producers and consumers of biofuels, have disposed an increase of the subsidies towards this sector: the United States, with the measures included in the 2008 Farm Bill, have increased the production of biofuels, created new refineries and encouraged research for alternative energy sources; Brazil has stimulated the production of biofuels through the proclamation of policies for agricultural development and of laws requiring the use of a minimum quantity of biofuel in blends.

Globally, the production and consumption of biofuels differs depending on the country considered. In the United States, the biofuels market is dominated by bioethanol obtained from corn processing: about 14 % of the corn crop in 2006 was used to produce bioethanol, which corresponds to 4 % of the total fuel used in the country (Kent Hoekman 2009). In Brazil, in 2011, 90 % of biofuels used consisted of bioethanol from sugar cane, and 10 % of biodiesel made mainly from soybean oil; currently about 45 % of the total energy and 19 % of the fuels used in Brazil comes from renewable sources (ANP 2012). In Europe, about 80 % of biofuels used is made up of biodiesel, 19 % of bioethanol and the remaining of vegetal oil and biogas; biodiesel is mainly (70 %) derived from rapeseed oil (Zezza, 2011).

Between the end of 2007 and the beginning of 2008, the world witnessed the first soaring of prices of agricultural commodities, in particular of corn, sugar and soybeans. In general, this increase involved cereals and oilseeds; both the price index passed, in a single year, from 150 to over 280 (Fanfani 2008). From the second half of 2008, the situation stabilized, thanks to the good results of global corn production, the appreciation of the dollar against the euro and the decrease in the price of oil. In 2010, however, a new rise occurred in prices of oil, commodities and food products, with the latter reaching a value even higher than in 2008. In the last 2 years (2011 and 2012), the prices have stabilized but are still relatively high compared to those of 2009 (see Fig. 1.1).

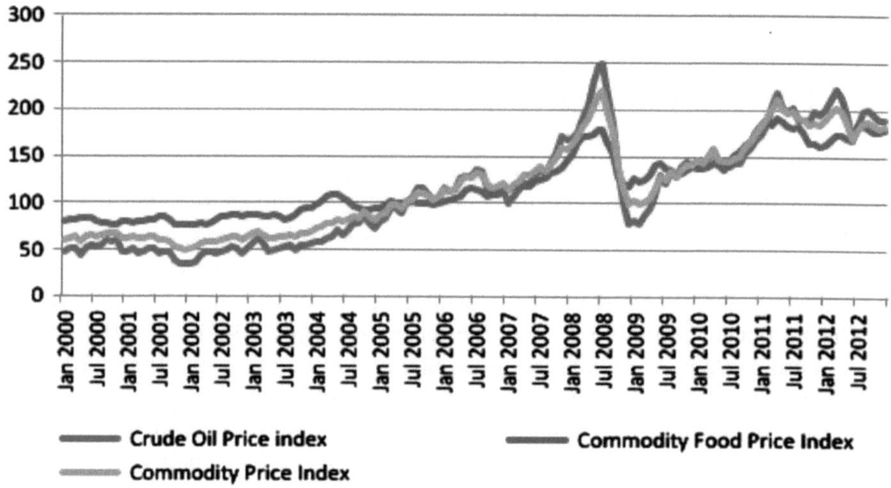

Fig. 1.1 Prices of commodities, food and fuel prices, 2000–2012 (2005 = 100)

Generally, a price increase can be attributed to several factors:

- On the supply side possible causes are adverse weather conditions, reduction of stocks, competition in the use of water resources and the rising costs of production and transport;
- On the demand side, possible causes are the increased demand for biofuels, the change in eating habits and financial speculation (FAO 2008).

In recent years, in particular, four factors have played an important role in determining the price increase: the growth in global demand for food goods, driven by rising disposable income of several large emerging countries (China, India, and Brazil), which has allowed improvements in the living conditions of millions of people, and changed their food habits; the sharp increase in the price of oil, which played a role in the increase in production costs (fertilizer) and transport; poor harvests in exporting countries (Australia, China, and many countries in Latin American); and finally financial speculation: in a situation of high global demand, reduced stock and absence of adjustment tools, speculators were attracted by high gains perspectives on the futures market, so they begun to bet on it, with heavy consequences in the real market of these goods (Maluf 2008).

1.2.2 Literature

The literature concerning this subject is very wide, and many of the most important studies use econometric models to analyze the relations between the prices and the production of agri-food commodities, and those of fuels. According to Hochman et al. (2012), the link between fuels and agricultural commodities depends on the market we focus on, the kind of commodity, the specification of the model, and the time series used. Kristoufek et al. (2011) analyzed the relations among a wide range of agri-food products and the prices of fuels in the United States and the European Union, and they found out that the interactions vary substantially if they consider time series collected weekly, monthly, or quarterly.

Some authors resort instead to time series analysis to study this problem: Serra et al. (2011) used auto-regressive models to identify the relations among the prices of corn, bioethanol, biodiesel, and oil on the United States market between 1990 and 2008, concluding that in the long run these prices are actually bounded, and that when the price of corn reaches high levels, it becomes the main factor on which the price of bioethanol depends. Zhang et al. (2009) adopted multivariate autoregressive estimators to study the volatility of the prices of corn, soy, biodiesel, and oil on the United States market between 1989 and 2007, discovering that the price of biodiesel influences the price of bioethanol, and that if the latter rises, the prices of agri-food commodities are influenced with short-term effects. Hertel et al. (2010) estimated that, on the United States market, the sharp rise in the prices of biofuels between 2001 and 2006 was mostly due to the increase in the price of oil; the same situation occurred in the European market, but it was mainly explained by the subsidies, and secondly the oil price trend.

Many studies were based on economic–mathematical models to assess the impact of biofuel production on commodity prices. Goldemberg et al. (2004), through an analysis of bioethanol production in Brazil, found that it increases both the demand and the supply for sugar cane. Mitchell (2008) used a multifactor analysis model to analyze the growth of prices of food commodities between 2002 and 2008 and concluded that this increase was attributable for 75 % to biofuels, together with other factors such as low levels of stocks, speculation and a halt to exports introduced in some countries.

Many, in addition to Mitchell, have attributed to biofuels a strong influence on the price of food commodities. The IMF has estimated that in 2008 the growth in the use of biofuels determined 70 % of the rise in the price of corn and 40 % of that of soy (Esposti 2008). For Trostle (2008), the increased share of corn and sugar cane produced for bioethanol is one of the main causes of the price boost of these commodities. According to the Farm Foundation (2008), the recent increase in oil prices, partly due to the depreciation of the dollar, has been the main cause of the growth of the demand for bioethanol in the United States, while before 2005 the demand for biofuels had a greater impact on the price of corn (because of the subsidies introduced by the U.S. government).

Conversely, Zilberman et al. (2012) argue that the price of ethanol is influenced both by the price of agricultural products and by that of fuel, but that the connection between the first two is weak. This is explained with two arguments:

- the studies on the relationship between fuel and food prices estimate *marginal* effects, while most of the literature on the impact of biofuels on the price of the agri-food commodities tries to evaluate the *total* effect on the price change, in the transition from food product to energy product;

- the impact of a change in the price of biofuels on the price of food products is not clear a priori.

1.3 Energy Market and Agri-Food Market: A VAR Analysis

1.3.1 VAR Models

The agricultural commodities market is characterized by complex dynamics, which require a multivariate approach. Economic variables often appear to be self-correlated and cross-correlated for several time lags. The need to build models that take into account the intertemporal structure of data arises from the complexity of the relations guiding the economic system.

In particular, for the analysis of time series, the use of vector autoregressive models, better known as VAR models is widespread. VAR approach was first proposed by Sims in 1980, as an alternative to Simultaneous Equations Models, which were the main instrument for macroeconomic analysis until that moment. VAR processes are the multivariate generalization of AR models: a VAR is actually a system where every variable is regressed on a set of deterministic variables on p lags, referring to every covariate in the model.

The lag operator is usually applied to numeric sequences and allows to transform the X_t sequence (both stochastic or not) in another sequence that has the same values present in X_t, with one lag (Podestà, 2011).

Therefore, the following form of the operator

$$\mathrm{LX}_t = X_{t-1} \tag{1.1}$$

becomes, after repeating n times the application of the lag,

$$L^n X_t = X_{t-n} \tag{1.2}$$

L is a linear operator, which means that if a and b are two constants, we will have

$$L(\mathrm{ax}_t + b) = \mathrm{aLx}_t + b = \mathrm{ax}_{t-1} + b \tag{1.3}$$

In general a VAR model of rank p will assume the following form:

$$y_t = \mu + A_1 y_{t-1} + \cdots + A_p y_{t-p} + u_t \tag{1.4}$$

where A_i are $k \times k$ coefficient matrices; $\mu = (\mu_1, \ldots, \mu_k)$ is a $k \times 1$ vector of intercepts and $u = (u_{1t}, \ldots, u_{kt})$ is a k-dimensional white noise process, where the variance-covariance matrix is non-singular for assumption.

The application of any methodology within VAR models requires as a necessary condition the stationarity of the autoregressive representation: a VAR process satisfies such a condition if all the eigenvalues of the *companion* matrix fall into the unit circle, that is, they are lower than one.

The *companion* matrix is $Kp \times Kp$ dimensional, composed by the A_i matrices, represented as follows:

$$A = \begin{bmatrix} A_1 & A_2 & \cdots & A_{p-1} & A_p \\ I_k & 0 & \cdots & 0 & 0 \\ 0 & I_k & \cdots & 0 & 0 \\ \vdots & \cdots & \ddots & \vdots & \vdots \\ 0 & \cdots & 0 & I_k & 0 \end{bmatrix} \tag{1.5}$$

VAR models, if stationary and well specified, allow analysis of many aspects of time series, and give substantial information about them. Examining the intertemporal links among different variables, these models are appropriate both for forecasting future values of the series, and for a dynamic analysis of present values, in particular for the existence of causality relations among the covariates. This particular analysis is based on the concept of Granger causality (1969).

1.3.2 Analysis of Granger Causality

In an empirical analysis of economic data, it is often of much interest to establish the cause–effect relationship, though it might also result in much difficulty. In general, if two variables X and Y show an important correlation, we might assume that they tend to follow the same trend, but in absence of further information, we cannot add more observations about the direction of the causality. Given an estimate VAR model, we can also take a test to verify the joint significance of the lag structure of y_{1t} in the equation referred to as y_{2t}. The test itself is built as a maximum likelihood ratio, or a simple F statistic.

The most appropriate way to interpret this kind of test is to see it as a graphic analysis that can show whether the trend of a variable follows or foreruns that of another variable.

Some interpretation problems often arise, mostly because the representation on a reduced form of VAR does not apply very well to draw general conclusions. These considerations are actually based, within VAR analysis, on the results obtained from the Impulse Response Functions.

The purpose of the analysis of Granger causality is to evaluate the predictive capacity of a single variable on the other ones in the system. If a variable, or a group of variables, y_{1t} fosters the forecasts of another variable y_{2t} or group, then y_1 Granger-causes y_2. Formally:

- let $y_t = \begin{bmatrix} y_{1t} \\ y_{2t} \end{bmatrix}$ be a multivariate time series where the K components are divided in two groups y_{1t} and y_{2t};
- let $F_t = \{y_t, y_{t-1}, \cdots\}$ be the set of all the observations until time t;
- let $y_{2,t+h/t}$ be the optimal predictor of $y_{2,t+h}$ based on F_t, and with $\sum_2 (h/F_t)$ its mean square error;

then y_{1t} Granger-causes y_{2t} if

$$\sum_2 (h/F_t) < \sum_2 (h|F_t/\{y_{1s}|s \leq t\}) \tag{1.6}$$

for at least one $h = 1, 2, \ldots$.

Considering, for example, the following stationary VAR(p):

$$\begin{bmatrix} y_{1t} \\ y_{2t} \end{bmatrix} = \begin{bmatrix} \phi_{11,1} & \phi_{12,1} \\ \phi_{21,1} & \phi_{22,1} \end{bmatrix} \begin{bmatrix} y_{1,t-1} \\ y_{2,t-1} \end{bmatrix} + \cdots + \begin{bmatrix} \phi_{11,p} & \phi_{12,p} \\ \phi_{21,p} & \phi_{22,p} \end{bmatrix} \begin{bmatrix} y_{1,t-p} \\ y_{2,t-p} \end{bmatrix} + \begin{bmatrix} u_{1t} \\ u_{2t} \end{bmatrix} \tag{1.7}$$

with $u_i \sim WN(0, \Sigma)$ and Σ non-singular, then

$$\sum_2 (h|F_t) < \sum_2 (h|F_t/\{y_{1s}|s \leq t\}), \; h = 1, 2, \ldots \Leftrightarrow \phi_{21,i} = 0 \tag{1.8}$$

then y_{1t} doesn't Granger-cause y_{2t} if $\phi_{21,i} = 0$, while if $\phi_{12,i} = 0$ for $i = 1, 2, \ldots p$, y_{2t} does not Granger-cause y_{1t}. It is important to note that if the VAR is stationary, the hypothesis "Y_1 does not cause Y_2" can be tested by a simple F-test: in fact, the hypothesis of absence of Granger causality is equivalent to the following linear restriction on the parameters:

$$H_0 : A_1 = A_2 = \cdots = A_p = 0 \tag{1.9}$$

1.3.3 Application of the VAR Model to Commodity Market

The objective of the research is to identify causality and connections between agri-food and energy markets, focusing in particular on prices: in the first part, the analysis was centered on the relationship among the prices, so we chose as variables time series of the prices of agricultural commodities and fuels, in particular the world current prices, recorded monthly, from January 2000 to December

2012[1]. This period was selected since it is large enough to show how the trend of the agricultural commodities prices, which until the beginning of 2000 had been of relatively slow and steady decline, has been reversed from 2003, assuming a positive trend but becoming extremely volatile (FAO 2011). This time, however, is not so extensive as to make the intertemporal comparison among prices prove meaningless without adjustments that take into account inflation and global economic trends: inflation, however, was not considered because it was difficult to assess, on a global level, in our model.

The products chosen as variables for our analysis are corn, rapeseed oil, sugar cane and soybean oil as regards agricultural commodities, and oil, bioethanol, and biodiesel as fuels. The choice of oil is almost obligatory, since it is the most widespread and used fossil fuel; the trend of its price is closely linked to that of other competitor energy goods, and also represents the main cost of production for most of the commodities, being the fuel for agricultural machinery and a component of nitrogen fertilizers. Bioethanol and biodiesel are, on the other hand, the two most widely used biofuels in the world. Corn, rapeseed oil, soybean oil, and sugar cane are the raw materials used to obtain these fuels. In particular, a geographic specialization of production is noted globally: corn is the main product for bio-ethanol in the United States, while Brazil uses sugar cane; biodiesel, rapeseed oil, and soybean oil are the most commonly used raw materials, especially in the EU. These three countries can be taken without any question as the target market for these biofuels: together they produce more than 90 % of the global share of bioethanol, and more than 80 % of biodiesel (Esposti 2008).

The first part of the analysis was conducted using as variables the time series of monthly world prices of corn, oil, biodiesel, soybean oil, sugar cane, and rapeseed oil[2]. The period covered goes from January 2000 to December 2012. The second part of the analysis has been restricted to the case of the U.S. market: this time we used as variables the share of corn and soybean oil intended, respectively, to produce bioethanol and biodiesel; the portion of corn and soybean oil destined for the domestic market; and world prices of corn, soybean oil, biodiesel and petroleum.

The vector autoregressive models have been particularly useful for the achievement of our objectives: the structure of the data provides indeed, through the VAR analysis, the generating process of the same, useful to forecast, and explain the links between economic variables. The analysis consists of the following phases:

- identification of the number of parameters p of the VAR;
- OLS estimation of the parameters of the VAR (p);

---

[1] The reference period applies to the prices of all commodities analyzed, with the only exception being the biodiesel prices, which are only available from July 2006, so VAR built with these data as variables were constructed considering only the period from 2006 to 2012.

[2] The time series used for the analysis of corn and ethanol ranged from January 2000 to December 2012, while the period considered for the analysis of soy was from July 2006 to December 2012. This choice is based on the availability of data.

- control of the adequacy of the estimated model through the diagnostic of residual;
- Granger causality.

In order to identify the number of parameters we used Akaike information criteria (AIC), Hannan-Queen (HQ), Schwarz (SC), and the final prediction error (FPE); for the diagnostic of residuals, we used the Portmanteau test, the Breusch-Godfrey and the Jarque-Bera test, to verify the absence of autocorrelation of residuals and normality distribution of the latter[3].

Initially, the analysis began with the construction of a correlation matrix which uses as variables the price of agricultural commodities involved in the production of biofuels (corn, canola, soybean oil, and sugar cane) and the price of oil. The observed correlations justify the construction of a VAR model, which results as a system of five equations with *m* indicating the price of corn, *s* the price of soybean oil, *r* that of rapeseed oil, *c* the sugarcane price, and *p* the price of oil. The constructed VAR is of the following type:

$$\begin{cases} p_t = \beta_1 p_{t-1} + \beta_2 m_{t-1} + \beta_3 r_{t-1} + \beta_4 b_{t-1} + \beta_5 s_{t-1} + \beta_6 c_{t-1} + \varepsilon \\ m_t = \beta_1 p_{t-1} + \beta_2 m_{t-1} + \beta_3 r_{t-1} + \beta_4 b_{t-1} + \beta_5 s_{t-1} + \beta_6 c_{t-1} + \varepsilon \\ r_t = \beta_1 p_{t-1} + \beta_2 m_{t-1} + \beta_3 r_{t-1} + \beta_4 b_{t-1} + \beta_5 s_{t-1} + \beta_6 c_{t-1} + \varepsilon \\ b_t = \beta_1 p_{t-1} + \beta_2 m_{t-1} + \beta_3 r_{t-1} + \beta_4 b_{t-1} + \beta_5 s_{t-1} + \beta_6 c_{t-1} + \varepsilon \\ s_t = \beta_1 p_{t-1} + \beta_2 m_{t-1} + \beta_3 r_{t-1} + \beta_4 b_{t-1} + \beta_5 s_{t-1} + \beta_6 c_{t-1} + \varepsilon \\ c_t = \beta_1 p_{t-1} + \beta_2 m_{t-1} + \beta_3 r_{t-1} + \beta_4 b_{t-1} + \beta_5 s_{t-1} + \beta_6 c_{t-1} + \varepsilon \end{cases} \tag{1.10}$$

Most of the parameters resulted were not significant; this model, however, allowed us to identify some relations, which we subsequently deepened through the construction of bivariate VAR. First or all, we built two separate models, one for the prices of food products and one for the prices of the two energy commodities, with the aim of identifying the relationships that bind intertemporally prices of products belonging to the same market.

Most of the VAR parameters built only with the prices of food commodities are not significant: this is an indication of a lack of intertemporal relationship between the prices of these products; the same results were obtained from the model containing the price of oil and biodiesel. This outcome shows that probably the prices of energy commodities and those of food commodities are inter-related, therefore it is appropriate to consider the construction of bivariate models using as a dataset the prices of products belonging to the food market and those belonging to the energy market.

The first bivariate model analyzed has been built using world prices of corn and oil. In Fig. 1.2, the two series are represented, and we can see that price trends are similar.

[3] The statistical software used is R 2.14.2; analysis was implemented using the R packages *Tseries* and *VAR*.

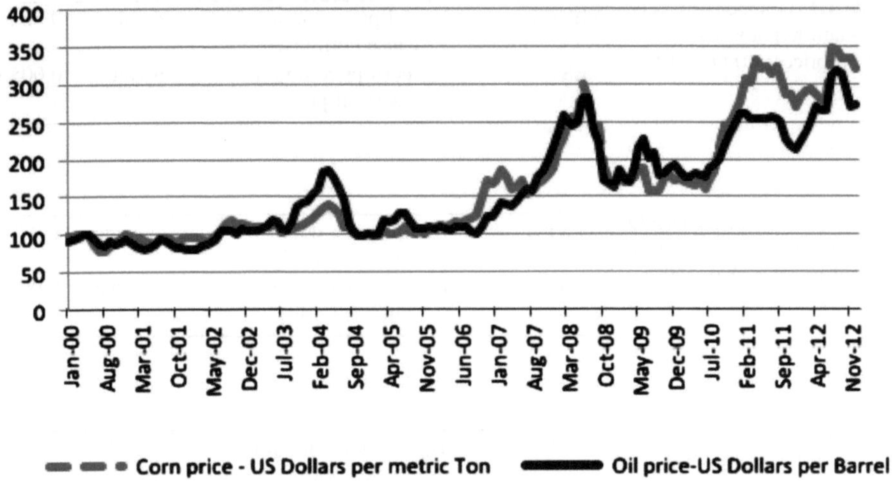

Fig. 1.2 Corn price and oil price, 2000–2012 (2005 = 100)

From this dataset, we esteemed a VAR with two lags that produced two equations; one that uses as a response variable the price of corn (m) at time t and the other using as a response variable the price of oil (p) at time t:

$$\begin{aligned} p_t &= 1.29p_{t-1} + 0.12m_{t-1} - 0.37p_{t-2} - 0.1m_{t-2} + 2.7 \\ m_t &= 0.02p_{t-1} + 1.12m_{t-1} - 0.04p_{t-2} - 0.16m_{t-2} + 2.7 \end{aligned} \quad (1.11)$$

All esteemed parameters are significant, i.e., there is an intertemporal relationship among these two prices. After that, we built the Granger causality test (Table 1.1) and found the direction of this relationship: corn price seems to Granger-cause oil price, and not vice versa.

We divided the period 2000–2012 in three-time series and made a VAR model for each one. This division is due to economic considerations: the first period (January 2000–December 2005) is the pre-crisis period without price rises; the second period (January 2006–December 2008) is the pre-crisis period with the first increase in the prices; and the third period (January 2009–December 2012) is the post-crisis period.

In the VAR model from 2000 to 2005, there is no statistically significant relationship between corn price and oil price, while the following models show a statistically significant intertemporal relationship, although with some differences. The VAR model with two lags from 2006 to 2008 introduces, in the equation with oil price as a response variable, an insignificant reported parameter to the corn price at time $t - 2$, while the VAR model with one lag from 2009 to 2012 has significant parameters. In the second VAR model (2006–2008), the Granger causality test results in a difficult interpretation: the p-value of the null hypothesis "*oil price doesn't cause, in the Granger sense, corn price*" is equal to 0.049, near to the threshold of significance; we can neither reject nor accept this hypothesis.

Table 1.1 Granger causality test—null hypothesis—Corn and oil prices (2000–2012)

TEST 1	H_0: oil price doesn't cause corn price	p-value = 0.7644
TEST 2	H_0: corn price doesn't cause oil price	p-value = 0.0056

Table 1.2 Granger causality test—null hypothesis—Corn and oil prices (2006–2008)

TEST 1	H_0: oil price doesn't cause corn price	p-value = 0.049
TEST 2	H_0: corn price doesn't cause oil price	p-value = 0.0049

Table 1.3 Granger causality test—null hypothesis—corn and oil prices (2000–2012)

TEST 1	H_0: oil price doesn't cause corn price	p-value = 0.049
TEST 2	H_0: corn price doesn't cause oil price	p-value = 0.051

The second test, instead, has a p-value equal to 0.0049, i.e., we cannot accept H_0, hence corn price causes, in the Granger sense, oil price (Table 1.2).

Threshold values come up in the two Granger causality tests of the last VAR model: p-value is 0.049, i.e., it is not possibile to reject or to accept the null hypothesis "*corn price doesn't cause, in the Granger sense, oil price*," while the second test allows us to say that "*oil price doesn't cause, in the Granger sense, corn price*" (Table 1.3).

The intertemporal relationship between corn price and oil price is recent and still changing. The importance of corn price to establish the direction of the causality is probably due to the convenience to produce bioethanol; besides, in the VAR models from 2006 to 2008 and from 2009 to 2012 the null-hypothesis "*corn price doesn't cause, in the Granger sense, oil price*" is rejected or is not entirely rejectable.

The same analysis has been made using world soybean oil and oil prices from January 2000 to December 2012 (see the trends in Fig. 1.3), dividing this series in the same three periods previously identified. The results obtained are similar to the results of the VAR models for corn price and oil price. Using as a response variable the price of soybean oil (s) and the price of oil (p) at time t, the model, with one lag, is:

$$\begin{aligned} p_t &= 0.69p_{t-1} + 0.02s_{t-1} + 1.79 \\ s_t &= -1.69p_{t-1} + 1.08s_{t-1} + 69.74 \end{aligned} \tag{1.12}$$

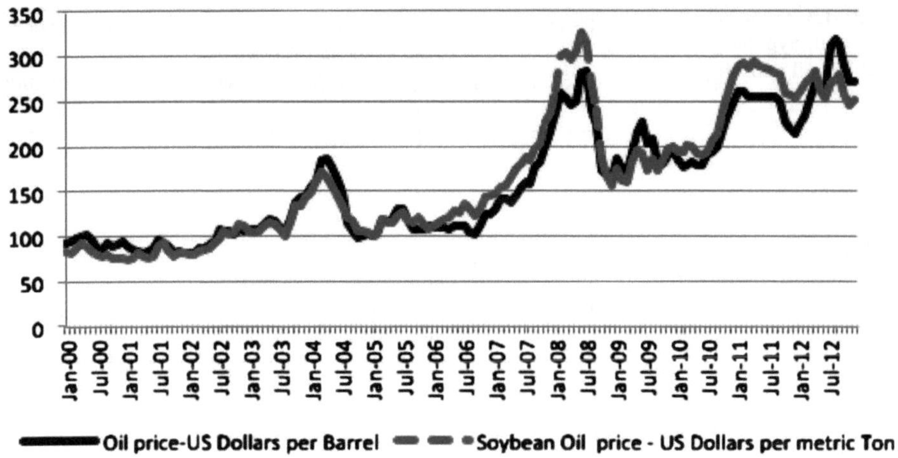

Fig. 1.3 Soybean oil price and oil price, 2000–2012 (2005 = 100)

All parameters were significant, a stationary VAR resulted and the residual-based test gave a negative result.

From the Granger causality test, it emerged that soybean oil price causes, in the Granger sense, oil price, and not vice versa. We could not draw any statistically significant intertemporal relationship from the first VAR model (from 2000 to 2005), while in the other periods the situation is different, and from 2006 we could observe a statistically significant intertemporal relationship. Particularly, there is a mutual causality in the Granger sense between soybean oil price and oil price; such mutual connection, however, does not exist in the following period: from January 2009 to December 2012, soybean oil price causes, in the Granger sense, oil price. We can say that the relationship between these prices is recent and the result "*soybean oil price causes, in the Granger sense, oil price*" has arisen in the last 4 years of the time series 2000–2012. The relationship can be explained by the increase of the soybean oil use for the production of biodiesel (Tables 1.4, 1.5 and 1.6).

With the monthly dataset of biodiesel price (b) and soybean oil price (s)[4], we finally built another VAR model:

$$\begin{aligned} b_t &= 0.71 b_{t-1} + 0.0007 s_{t-1} + 0.02 \\ s_t &= -58.91 b_{t-1} + 1.08 s_{t-1} + 70.03 \end{aligned} \tag{1.13}$$

From the Granger causality test, we observed again a mutual causality among these two prices: it means that soybean oil price depends on the energy market.

[4] Time series is from July 2006 to December 2012, due to the availability of data.

Table 1.4 Granger causality test—null hypothesis—soybean oil and oil prices (2000–2012)

TEST 1	H_0 : oil price doesn't cause soybean oil price	p-value = 0.065
TEST 2	H_0 : soybean oil price doesn't cause oil price	p-value = 0.0006

Table 1.5 Granger causality test—null hypothesis—soybean oil and oil prices (2006–2008)

TEST 1	H_0 : oil price doesn't cause soybean oil price	p-value = 0.007
TEST 2	H_0 : soybean oil price doesn't cause oil price	p-value = 0.0219

Table 1.6 Granger causality test—null hypothesis—soybean oil and oil prices (2009–2012)

TEST 1	H_0 : oil price doesn't cause soybean oil price	p-value = 0.2126
TEST 2	H_0 : soybean oil price doesn't cause oil price	p-value = 0.0017

1.3.4 The United States Market

The choice to analyze the United States market depends on the central role that the country has assumed in the bioenergy market, and in particular in the production of corn for bioethanol. The variables considered in this model are not only related to the prices of corn, soybean oil and energy products, but also to the quantities produced and the share of the crops of these two cereals intended for food use and energy use.

Once again it was decided to deal with the problem by building bivariate VAR. It is interesting to observe Fig. 1.4 where, in addition to the price, we reported that in the amount of corn produced, it was intended for the production of bioethanol and it was destined for domestic use from January 2000 to December 2012: all these quantities are characterized by an increasing trend, but that of the amount of corn for the production of bioethanol followed a different evolution compared to the other variables.

In January 2000, the percentage of corn destined to the energy market was 8 %, while at the end of 2012, the share used for ethanol was 59 %, and in general the volume of corn production increased by 56 % from 2000 to 2012. The most important growth in the amount of corn for the production of bioethanol happened in January 2011: in December 2010, the percentage of corn destined for the energy market stood at 9 %, while in the following month it jumped to 67 %. This is probably due to the approval, in February 2010, of the RFS2 program (Renewable Fuel Standard 2), which has set ambitious targets for the reduction of greenhouse gases through the use of biofuels and the development of the alternative energies

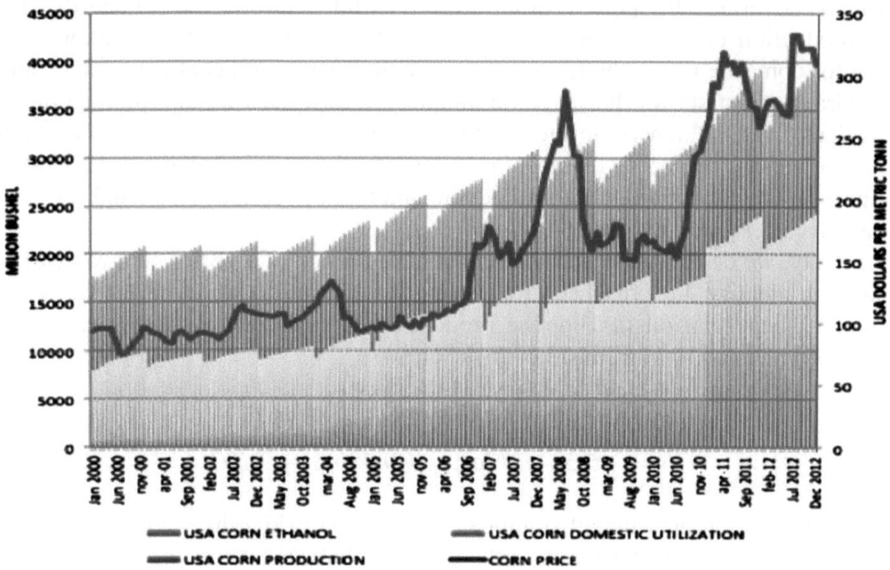

Fig. 1.4 Corn: price, total production, domestic and fuel destination, 2000–2012

sector: This caused a strong increase in the demand for corn for the production of biofuels. In the same period, the price of corn grew sharply, both thanks to the policy factor, and because of forecast errors in the estimation of crops and reduction of stocks of corn. The scarcity of production might have prompted many farmers to devote the majority of their crop for bioethanol production, since it was more convenient than the food destination.

The first VAR considered in this context is one that uses as variables the price of corn (m) and the amount of corn (qm) intended for bioethanol:

$$\begin{aligned} \mathrm{qm}_t &= 0.91\mathrm{qm}_{t-1} + 4.12m_{t-1} - 189.7 \\ m_t &= 0.001\mathrm{qm}_{t-1} + 0.94m_{t-1} + 4.05 \end{aligned} \tag{1.14}$$

The model was built with one lag and fits the data well; it is important, however, to note that in the equation, with the price of corn as a response at time t, the parameter related to the amount of corn for the production of bioethanol has a significance level of 1 %; therefore, it cannot be considered significant. The Granger causality test, indeed, shows that it is not the amount of corn used for the production of biofuel to cause, in the Granger sense, the price of corn, but rather the contrary. The direction of this relation explains how the greater or lesser convenience in producing bioethanol determines the grown production, causing variations in the portion of corn intended for the energy market rather than for food.

Always looking at Fig. 1.4, we can see that the corn produced for domestic use and employed for bioethanol has increased significantly over time. Therefore, using as covariates the portion of corn destined for domestic use (*d*) and that for the production of bioethanol, interesting results emerge: only the equation with corn for domestic use as the response variable presents significant parameters, and the Granger test shows that it is the amount of corn used for the production of bioethanol that influences, in the Granger sense, the quantity for internal use and not vice versa. This means that the growing demand for biofuels has a considerable impact on the allocation of the production, and that the domestic demand for corn was affected by this influence.

Finally, we decided to proceed with the construction of a vector auto-regressive model to identify the existence of a linear intertemporal relationship between the amount of soybean oil (qs) for the production of biodiesel and the price of the latter (*b*). The VAR was constructed as follows:

$$\begin{aligned} \mathrm{qs}_t &= 0.51\mathrm{d}s_{t-1} + 544.82b_{t-1} + 153.97 \\ b_t &= -5.544e^{-0.5}\mathrm{qs}_{t-1} + 9.78b_{t-1} + 2.697 \end{aligned} \quad (1.15)$$

The first equation presents all significant parameters, while the second equation's only significant parameter refers to the price of biodiesel at time $t-1$. Thus, the Granger causality test has allowed us to identify the direction of this relationship: it is the price of biodiesel that determines, in the Granger sense, the amount of soybean oil to be allocated to the production of biofuel. Again, therefore, it is the greater or lesser convenience in producing biodiesel that affects heavily the amount of soybean oil produced.

1.4 Conclusions

We have analyzed the relationship among the principal agri-food commodities used for the production of biofuels, crude oil and biofuels themselves. Most of these products are made with the same input destined to human and animal nutrition, and this creates a competition in the allocation and the use of land and raw materials between energy and the food market. Through VAR models we tried to understand how agri-food commodities' prices have been influenced by the energy market rather than the food market.

The first conclusion is methodological: the vector auto-regressive models that resulted were particularly appropriate for assessing prices; in particular the data are well described by the VAR model which, in all cases, showed very high indices of goodness of fit. The VAR methodology has been particularly interesting since it has allowed us to detect the presence or absence of statistically significant relationships between the prices used as variables and, most of all, through the Granger causality; it let us deepen these relationships by identifying the direction of the causality. It is important, however, to emphasize that this kind of test does not claim to identify ever-valid relationships, but the ties apply in the context we analyzed, which is in the VAR we built, considering only two variables. In any case, the results obtained allow us to draw some interesting conclusions from an economic point of view.

Considering the global level, corn price and soybean oil price cause, in the Granger sense, the crude oil price. It means that, through VAR models, it has been observed that "*corn price*" and "*soybean oil price*" past observations are useful to predict the trend of the "*oil*" variable. We found this causality between 2009 and 2012: this is probably due to the increasing employment of corn and soybean oil as factors of production for biofuels, and to the fact that oil is used both for production of these cereals and a good substitute for biofuels. The recent strengthening of this relationship has also been observed by Wisner (2009), who found a weak relationship between oil price and corn price until 2007, whereas from 2007 to 2009 he noticed a consolidation of this link.

We also observed a relationship of mutual causality between biodiesel price and soybean oil price. This relationship can be explained by the use of soybean oil for the production of biodiesel in Brazil and USA, which are both the greatest producers of soybean in the world and the main consumers for biodiesel. In Brazil, the National Program of Production and Use of Biodiesel (PNPB) has stimulated the use of soybean oil as a primary input to produce biodiesel (among 77 and 86 % from 2007 to 2011): currently 14 % of Brazilian soybean (ANP 2012) is destined for the production of biofuels. Also in the USA about 14 % of soybean production is destined for the production of biodiesel (in 2011) and 65 % of biodiesel is composed mainly of soybean oil (Wisner 2013).

The analysis of the U.S. market has shown that the price of corn causes (always in the Granger sense) the amount of corn used to produce ethanol, which in turn causes the amount of corn for domestic use. The growing importance of the biofuel market clearly emerges here, and evidence in current and future trends in the consumption of biofuels confirms this hypothesis [in the United States, the amount of corn used for ethanol production increased from 5 % of the total in 2001 to 30 % in 2010 (Hertel and Beckman, 2010)]. This growth was influenced by several factors:

- the increase in state subsidies given over the past 10 years;
- the future prospect of an increase in profits related to the production of corn;
- the promotion of "green" policies that fostered the development of biofuels.

Food production destined for human consumption and commodity prices has been influenced by an increase in the importance of bioenergy. Allocation of goods for energetic use interferes in the available quantity and price of food products, and therefore it influences food safety, especially in developing countries (Diouf 2008). On the other hand, enhancing the use of renewable energies could be a good practice to prevent the negative impact of fossil fuels in the environment. It is important to discuss such problems and renew the debate on the sustainability of energy policies, to avoid a lack of balance between food production and biofuel production.

Acknowledgments All authors would like to acknowledge Professor Roberto Fanfani of the Department of Statistical Sciences "P. Fortunati" (UNIBO) for his precious advice and help, and mostly for being a mentor for all of us.

References

ANP—Agencia Nacional de Petróleo, Gás Natural e Biocombustíveis, 2012. "Biocombustíveis".

Diouf, J. (2008). *La sicurezza alimentare in tempi di cambiamento climatico e produzione bioenergetica*. Roma: ENEA Magazine.

Esposti, R. (2008) Food, feed and fuel: biocarburanti, mercati agricoli e politiche, *Working paper* n. 10, Gruppo 2013.

Fanfani, R. (2008). *L'aumento dei prezzi e il complesso sistema agroalimentare mondiale*, Bologna, Il Mulino, vol. 5, pp. 919–938.

FAO. (2008). *Food outlook. Global market analysis—June 2008*, Rome.

FAO. (2011). *The state of food insecurity in the world, 2011*. Rome: Food and Agricolture Organization of the United Nations.

Farm Foundation. (2008). *What's driving food prices?* Issue Report, Oak Brook, Illinois

Frondel, M., & Peters, J. (2007). Biodiesel: A new Oildorado? *Energy Policy, 35*(3), 1675–1684.

Goldemberg, J., Teixeira Coelho, S., Nastari, P. M., & Lucon, O. (2004). Ethanol learning curve—the Brazilian experience. *Biomass and Bioenergy, 26*, 301–304.

Granger, C. W. J. (1969). Investigating causal relations by econometric model and cross-spectral methods. *Econometrica, 37*, 424–438.

Hertel, T. W., Beckman, J. (2010). Commodity price volatility in the biofuel era: An examination of the linkage between energy and agricultural markets, GTAP *Working Paper*, n. 60, Paper Prepared for the NBER Agricultural Economics Conference, March 4–5, 2010, Cambridge, Massachusetts.

Hertel, T. W., Tyner, W. E., Birur, D. K. (2010). The global impacts of biofuel mandates. *The Energy Journal, 31* (1).

Hochman, G., Kaplan, S., Rajagopal, D., Zilberman, D. (2012). Biofuel and food-commodity prices. *Agriculture*, (2), 272–281

Kent Hoekman, S. (2009). Biofuels in the U.S.—challenges and opportunities. *Renewable Energy, 34*, 14–22.

Kristoufek, L., Janda, K., Zilberman, D. (2011). Topological properties of biofuels networks, *Working Paper*, University of Praga.

Maluf, R. (2008). Evolução no preço dos alimentos e o sistema alimentar global, *Boletins OPPA*, 18.

Mitchell, D. (2008). A note on rising food prices, *Technical Report n. 4682 for World Bank Policy Research*, Washington DC, USA.

Podestà, F. (2011). *Tecniche di analisi per la ricerca comparata trans-nazionale*. Milano: FrancoAngeli Editore.

Rosegrant, M. W. (2008). *Biofuels and grain prices: Impacts and policy responses*. Washington DC, USA: International Food Policy Research Institute.

Serra, T., Zilberman, D., Gil, J. M., & Goodwin, B. K. (2011). Nonlinearities in the U.S. corn-ethanol-oil-gasoline price system. *Agricultural Economics, 42*(1), 35–45.

Trostle, R. (2008). *Global agriculture supply and demand: Factors contributing to the recent increase in food commodity prices*, WRS-0801, Washington DC, USA: United States Department of Agricultural (USDA—ERS).

Wisner, R. (2009). Corn, ethanol and crude oil prices relationship—implications for the biofuels industry. *AgMRC Renewable Energy Newsletter*.

Wisner, R. (2013). Soybean oil and biodiesel usage projections & balance sheet, *USDA Report*, http://www.extension.iastate.edu/agdm/crops/outlook/biodieselbalancesheet.pdf

Zezza, A. (2011) Le politiche per I biocarburanti nei principali paesi produttori, *Agriregionieuropa*, Year 7, *24*.

Zezza, A. (2007). Sostenibilità economica e ambientale della produzione dei biocarburanti. *QA—Rivista dell'Associazione Rossi-Doria, 4*.

Zhang, Z., Lohr, L., Escalante, C. E., & Wetzstein, M. E. (2009). Ethanol, *corn and soybean price relations in a volatile vehicle-fuels market. Energies, 2*, 320–339.

Zilberman, D., Hochman, G., Rajagopal, D., Sexton, S., & Timilsina, G. (2012). The impact of biofuels on commodity food prices: Assessment of findings. *American Journal of Agricultural Economics, 95*(2), 275–281.

Additional Bibliography

Beidas-Strom, S., Shik Kang, J., Loungani, P., Matsumoto, A., Rousset, M. (2012). Commodity market review. *World Economic Outlook*, IMF, Washington, DC.

FAO. (2012). *Impacts of bioenergy on food security—guidance for assessment and response at national and project levels*. Rome: Food and Agricolture Organization of the United Nations.

Fortenbery, T. R., Park, H. (2008). The effect of ethanol production on the U.S. National corn price, *Staff Paper* n. 523, Department of Agricultural & Applied Economics, University of Wisconsin—Madison.

Hochman, G., Rajagopal, D., Timilsina, G., Zilberman, D. (2011). The role of inventory adjustments in quantifying factors causing food price inflation. *Policy Research Working Paper Series*, no. 5744, The World Bank, Washington DC, USA.

OECD-FAO. (2010). *Agricultural outlook 2010–2019*.

Piesse, J., & Thirtle, C. (2009). Three bubbles and a panic: An explanatory review of recent food commodity prices events. *Food Policy, 34*(2), 119–129.

Sims, C. A. (1980). Macroeconomics and reality. *Econometrica, 48*(1), 1–48.

Author Biographies

Anna Caterina Leucci Ph. D. student in Statistical Science at the Department of Statistics "Paolo Fortunati", University of Bologna. She graduated in Statistics and Economy with a thesis about the use of VAR models to forecast commodities prices. Currently she is analyzing the gluten-free market and celiac consumer behavior. Since 2013 she is working for ECONAG, a spin-off of University of Bologna, on the "Commodities prices observatory" project.

Stefano Ghinoi Ph. D. candidate in Statistical Sciences at the University of Bologna. He has a Master's degree in Economics and Public Policy Evaluation and his fields of research are environmental economics, innovation economics and local development. He worked in a consulting firm and as a tutor of microeconomics for the University of Modena e Reggio Emilia. Since 2013 he is collaborating with ECONAG, a spin-off of University of Bologna, on the "Commodities prices observatory" project.

Daria Sgargi Ph. D. student in Statistical Sciences at the Department of Statistics "Paolo Fortunati," University of Bologna, Italy. She graduated in Cooperation and Development, with a thesis on Fome Zero, a Brazilian program for food security. She has studied volatility and interactions between food and commodities prices. She is currently working on carcinogenicity and toxicity tests of food additives and pesticides. Since 2013 she is collaborating with ECONAG, a spin-off of University of Bologna, on the "Commodities prices observatory" project.

Valdemar João Wesz Junior Ph. D. student in Development, Agriculture and Society (CPDA/UFRRJ) with a scholarship from CNPq (Brazil) and visitor Ph. D. student in Agri-Food Economics and Statistics (Unibo-Italy) with a scholarship from the Faperj Foundation (Brazil).

Chapter 2
Irrigation Water Resource in a Rice-Growing Area: Economic Evaluation under Different Pricing Conditions

Guido Sali and Federica Monaco

Abstract Water scarcity is an increasing phenomenon affecting all sectors of economic interest. This problem is stressing agriculture as well, and in particular primary activities that use huge amounts of the resource to maintain their productions at sufficiently high levels. A way to contrast scarcity is the improvement of the efficiency of water allocation and the reduction of its losses, through the adoption of political instruments and pricing aimed to a more aware use of the resource itself. In this context, the Water Framework Directive, in order to assign an appropriate cost to irrigation water, urges member states to introduce the concept of full cost, and to apply a volumetric supply fee promoting the rationalization of the resource, thus playing a role in addressing emerging and future problems of water scarcity. However, several studies have already demonstrated through modeling approaches that these interventions could strongly affect farms' choices and performances, resulting in consequences that would have repercussions on the whole agricultural system. The study aims to evaluate economic performances of farms in a typical rice-cultivated area in Lombardy, Northern Italy, under different supply tariff levels. A simple programming model has been used to run a scenario analysis. Structural features of farms, their productive inputs and performances are reported in current conditions, under different pricing and progressively increasing fee levels, in order to evaluate their effects on farms' economic performances and operative strategies. The obtained results allow for a first identification of critical points in the water management of the area and hypothesize interventions for a better resource allocation, as a useful instrument for supporting future policies on water resources.

G. Sali · F. Monaco (✉)
Department of Agricultural and Environmental Sciences, University of Milan, Milan, Italy
e-mail: federica.monaco@unimi.it

C. Zopounidis et al. (eds.), *Agricultural Cooperative Management and Policy*, Cooperative Management, DOI: 10.1007/978-3-319-06635-6_2,

2.1 Introduction

Water represents a fundamental element for all sectors of economic, social, and environmental interest. Particularly in agriculture, it undoubtedly plays a key role as a fundamental productive input for the conduction of all the related activities, in arid and semi-arid regions, as well as temperate ones. In the former, water allows to obtain a sufficient crop production, while in the latter it maintains yields at high levels, reducing the risk of loss of the product (Tarimo et al. 1998; Iglesias et al. 2005; IPPC 2012). However, in relation to several emerging issues, its importance is increasing, even in such areas where water availability for the primary sector has not traditionally been limiting. Also for irrigated agriculture, in fact, a quantitative reduction of the resource is occurring, due to the global phenomena of climate change (Fischler et al. 2007), an increasing population and rapid urbanization, which are emphasizing the conflict of water use among different sectors, as a result of an increasing demand on the part of each one, at the same time exacerbating the effects of decreased usability (UNEP 1999). Water scarcity in agriculture is becoming a significant issue and it inevitably has repercussions both on the productive and economic performances of farms, modifying in the long-term period their competitiveness, and burdening the possibility of continuing the activity. Along with water scarcity, and as a possible strategy to face it, the need for reducing the wastes of the resource also has to be considered. Water as an economic asset with limited availability (ICWE 1992) is to be protected through promoting its efficient and equal, which is possible only by the attribution of a fair price. The estimation of water irrigation costs is then a significant topic with an important role in supporting water regulations, allowing decision makers to make aware choices to face water shortages.

2.2 The Cost of the Resource

The Water Framework Directive (WFD) 60/2000/EC (European Parliament and the Council of EU 2000) emphasizes the allocation of a fair price for irrigation water and calls on member states for the introduction of the so-called "*full cost*" (Fig. 2.1), which, taking into account financial, opportunity and environmental costs, could represent the practical application of the "polluter-pays principle": it ensures that the end user pays a price high enough to recover all the costs arising from the use of water, and its adoption reduces wastes and nonvirtuous behaviors caused by an underestimation of the resource.

In agriculture, applied fees are much lower than those hypothesized by regulatory bodies which could lead to an increase in irrigation costs; paradoxically, the farmer, as the end user of the resource, would then be in the condition of having less water at a higher cost; therefore this situation would not be sustainable from the farmers' point of view. In order to achieve a sustainable use of the resource, the

Fig. 2.1 Structure of the full cost (WATECO 2003, modified)

<table>
<tr><td>Environmental costs not related to water</td><td rowspan="2">Environmental costs (external)</td><td rowspan="6">Economic costs</td></tr>
<tr><td>Environmental costs related to water</td></tr>
<tr><td>Opportunity costs (scarcity)</td><td>Cost of the resource (external)</td></tr>
<tr><td>Other direct costs</td><td rowspan="3">Financial costs (including environmental and opportunity costs already internalized)</td></tr>
<tr><td>Administrative costs</td></tr>
<tr><td>Capital, operating and maintenance costs</td></tr>
</table>

suppliers can adopt different modalities for the delivery of water service. Pricing and fees differ according to their efficiency in promoting a more rational use of irrigation water. A fixed fee set per irrigated or irrigable hectare tends not to encourage such practices, but is relatively easier to adopt and may in some cases represent the most recommended solution (Giannoccaro et al. 2007); volumetric fees, instead, determine a more aware use of water, but could have unit costs much lower than the actual cost of the resource. WFD suggests preferentially using a volumetric rate, as it would represent an economic instrument able both to reduce water consumption and cover all the costs of water service. It represents a more transparent and efficient (Tsur et al. 2003) pricing method, since it is based on the water quantity actually supplied. As several studies have already demonstrated (Dono et al. 2006; Giannoccaro et al. 2007; Bartolini et al. 2007), a different tariff level, a different pricing and the increase of irrigation water costs influence farmers' choices, and lead to a significant reduction in water consumption, at the expense of withdrawals from wells and private water sources, as well as the need for management and/or productive changes; but these strategies, such as a reduced irrigated area, crop diversification toward less water-demanding crops, an increase in the efficiency of distribution and a different method of water application, can finally result in a significant decrease in farm income.

Moreover, some authors consider the use of incentives to be not so encouraging of good behavior and the assignment of a political price to water service supply to be an inefficient management system, not stimulating proper use (Rogers et al. 2002), but efficient pricing, which may determine undesirable effects on farmers' decisions or environmental implications not immediately anticipated. In the fields of ancient irrigation, such as rice-cultivated areas in Northern Italy, environmental aspects also related to multiple use of the resource must be considered (Cadario and Bischetti 2006): even though water distribution techniques are technically inefficient and characterized by huge losses due to filtration, the complex system, and water network developed over the centuries has allowed the creation of valuable paranatural aquatic environments. Even in these areas with high natural and environmental value the quantification of environmental costs is something difficult (EEB 2001), leading to an uncertain estimation of full cost.

Finally, it must be considered that irrigation water value is strictly linked to that of the agricultural production it contributes to. Consequently, a higher water cost inevitably reflects on water use efficiency and productivity (Molden 1997; Seckler et al. 1998; Kassam and Smith 2001), an increase of which could represent a further way to achieve an efficient use of water.

2.3 Modeling for Irrigation Water Management

A valid support to policy makers and to decisional processes lies in the results of appropriate tools, such as mathematical programming models. They provide information not directly observable and allow simulations of different scenarios related to changes in agricultural policies, resource management or market development, and can guide decision makers toward the identification of the most suitable interventions to achieve economic and environmental targets of water policies.

Economic analyses of irrigation water are based on the formalization and implementation of both econometric and programming models, at different scales and levels (farm, local, regional). Among them the regional level is able to answer the requirements of the WFD, which states that the catchment area is the unit for the analysis and the integrated management of water resources.

The econometric approach, based on less informative inputs, has demonstrated on several occasions the possibility to estimate a function of operating costs of water distribution, in irrigation districts and consortia (Dono 2003; Dono and Giraldo 2010; Giraldo 2011; Dono et al. 2011); more often the economic analysis of irrigated agriculture is realized through the application of linear programming models (mono-objective, multicriteria, stochastic discrete) to evaluate the impacts derived from alternative conditions, both internal and external to the system: each simulation generates a new solution showing the effects of the changes themselves on crops, technological choices, use of productive inputs, and economic performances of farms (Dono 2003; Bazzani et al. 2005; Dono et al., 2008; Giannoccaro et al. 2008; Bazzani and Zucaro 2008; Bazzani and Scardigno 2008; Dono and Giraldo 2010; Giraldo 2010; Dono et al. 2011). However, these models require the collection and processing of a large amount of economic and productive data and information; even though they are useful to understand the features of the agricultural system by identifying relationships between the use of inputs and productive levels, their results strongly depend on the constraints imposed on the model.

In the same context, the use of Positive Mathematical Programming (PMP) (Howitt 1995; Paris and Howitt 1998) is recently spreading. This new approach requires a limited amount of data to perfectly calibrate the model for the reference period, according to three main phases: specification of a linear programming model that uses all the information available, reconstruction of a total variable cost (Arfini and Paris 1995), and formulation of a nonlinear programming model to be used to perform simulations. Its application for water resource analyses is, however, currently underdeveloped. In this regard, it recalls the work of Blanco et al. (2004) which considers the impact of pricing policies on two irrigation districts in Spain by specifying a cost function for each one, what Cortignani and Severini (2008, 2009) have developed in relation to territorial analysis, also following the introduction of tariffs differentiated depending on the season.

These models can be used to face issues related to the variation in the cost of the water and its availability, but the possibility of analyzing future scenarios is limited, since they do not allow the consideration of new and different production activities compared to the reference situation.

2.4 Aims and Analysis: Methodology

The paper aims to simulate possible changes in water management and water use, if different types and levels of payment were introduced. In many parts of Northern Italy irrigation consortia apply to the supply of water a fee based on the served surface, rather than according to the distributed volume. In order to simulate the farmers' behavior in the adaptation to face a different basis for water payment, a mathematical programming model has been implemented. The analysis has been carried out in a rice-cultivated area in Lombardy, Northern Italy, characterized by peculiar uses of the resource itself and particularly suited for this analysis.

Data collection has been carried out through direct surveys at sample farms, using results of *ad hoc* experimentations conducted in an experimental farm in the same area.

The selection of rice-growing farms operating in the district started from their extraction from the regional database *Sistema Informativo Agricolo della Regione Lombardia* (SIARL), their classification on the basis of Utilized Agricultural Areas (UAA) of rice, and the sampling within each class. To each farm a specific questionnaire requiring information about the crop year 2010–2011 was submitted and filled through direct surveys to farmers, for a total of 19 surveys carried out and a total rice-cultivated area of 730 ha. The cultivated area is dedicated to four main cultivars, namely *Gladio, Loto, Baldo*, and *Selenio.*

Data were then elaborated to describe the features of the system, and used for the identification and implementation of a model, returning current economic and productive conditions of farms. In order to evaluate the effects of new managerial and/or productive strategies on cultivated areas (possible reduction of the irrigated area, crop diversification, increase in the distribution efficiency and different method of water provision), it has also been used to make scenario analysis, related to a different pricing system and levels.

2.4.1 Case Study Area: Main Features

The case study area is located in a typical rice-growing district in South-Eastern Lombardy, i.e., the so-called Lomellina, with a particular focus on the area of San Giorgio di Lomellina (PV). Agriculture in the district is mainly dedicated to rice, with a marginal portion for other arable crops, such as corn, soybean and poplar. The consortium supplier (*Associazione Irrigazione Est Sesia*) provides water to farms, deriving it from Cavour Canal, Arbogna River and leakages, even though supplies from private sources also exist. The distribution of water is mostly continuous and, for a lesser part, it refers to pre-established rotating shifts. Combinations between water dispensation and cultivation strategies return in different typologies for the conduction of rice-fields, as shown in Table 2.1.

According to conducted experimentations, crop production is linked to water and agronomic management, since differences among yields exist.

The estimation of distributed water indicates the traditional method as the most water-requiring, while the differentiation of sowing techniques shows a lower overall water distribution for soil-seeding (Table 2.2). At the same delivery typology, the determining factor increasing its resource management typology during the growing season. Water quantity seems, then, to affect yields, suggesting that lower provision and availability cause a lower production.

2.4.2 The Implemented Model

For an economic evaluation of irrigation water in the district, a simple nonlinear programming model was developed. A decisional variable set in simulations is the rice-growing area ($xcrop_{f,c}$) in each farm (*f* index) subject to irrigation according to the different methods of water supply and agronomic management (*c* index).

The objective function *Z* aims to maximize gross margin of the group of farms, as a difference between obtainable revenues (*R*) and costs supported during the whole growing season (*C*), including, along with production costs, water supply costs and water management costs (Castellani et al. 2008) (see also Table 2.3). It takes the following synthetic form:

$$Z = \sum_{f,c} (R_{f,c} - C_{f,c})$$

Revenues encompass those from CAP subsidies for rice-growing activity and those from the sale of paddy rice. In detail, the former amount to an average premium of 850€/ha, with a reduction of 8 % for the part exceeding 5,000€, according to European guidelines: each farm can obtain an average more than 30,400€, which currently represents 20 % of total revenues; the remaining 80 % is due to the sale of paddy rice at market prices in 2011, equal to 331€/ton (Camera di Commercio di Pavia 2011).

Table 2.1 Different conductions of rice-fields: general characteristics

Irrigation type code	Water dispensation	Water management		Agronomic management	Farms (n.)	% UAA	Yield (tons/ha)
CFW	Continuous	Continuous flooding. Water flows continuously for the whole duration of the crop cycle. Submersions are interrupted by 3 or 4 dries in correspondence with certain phases of the cycle or treatments with herbicides or fertilizers		Water-seeding after the submersion of the field	10	50.63	9.72
CFS	Continuous	Continuous flooding. Water flows continuously for the whole duration of the crop cycle. Submersions are interrupted by 3 or 4 dries in correspondence with certain phases of the cycle or treatments with herbicides or fertilizers		Soil-seeding; the ground remains dry until the rice has reached the stage of the 4th-5th leaf, then the normal regime of submersion is restored	5	8.95	8.33
SCFW	Rotating shifts	Intermittent flooding. Water is available continuously only during predefined shifts	Continuous flooding during the shift	Water-seeding after the submersion of the field	1	8.02	9.72
SCFS	Rotating shifts	Intermittent flooding. Water is available continuously only during predefined shifts	Continuous flooding during the shift	Soil-seeding before the first irrigation	2	10.02	8.33
SIS	Rotating shifts	Intermittent flooding. Water is available continuously only during predefined shifts	Flowing irrigation, trying to maintain water on the ground until the next shift	Soil-seeding before the first irrigation	6	21.91	7.81

Table 2.2 Seasonal water dispensation and flow for each crop type

Irrigation type	Distributed water (m^3/ha)	Water flow (i_c) ($l * s^{-1} ha^{-1}$)
CFW	22,712 ± 1,696	2.4
CFS	20,842 ± 114	2.4
SCFW	17,075[a]	1.5
SCFS	13,073 ± 84	1.5
SIS	5,476 ± 6,344	1.5

[a] one data available only

Table 2.3 Elements of the implemented model

Total revenues (R)	CAP subsidies	Contribution for single payment with reduction for the modulation
	Sale of paddy-rice	Revenues from selling rice to processing industries
Direct costs (C)	*Water supply cost*	
	Water supply cost	Payment to Irrigation Consortium for water supply during watering season
	Water management costs	
	Maintenance and repair of technical means used for irrigation	Managing costs for irrigation structures inside the farm (maintenance, repair and operations)
	Costs for energy and consumables	Fuel, oil, electricity for pumping, lifting and distributing water
	Labor costs	Manpower for water management
	Amortization of machines	Share of deterioration of the machines used for irrigation
	Other production costs	
	Farm-level operations, from sowing to harvest	

As summarized in Table 2.3, various expenses are traced back to three main cost categories. The expenses related to water supply costs refer to the current tariff condition set by the consortium and equal 278.62€/ha, or the volumetric rates introduced in different scenarios. 15 % of total costs are due to this aspect.

Water management costs, as suggested by Lazzari and Mazzetto (2005), take into account various economic aspects linked to irrigation practices. Some technical elements needed for the estimation of this cost category have been directly surveyed at farms (working capacity of the pump and power of the tractor used for irrigation, number of irrigations during watering season), while others have been assumed as starting points (hourly labor cost, value of a new machine, its economic and physical life, repair and maintenance factors and coefficient, and depreciation rate). The estimation of these costs reveals that they represent 20 % of direct costs, and in particular 12 % are linked to labor, 7 % to the management of technical means, and 4 % to consumables.

Finally production costs, or costs for operations at the farm level from sowing to harvest, are estimated to be 1,200€/ha, returning almost one third (65 %) of the total expenses that farms support.

Z is subjected to two main farm-level and district-level constraints regarding land and water. Land balance ensures that no more land than the total available in each farm (land_f) is cultivated (2.1) and that cultivated areas ($a_{f,c}$) still maintain the same water dispensation, continuous (2.2a) or not (2.2b):

$$\sum_c \text{xcrop}_{f,c} \leq \text{land}_f \tag{2.1}$$

$$\text{xcrop}_{f,\text{CFW}} + \text{xcrop}_{f,\text{CFS}} \leq a_{f,\text{CFW}} + a_{f,\text{CFS}} \tag{2.2a}$$

$$\text{xcrop}_{f,\text{SCFW}} + \text{xcrop}_{f,\text{SCFS}} + \text{xcrop}_{f,\text{SIS}} \leq a_{f,\text{SCFW}} + a_{f,\text{SCFS}} + a_{f,\text{SIS}} \tag{2.2b}$$

Water balance ensures that water flow resulting from the model is not higher than that currently provided by the consortium (i_c), differing for each water dispensation (see Table 2.2):

$$\sum_c i_c * \text{xcrop}_{f,c} \leq \sum_c i_c * a_{f,c}$$

2.4.3 Scenario Analysis

A scenario analysis has then been performed. The first condition (scenario #0) applies the maximization to the current situation, characterized by a water payment per irrigated hectare. In further scenarios, a volumetric fee replaces the current one, *ceteris paribus*. In scenario #1 the fee is calculated so as to return the same expenditure, deriving from the fixed rate per hectare.

Prices introduced in scenarios #2 and #3 allow us to understand which types of water management are chosen in order to maximize the gross margin, and at the same time, how much water is saved. This information is synthesized in economic and productive parameters. In particular the following have been considered:

- *Total costs and revenues*;
- *Water cost*, or the price of irrigation water (PU, in €/m^3), as the ratio between costs of irrigation and water available to the farm, and distributed:

$$PU\,[€/m^3] = \text{costs of irrigation/distributed water}$$

- *Water productivity,* defined as the ratio between total yield (in tons) and its water consumption (m^3) during the season due to evapotranspiration (Teixeira et al. 2008; Vazifedoust et al. 2008); we have instead calculated it as yield compared to total amount of water used during the watering season, not considering line losses, namely the amount potentially distributed each year according to the available resource; this productivity can be named *Irrigation Water Productivity*:

$$\text{IWP}\,[\text{g/kg}] = \text{total yield/distributed water} * 1000$$

In addition, the *Economic Water Productivity (EWP)* has also been considered based on the market value of the crop (Igbadun et al. 2006; Palanisami and Suresh Kumar 2006; Teixeira et al. 2008; Vazifedoust et al. 2008):

$$EWP[\text{€/m}^3] = \text{crop economic value/distributed water}$$

2.5 Results and Comments

The model was solved through the software GAMS (*General Algebric Modeling System*) (Brooke et al. 1988; Rosenthal 2007) and has allowed the generation and display of several data output.

The model returned information about current structural features of farms, their productive inputs, as well as the productive and economic performances of each one and for every type of culture, allowing comparisons between farm and cultural types, homogeneous or not.

Optimal management of cultivated areas in comparison with current conditions is given in Table 2.4. The maximization of overall margin leads in any case to managerial and agronomic choices quite far from what really applied. In the current situation, the fee for water supply in the area analyzed is equal to 278€/ha, compared to the circulated volumes during watering season, which correspond to 0.017€/m^3.

Fixed fees per hectare do not seem to encourage water saving, since areas with continuous supply are suggested to be cultivated according to water seeding, which is more water-demanding than soil-seeding, and those provided periodically shift to the most demanding method within the category (SCFW). In this case, water management costs are brought down, rather than those relating to water supply. On the contrary, the adoption of different pricing has more evident effects both on typology of water management and agronomic strategies: in relation to periodic irrigations, water saving techniques are preferred. The opportunity to adopt dry or semi-dry cultivation is confirmed by previous surveys carried out in the same area: during the season 2004–2005, 5.4 % of the denounced rice-fields in the S. Giorgio di Lomellina area were soil-seeded, and from 2008 so far this percentage is passed to almost 30 %, with peak values of 37 %.[1] The volumetric

[1] These results derive from a study carried out at the experimental farm of Centro Ricerche sul Riso (*Rice Research Center*) in Castello d'Agogna (PV), pertaining to Ente Nazionale Risi (www.enterisi.it).

Table 2.4 Cultivated areas (%UAA) per irrigation type in different scenarios

Irrigation type	Currently Fee = 278€/ha	Scenario #0 Fee = 0.017€/m^3	Scenario #1 Fee = 0.03€/m^3	Scenario #2 Fee = 0.05€/m^3	Scenario #3 Fee = 0.11€/m^3
CFW	50.5	59.0	31.0	28.0	15.1
CFS	8.5		28.1	31.0	44.0
SCFW	8.2	41.0	12.3		
SCFS	10.3		28.6	41.0	40.1
SIS	22.5				0.8

fees hypothesized to be adopted have also allowed to identify cost levels favoring different irrigation techniques: with a tariff of 0.03€/m^3, part of the surface served with continuous dispensation is managed with delayed flooding (CFS). Similarly, most of the surface served by rotating shift is converted to SCFS type (seeding before the first irrigation). The more the fee increases, the more water-saving methods are preferred. With a tariff equal to 0.05€/m^3, CFS and SCFS are the most used methods, whilst in the presence of a fee equal to 0.11€/m^3, the SIS irrigation type begins to be chosen. At this level of price, in fact, the gross margin begins to be more favourable than that returned by irrigation systems ensuring higher yields.

2.5.1 Costs Analysis

The fee type currently adopted links proportional water supply cost to irrigated areas, independently from the amount of available water. Irrigation water supply costs expressed in a volumetric rate show that the lower costs are, the lower water distribution is. Actually, however, the adoption of periodic irrigation is independent of farmers' will: since the fee is set by a consortium, whenever it is modified, it would consequently affect these aspects.

Inevitably higher fee levels would mainly affect supply costs (Tables 2.5 and 2.6). For continuous dispensation, water supply costs rise proportionally to the fee introduced. Volumetric fees higher than 0.05€/m^3 lead to supply costs higher for a unit size than those deriving from a fixed fee per hectare. In these unfavourable conditions, farmers are driven to choose irrigation systems that, using less water, ensure a minor expense for the supply of the resource.

Total costs related to irrigation (Table 2.7) differ from each other according to the operative procedures adopted by each farm for the irrigation practice. However, farm water management costs remain quite constant amongst irrigation typologies, with an average value around 700€/ha, since irrigation is essentially by gravity and energy consumptions are negligible. The relevant elements in this sense seem to be the components related to the maintenance of irrigation network and labor costs, very little depending on volumes and not very compressible with a

Table 2.5 Water supply cost (€/ha), for each irrigation type in different scenarios

Irrigation type	Currently	Scenario #1	Scenario #2	Scenario #3
CFW	278	673	118	2,391
CFS		627	1,108	2,401
SCFW		244		
SCFS		133	266	578
SIS				78

Table 2.6 Water cost (€/m^3) in different scenarios

Irrigation type	Currently	Scenario #1	Scenario #2	Scenario #3
CFW	0.04	0.06	0.09	0.15
CFS	0.05	0.06	0.09	0.15
SCFW	0.10	0.11		
SCFS	0.17	0.18	0.19	0.25
SIS	0.13			1.05

Table 2.7 Costs (€/ha) related to farm water use (supply and management) in different scenarios

Irrigation type	Currently	Scenario #1	Scenario #2	Scenario #3
CFW	1,009	1,412	1,923	3,128
CFS	1,003	1,346	1,826	3,113
SCFW	963	928		
SCFS	979	816	949	1,226
SIS	953			717

quantitative reduction of distributed water. In these cases, higher outputs are due to an increased labor for periodic irrigations, as confirmed by water use cost: a higher increase is, in fact, observable in correspondence of periodic irrigations.

2.5.2 Gross Margin Analysis and Water Productivity

Since different pricing and pricing levels do not affect total revenues, as they depend only on the amount of cultivated area, economic performances in terms of gross margins mostly depend on costs. However, as observed in Table 2.8, they do not show significant differences among irrigation typologies and scenarios as well; this is due to the specific objective function utilized that imposes the maximization of the overall margin of the group of farms, and not the single margins of each individual farm. However, if considered in purely economic terms (€), different scenarios lead to a decrease in the overall margin, with a diminution in comparison to the current condition ranging from −5 to −77 %.

Table 2.8 Gross margin (€/ha) in different scenarios

Irrigation type	Currently	Scenario #1	Scenario #2	Scenario #3
CFW	4,021	4,021	4,022	4,023
CFS	3,933	3,929	3,929	3,928
SCFW	4,026	4,024		
SCFS	3,932	3,929	3,929	3,928
SIS	3,436			3,495

Table 2.9 Irrigation water productivity (g/kg) in different scenarios

Irrigation type	Currently	Scenario #1	Scenario #2	Scenario #3
CFW	0.42	0.44	0.44	0.47
CFS	0.46	0.45	0.45	0.45
SCFW	1.02	1.20		
SCFS	2.13	2.14	1.89	1.89
SIS	1.36			11.76

Water productivity expresses at what extent different irrigation typologies contribute to the productive and economic performances of farms. It does not directly depend on the imposed tariff level, as essentially based on seasonal water distribution, but rather from the amount of distributed water; however, it is affected by the effects an increased tariff can produce on the management of cultivated areas: dissimilar values then result according to different scenarios, to which diverse amounts of distributed water correspond.

Irrigation Water Productivity (Table 2.9) may be intended as a proxy for water use efficiency, not from an agronomic point of view but rather in terms of technical-management efficiency. A lower provision to the field still allows for quite uniform yields, despite being lower than those from traditional conduction. Thus it would derive a higher value in correspondence to a minor use of resource, i.e., alternative irrigation techniques. These deviations are not immediately identifiable by analyzing each single irrigation typology, but IWP values are higher if the dispensation is not continuous, particularly evident in the case of SIS, while a more traditional conduction (CFW and CFS) does not show significant variations despite increasing fees. On the other hand, if rising tariffs lead to more water-saving methods, a slight change in the overall productivity along scenarios occurs (respectively +1 % from #1 to #2, +4 % from #1 to #3 and +12 % from currently to #3).

Economic Water Productivity, meant as the ratio between the value of obtained production and distributed water, represents the remunerativeness of the resource and shows the same trend of IWP (Table 2.10), as they are directly linked. A higher value indicates a better capability in deriving a certain revenue from crop production, even in combination with a more efficient management of water.

The progressive rise of the tariff leads to not particularly significant improvements, passing from 0.58€/m^3 in scenario #1 to 0.66€/m^3 in scenario #3. This means that while the tariff increases ten times (from 0.017 to 0.11€/m^3), economic

Table 2.10 Economic water productivity (€/m^3) in different scenarios

Irrigation type	Currently	Scenario #1	Scenario #2	Scenario #3
CFW	0.14	0.14	0.15	0.16
CFS	0.15	0.15	0.15	0.15
SCFW	0.40	0.40		
SCFS	0.71	0.71	0.63	0.63
SIS	0.45			3.90

productivity rises in the order of 12 %, suggesting, at least for this aspect, that higher supply costs do not cause particularly negative effects on economic performances.

However, it must be considered that productivity values represent the lowest possible ones (minimum benchmark), as the starting assumption for the definition of productivity itself has stated that water volumes are gross volumes, not considering overall losses of the resource (along line losses and water flows).

2.6 Conclusions

In rice–paddy fields, the adoption of nontraditional managerial and agronomic techniques allows the achievement of positive targets in terms of water saving and use efficiency, expressed by water productivity. From an economic point of view, they do not substantially modify revenues of farms but affect their costs; in particular for dry cultivation, it could be necessary to increase workforce or labor per worker, which could lead to higher costs for manpower. The increase in water supply cost could also determine a better allocation of the resource.

The adoption of a volumetric rate appears as a solution with contrasting effects. It is a valid incentive for diversification of irrigation techniques toward more water-saving methods, but it is also inevitably accompanied with negative economic effects, such as lower margins, due to the need to apply tariff levels to the limitation of water quantities distributed. This leads also to the reconsideration of the concept of water use efficiency, which nowadays appears relatively high, given the ability of the system to handle huge volumes with modest costs. The model shows that such costs are also slightly compressible, as, with the reduction of distributed volumes, costs of water management vary little.

It should however be noted that the introduction of volumetric rates must be accompanied by accurate assessments about two important aspects. The first concerns the need to overcome the rigidity of supply still practiced by consortia. The possibility for suppliers in reducing the amount of water to farms, or increasing its cost (and then a decrease in demand), as well as a factor changing their managerial aspects and their farming systems, could determine a less efficient allocation of the resource, affecting hydrological cycles on a local scale, interfering and changing the water returns to farms, surface water bodies and groundwater. In this sense, a different irrigation method may result in a delay in the loading of the water table and a lowering in the water table itself can occur (in particular for dry cultivation). Similarly, a higher technical and infrastructural efficiency able to reduce distribution losses can have implications in the recharging and supplying of water sources, eliminating the potential benefits of reallocation, even if in many cases a large part of the water flow available to farms comes from internal recirculation, as a means to contrast the reduction of the water demand. A dry cultivation could finally affect the created paranatural aquatic environments. In fact the environmental role of rice fields and their irrigation systems must be considered, not just in the study area but throughout the rice-growing area of Lombardy and Piedmont. The circulation of very high volumes of water has significant effects on habitats constituted over time, becoming important ecosystems, even recognized at the Community level (SPAs Rice fields of Lomellina). For this reason, water-saving should be carefully evaluated according to the environmental functions that traditional irrigation systems perform in large parts of the territory.

These important considerations must be properly considered in order to make a complete economic evaluation of water resources. In this sense, it is then important to identify the best method for the estimation of environmental costs, since this step plays a key role as a starting point toward the quantification of the *full cost*, which represents itself as a crucial instrument in order to strengthen decisional support to policy makers.

Acknowledgments Results presented in this chapter are part of the activities carried out for the research funded by Regione Lombardia, fondo per la promozione di Accordi Istituzionali, project BIOGESTECA 15083/RCC.

References

Arfini, F., & Paris, Q. (1995). *A positive mathematical programming model for regional analysis of agricultural policies. EAAE Proceedings, 40th Seminar of the of the EAAE, Italy, 1995* (pp. 17–35).

Bartolini, F., Bazzani, G. M., Gallerani, V., Raggi, M., & Viaggi, D. (2007). The impact of water and agricultural policy scenarios on irrigated farming systems in Italy: An analysis based on farm multi-attribute linear programming models. *Agricultural Systems, 93*, 90–114.

Bazzani, G. M., & Scardigno, A. (2008). Un modello di simulazione territoriale per l'analisi economica dell'uso dell'acqua e della riforma della PAC: una proposta metodologica e prime applicazioni all'agricoltura pugliese. Paper presentated at the XLV Convegno di studi SIDEA, 25-27 September 2008, Portici, Italy. http://www.depa.unina.it/sidea2008/ris_idriche/Bazzani_et_al.pdf

Bazzani, G. M., & Zucaro, R. (2008). Scarsità idrica e Direttiva Acque. Politiche e metodologie di analisi: un caso di studio in Italia centrale. In L. Casini, V. Gallerani, D. Viaggi (Eds.), *Acqua, agricoltura e ambiente nei nuovi scenari di politica comunitaria* (pp. 67–84). Milano, Italy: Franco Angeli Editori. ISBN 978-88-568-0488-1.

Bazzani, G. M., Di Pasquale, S., Gallerani, V., Morganti, S., Raggi, M., & Viaggi, D. (2005). The sustainability of irrigated agricultural systems under the water framework directive: First results. *Environmental Modelling Software, 20*, 165–175.

Blanco, M., Iglesias, E., & Sumpsi, J. M. (2004). *Environmental and socioeconomic effect of water pricing policies: key issues in the implementation of the Water framework Directive. 13th Annual EAERE Conference, Budapest*, Hungary.

Brooke, A., Kendrick, D., & Meeraus, A. (1988). *GAMS: A user's guide*. Redwood City, CA, USA: The Scientific Press.

Cadario, D., & Bischetti, G. B. (2006). Caratteri e funzioni del reticolo idrografico della pianura lombarda. *Valutazione Ambientale*. ISSN 1826-2201.- 10, *July 2006*, 58–61.

Camera di Commercio di Pavia. (2011). Rilevazioni sala contrattazione merci di Mortasa—anno 2011. Retrieved December 15, 2012, from http://www.pv.camcom.it/index.phtml?Id_VMenu=561&daabstract=2769

Castellani, L., Mancuso, T., & Massaglia, S. (2008). Aspetti normativi ed economici in materia di gestione e tutela delle risorse idriche ad uso irriguo in Piemonte. In L. Casini, V. Gallerani, D. Viaggi (Eds.), *Acqua, agricoltura e ambiente nei nuovi scenari di politica comunitaria* (pp. 31–48). Milano, Italy: Franco Angeli Editori.

Cortignani, R., & Severini, S. (2008). *Introducing deficit irrigation crop techniques derived by crop growth into a positive mathematical programming model*. Paper presented at the XII Congress EAAE, Ghent, Belgium, 26–29 August 2008. http://ageconsearch.umn.edu/bitstream/44010/2/380.pdf

Cortignani, R., & Severini, S. (2009). L'impatto delle riforme OCM zucchero ed ortofrutta sull'uso dell'irrigazione: un'analisi condotta mediante un modello di PMP senza prima fase. Communication, XLVI Convegno di studi SIDEA, 16–19 September 2009, Piacenza, Italy. http://www.sidea.org/Archivio_Convegni.html

Dono, G. (2003). Costi della distribuzione idrica per l'irrigazione nell'Italia meridionale e problemi della formazione dei prezzi dell'acqua per l'agricoltura. *Rivista di Economia Agraria, 58*, (1).

Dono G., & Giraldo, L. (2010). Un'analisi dei costi per la distribuzione dell'acqua in agricoltura. *Rivista di Economia Agraria, 64*, (3–4).

Dono G., Giraldo, L., & Severini, S. (2011). I costi della distribuzione irrigua: un tentativo di riconciliare i concetti di copertura dei costi. *Economia e Diritto Agroalimentare, XVI* 1/2011.

Dono G., Marongiu, S., Severini, S., Sistu, G., & Strazzera, E. (2008). *Studio sulla gestione sostenibile delle risorse idriche: analisi dei modelli di consumo per usi irrigui e civili*. ENEA editore. Collana Desertificazione—Progetto RIADE.

Dono, G., Marongiu, S., & Severini, S. (2006). Gli effetti della Riforma della PAC sull'uso delle risorse idriche di falda e di superficie. Un modello di analisi territoriale. In L. Casini, V. Gallerani, D. Viaggi (Eds.), *Acqua, agricoltura e ambiente nei nuovi scenari di politica comunitaria* (pp. 85–104). Milano, Italy: Franco Angeli Editori.

European Environmental Bureau (EEB). 2001. Proposals to strengthen the Environmental Action programme. http://www.eeb.publication/6th%20EAP%20final.pdf.

European Parliament and the Council of EU. 2000. DIRECTIVE 2000/60/EC establishing a framework for Community action in the field of water policy. OJEC L 327, 22/12/2000. http://eur-lex.europa.eu/LexUriServ/LexUriServ.do?uri=OJ:L:2000:327:0001:0072:IT:PDF

Fischler, G., Tubiello, F. N., van Velthuizen, H., & Wiberg, D. A. (2007). Climate change impacts on irrigation water requirements: Effects of mitigation, 1990–2080. *Technological Forecasting and Social Change, 74*, 1083–1107.

Giannoccaro, G., Prosperi, M., & Zanni, G. (2007). *Analisi dell'efficienza tecnica ed ecologica dei metodi di tariffazione diretti ed indiretti dell'acqua irrigua*. Paper presented at XLIV congress SIDEA, Taormina, Italy. http://www.sidea.org/Archivio_Convegni.html

Giannoccaro, G., Prosperi, M., & Zanni, G. (2008). DEA Application to Evaluate the Technical and Ecological Efficiency of Water Pricing Policies. In: L. Bartova, R. M'Barek, T. Ratinger (eds.) *Modelling agricultural and rural development policies. Proceedings of the 107th*

Seminar of the EAAE, 29 January–1 February 2008, Sevilla, Spain. European Communities. ISBN 978-92-79-08068-5.

Giraldo, L. (2010). I costi operativi per la distribuzione dell'acqua a scopo irriguo. Un'analisi empirica con una forma funzionale flessibile. Ph. D. Thesis, DEAR, Università della Tuscia. http://dspace.unitus.it/bitstream/2067/1005/1/lgiraldo_tesid.pdf

Howitt, R. E. (1995). Positive mathematical programming. *American Journal of Agricultural Economics, 77*, 329–342.

ICWE. (1992). *The Dublin statement and report of the conference*. International Conference on Water and the Environment: Development Issues for the 21st century. 26–31 January, Dublin, Ireland. Geneva: World Meteorological Organization. http://docs.watsan.net/Scanned_PDF_Files/Class_Code_7_Conference/71-ICWE92-9739.pdf

Igbadun, H., Mahoo, H., Tarimo, A., & Salim, B. (2006). Perfomance of two temp-based reference evapotranspiration models in the Mkoji subcatchment in Tanzania. *Agricultural Engineering International: CIGR Journal*. Manuscript LW 05 008. *VIII*.

Iglesias, A., Estrela, T., & Gallart, F. (2005). Impacts on hydric resources. A preliminary general assessment of the impacts in Spain due to the effects of climate change. Ministerio de Medio Ambiente, Spain. http://www.magrama.gob.es/es/cambio-climatico/temas/impactos-vulnerabilidad-y-adaptacion/full_report_tcm7-199440.pdf

IPPC. (2012). *Renewable energy sources and climate change mitigation—special Report of the Intergovernmental Panel on Climate Change* (p. 1075). United Kingdom and New York, NY, USA: Cambridge University Press. Export Citation (RIS format).

Kassam, A., & Smith, M. (2001). "FAO methodologies on crop water use and crop water productivity". Expert meeting on water productivity, FAO, Rome 3–5 December 2001.

Molden, D. (1997). Accounting for water use and productivity. International Irrigation Management Institute—System Wide Initiative for Water Management (SWIM) paper 1. http://www.iwmi.cgiar.org/Publications/SWIM_Papers/PDFs/SWIM01.PDF

Lazzari, M., & Mazzetto, F. (2005). *Prontuario di meccanica agraria e meccanizzazione*. Torino, Italy: Reda Edizioni.

Palanisami, K., & Suresh Kumar, D. (Eds.). (2006). *Challenges in impact assessment of watershed development: methodological issues and experiences*. New Delhi, India: Associated Publishing Company Ltd.

Paris, Q., & Howitt, R. E. (1998). An analysis of Ill-posed production problems using maximum entropy. *American Journal of Agricultural Economics, 80*(1), 124–138.

Rogers, P., de Silva, R., & Bathia, R. (2002). Water in an economic good: how to use prices to promote equity, efficiency and sustainability. *Water Policy, 4*, 1–17.

Rosenthal, R. E. (2007). *GAMS: a user's guide*. Washington, DC, USA: GAMS development corporation.

Seckler, D., Amarasinghe, U., Molden, D., de Silva, R, & Barker, R. (1998). World Water Demand and Supply, 1990 to 2025: Scenarios and Issues. International Water Management Institute Research Report 19, Sri Lanka. http://www.iwmi.cgiar.org/Publications/IWMI_Research_Reports/PDF/PUB019/REPORT19.PDF

Tarimo, A. K. P. R., Mdoe, N. S., & Lutatina, J. M. (1998). Irrigation water prices for farmer-managed irrigation systems in Tanzania: a case study of lower Moshi irrigation scheme. *Agricultural Water Management, 38*, 33–44.

Teixeira, A. H., de C.,. Bastiaanssen, W. G. M., Moura, M. S. B., Soares, J. M., Ahmad, M. D., & Bos, M. G. (2008). Energy and water balance measurements for water productivity analysis in irrigated mango trees, Northeast Brazil. *Agricultural Forest Meteorology. 148*, 1524–1537.

Tsur, Y., Roe, T., Doukkali, R., & Dinar, A. (2003). *Pricing irrigation water: Principles and cases from developing countries* (p. 319). Washington DC, USA: Resources for the Future.

United Nations Environment Program (UNEP). (1999). *Global environmental outlook 2000*. London, England: Earthscan Publications Ltd. http://www.grid.unep.ch/geo2000/index.html

Vazifedoust, M., van Dam, J. C., Feddes, R. A., & Feizi, M. (2008). Increasing water productivity of irrigated crops under limited water supply at field scale. *Agricultural Water Management, 95*, 89–102.

WATECO. (2003). *Common implementation strategy for the water framework directive* (2000/60/CE)—Guidance Document no. 1, Bruxelles, Begium: European Commission. http://www.enorasis.eu/uploads/files/Water%20Governance/9.Guidancedoc01_Economics_WATECO_WG2-6.pdf

Author Biographies

Guido Sali is a Professor of Agricultural and Environmental Economics at the University of Milan. He is a member of the Research Centre for Appraisal and Land Economics. His research focuses on the economics of agricultural and agro-food policies, rural development and multifunctional agriculture. He is the scientific coordinator of several projects on the use of natural resources in agriculture and on the relationship between urban and rural areas.

Federica Monaco is a Ph.D. student at the Department of Agricultural and Environmental Sciences of the University of Milan. Her research interests are economic evaluation of natural resources and European policies on rural development, sustainability and agro-food systems.

Chapter 3
Stochastic Partial Equilibrium Modelling: An Application to Crop Yield Variability

Siyi Feng, Julian Binfield, Myles Patton and John Davis

Abstract Market volatility is of increasing concern within the agricultural sector of the EU and consequently the FAPRI-UK and EU-GOLD partial equilibrium modelling system is being developed to produce stochastic projections, which reflect underlying uncertainties. This chapter describes the underlying methodology. In the initial simulations, only one source of uncertainty, namely crop yield variability within the EU, is examined. Positive correlations among crop yield deviates amplify variations in crop production. This is supported by our results of the experimental simulations in which correlations among crop yield deviates are taken into account and compared to results where they are not addressed. Variability in output value varies widely depending on the aggregation level under consideration. In particular, at levels below where prices are determined, output value variability is larger than either production variability or price variability alone, supporting the potential need for risk management policy.

3.1 Introduction

Risk is an inherent aspect in the agricultural sector and has important implications on both private decision making and public policy. There are mainly two types of risks: production risk and price risk. Production risk is a distinctive risk in the agricultural sector where nature is an important input. Price risk stems from the uncertainties in supply and demand. As supply is mostly determined by production, production risk and price risk are closely interrelated to each other.

S. Feng (✉) · M. Patton · J. Davis
Agri-Food & Biosciences Institute, Belfast, UK
e-mail: Siyi.Feng@afbini.gov.uk

J. Davis
Queen's University Belfast, Belfast, UK

J. Binfield
FAPRI, University of Missouri, Columbia, MO, USA

C. Zopounidis et al. (eds.), *Agricultural Cooperative Management and Policy*, Cooperative Management, DOI: 10.1007/978-3-319-06635-6_3,

Nevertheless, the links between the two may be weakened in the presence of demand management policy.

Within the agricultural sector of the European Union (EU), risk is of increasing concern in recent years. This is a result of the progressive reforms of the Common Agricultural Policy (CAP), coupled with substantial increases in world agriculture prices. The reforms and developments on the global market have allowed internal prices to vary over a wider range and strengthened the integration between the regional and world market. Since the MacSharry Reforms in 1992, the CAP has changed significantly with the focus switching from product support to producer support. Intervention prices were initially reduced under the MacSharry Reforms. This was accompanied with the provision of direct payments to compensate for the reduced support prices. These payments were, however, linked to production and thus, the markets continued to be distorted. The Agenda 2000 reforms deepened the reduction in intervention prices. The replacement of the coupled direct payments with the decoupled Single Farm Payment (SFP) under the Fischler CAP reforms (agreed 2003, implemented 2005–2007) separated financial support from the level of production. The reforms have resulted in a less administered market and consequently variability has become more prominent. At the same time, because of the reductions in the intervention prices, the EU market is less isolated from the world market than it used to be and therefore more susceptible to the volatilities on the world market. The implications of these developments on policy analysis are at least twofold: first, although individuals in the sector will undoubtedly need to adapt to the changes and be more proactive in managing risks, there may still be some risks that they are not able to cope with and thus new policies may be needed to provide assistance; secondly, traditional policies need to be evaluated in the light of the new environment.

The FAPRI-UK model, which is integrated with the EU-GOLD model, is a deterministic partial equilibrium model on the EU agricultural sector for ex-ante policy analysis. However, within a volatile market environment, a deterministic model is likely to miss important implications of potential policy changes, especially when the policy actions are dependent on market outcomes. In view of this, a stochastic modelling framework is being developed. Stochastic modelling provides a means to capture some of the inherent uncertainty associated with agricultural production systems. By varying assumptions about certain exogenous variables, stochastic models can be used to examine the different ways markets may behave. In the past, volatile world prices have been incorporated into the FAPRI-UK model (Moss et al. 2010). This chapter represents modelling advancements of introducing internal uncertainties to the modelling system, with an initial focus on crop yields. Variability in crop yield affects producers' welfare through multiple ways. On the one hand, the market mechanism to some extent provides a 'natural hedge' to producers' income as yields and prices generally move in opposite directions; on the other hand, uncertainties in crop yields tend to be positively correlated and the positive correlations essentially amplify rather than mitigate uncertainties. The positive correlations hold particularly for different crops in the same region and for crops in neighbouring regions that experience similar weather conditions. This arises because if there is good (bad) weather in a particular year, it will be favourable (unfavourable) to all the affected crops. In other words, a below average yield for one crop is likely to be accompanied by a below average yield of another crop in the same year and vice versa. The positive correlations lead to large variability in production at the farm level and regional level, which may cause adverse consequences to producers. Crop yield is found to be the most important source of price volatility in a recent study by OECD using the AGLINK-COSIMO model (Taya 2010). Nevertheless, in this study the EU-27 is divided into only two blocks, namely EU-15 and EU-12, and it is therefore not feasible to assess the impacts of uncertainties at the individual country level. This chapter describes a methodology for introducing stochastic modelling to the FAPRI-UK and EU-GOLD modelling system using crop yields as an example and further examines the contribution of positive correlations among crop yields to volatility of the sectoral outcomes.

3.2 Methodology

This section describes the methodology.

3.2.1 The Stochastic Modelling Framework

The stochastic modelling framework of the FAPRI-UK and EU-GOLD model is shown in Fig. 3.1. Generation of the stochastic crop yields is explained in the following sections.

In terms of geographical scope, the FAPRI-UK model covers the main agricultural sectors in England, Wales, Scotland and Northern Ireland separately. The EU-GOLD includes six individual member states (i.e. France, Germany, Italy, Ireland, Poland and Hungary) and three member state groups: the Balkan countries (i.e. Romania and Bulgaria), the rest of the EU-15 and the rest of the EU-10 (i.e. EU-12 minus the Balkan countries).

Crops considered include wheat, barley, maize, rye, rapeseed, sunflower seed and soybean. Restricted by climatic conditions, some of these crops are planted only in a subset of member states. Together with the grouping of the smaller member states, this results in 63 individual crops being modelled.

3.2.2 Generation of Stochastic Inputs

In the FAPRI-UK and EU-GOLD modelling system, crop yields are modelled as follows:

$$y_{\text{it}} = a_i + b_{i1}\text{Trend}_t + b_{i2}\text{HA}_{\text{it}} + b_{i3}f(\bar{p}_t) + e_{\text{it}} \tag{3.1}$$

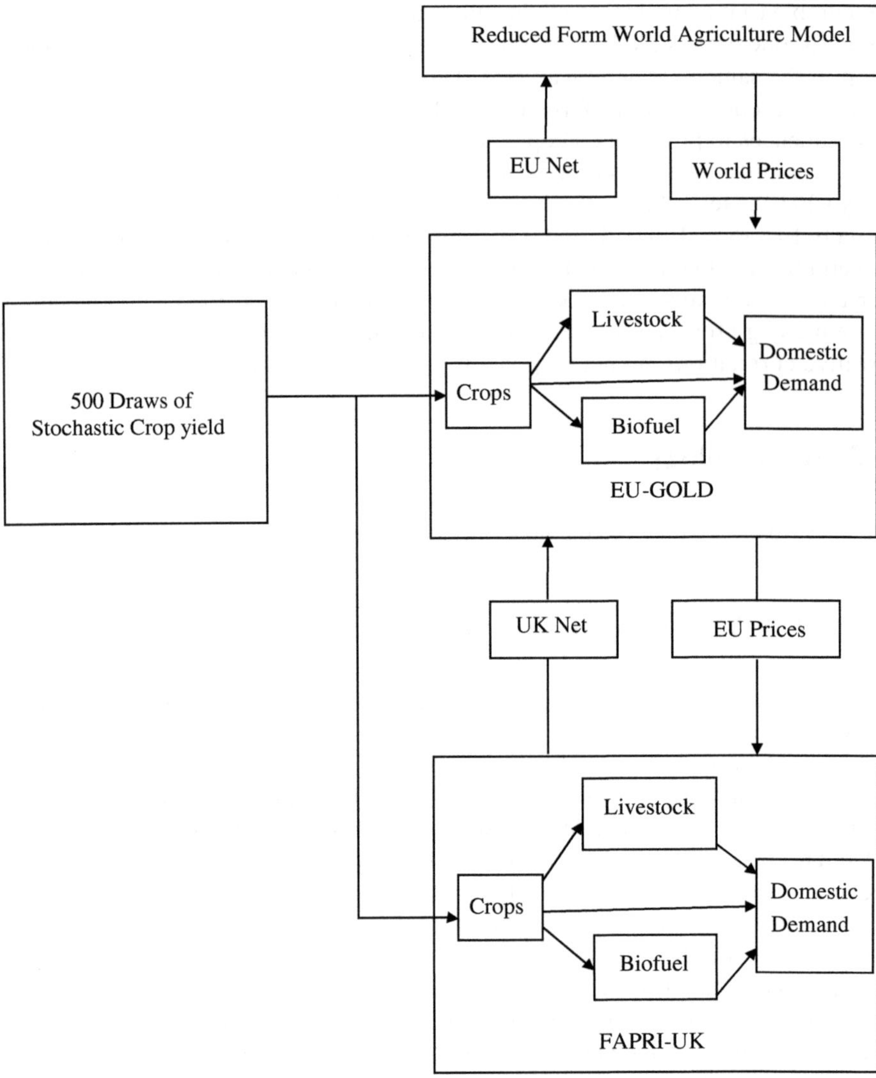

Fig. 3.1 Stochastic modelling framework

where i denotes a specific crop in a particular country, t is the year, a_i, b_{i1}, b_{i2} and b_{i3} are the equation coefficients, y_{it} is the realised yield in year t, $Trend_t$ is the trend term, HA_{it} is area of the crop in year t and $\bar{p}_t$ is the vector of crop prices and input prices that affect the yield of crop i. Deviates e_{it} represent the variations not accounted for by either the trend or the price effects each year, a large part of which is a result of the variation in weather conditions. In the deterministic model, normal weather conditions are assumed and the deviates are held constant. In the stochastic model, the distributions of the deviates are estimated based on historical data, from which stochastic inputs are subsequently drawn. Historical data for all the crops in the EU-15 countries include 22 observations from 1990 to 2011, while that for some of the crops in the new member states include only 12 observations from 2000 to 2011. Summary statistics of the deviates are given in Table 3.1. Since the statistics are based on the deviates rather than the original data, it is somewhat difficult to compare the variabilities of the crop yields from the table. For example, although the table indicates that standard deviates of oil crops are smaller than the grains, the means of oil crop yields are smaller than grains. As a result, oil crops exhibit more yield variability. The skewness statistics are also of specific interest. It is suggested in the literature, most of which are American studies and use maize data, that crop yields are more likely to be negatively skewed (Sherrick et al. 2004). The explanation is that there is a limit on how much good weather can bring some improvement to the yields while bad weather can potentially cause substantial loss. This negative skewness holds for the majority of crops (Table 3.1). In particular, the crop yield deviates for maize are negatively skewed in each of the countries modelled. Negative skewness, however, does not necessarily apply to all the crops. The skewness statistics have important implications for the distribution estimation step of the stochastic modelling, which will be further explained in the next section.

Following the estimation of distributions, discussion will focus on correlation among the crop yields, another crucial factor underlying the stochastic modelling methodology.

3.2.2.1 Probability Distributions of the Random Inputs

To estimate the distributions of the deviates, two questions are involved: (1) the type of the distribution and (2) the values of their characterising parameters. With regards to the first question, economic literature carefully distinguishes between situations with a 'yes' answer and those with a 'no' answer. This can sometimes have important implications on the question under investigation. In our stochastic analysis, the probability distributions are never known and have to be assumed and estimated. Estimation can start either with or without a parametric distribution assumption. With a distribution assumption, sample data are used to estimate the values of the distribution parameters. Alternatively, a distribution can be estimated based on the sample data solely using nonparametric methods (the empirical distributions). In this process, there are two closely related issues that require careful treatment: (1) whether the extremes (the minimum and the maximum) represent the range of all the possible outcomes, that is, in the case of crop yields, the worst in the historic period that we could ever have; (2) whether the probability densities assigned to the extremes are their true probabilities. They are manifestations of the first question to some extent and cannot be answered in the estimation process. There are cases in which boundaries can be estimated; however, the estimations are not necessarily good ones. These issues mean that modeller judgement is required. In the analysis presented below, the range of possible

Table 3.1 Summary statistics of historical deviates of all the crops

	Min	Median	Max	Mean	St Dev	Skewness		Min	Median	Max	Mean	St Dev	Skewness
Grains							Oil crops						
Soft Wheat							*Rapeseed*						
EN	−0.76	0.00	0.78	0.00	0.41	−0.16	EN	−0.64	0.01	0.41	0.00	0.28	−0.45
WA	−0.88	0.08	0.96	0.00	0.50	−0.06	FR	−0.52	−0.02	0.48	0.00	0.29	−0.10
SC	−0.62	−0.09	0.83	0.00	0.43	0.35	DE	−0.90	0.10	0.64	0.00	0.41	−0.68
NI	−1.41	0.06	0.80	0.00	0.57	−0.86	OE	−0.53	−0.10	0.63	0.00	0.37	0.53
FR	−0.91	−0.07	0.81	0.00	0.47	0.05	PL	−0.60	0.05	0.56	0.00	0.33	−0.13
DE	−0.84	0.01	0.95	0.00	0.47	0.19	HU	−0.59	0.00	0.66	0.00	0.35	0.16
IT	−0.37	0.07	0.35	0.00	0.20	−0.42	OC	−0.84	−0.02	0.56	0.00	0.29	−0.61
OE	−0.47	−0.02	0.53	0.00	0.28	0.37	BK	−0.87	0.09	1.05	0.00	0.54	0.04
PL	−0.69	−0.05	0.61	0.00	0.31	0.04	*Sunflower*						
HU	−1.37	0.08	1.13	0.00	0.66	−0.16	FR	−0.17	−0.02	0.29	0.00	0.12	0.56
OC	−0.73	−0.13	0.83	0.00	0.39	0.41	DE	−0.48	−0.02	0.37	0.00	0.25	−0.29
BK	−1.27	0.02	1.15	0.00	0.59	−0.59	IT	−0.62	0.01	0.26	0.00	0.17	−2.34
Durum Wheat							OE	−0.37	0.01	0.44	0.00	0.20	0.11
FR	−1.44	0.10	0.57	0.00	0.51	−1.19	HU	−0.73	0.03	0.48	0.00	0.29	−0.77
DE	−0.63	0.02	0.78	0.00	0.40	0.11	OC	−0.42	−0.03	0.44	0.00	0.23	0.37
IT	−0.56	−0.01	0.54	0.00	0.32	0.02	BK	−0.60	0.01	0.37	0.00	0.23	−1.28
OE	−0.61	0.01	0.93	0.00	0.37	0.30	*Soybeans*						
Barley							FR	−0.79	0.07	0.27	0.00	0.26	−1.72
EN	−0.66	0.10	0.64	0.00	0.30	−0.24	IT	−0.91	0.05	0.36	0.00	0.30	−1.32
WA	−0.46	0.00	0.70	0.00	0.29	0.45	OE	−0.48	−0.02	0.79	0.00	0.31	0.50
SC	−0.78	0.01	0.58	0.00	0.35	−0.51	HU	−0.61	0.04	0.58	0.00	0.32	−0.49
NI	−0.98	0.01	0.64	0.00	0.42	−0.88	OC	−0.55	0.02	0.85	0.00	0.32	0.50
FR	−0.80	0.07	0.58	0.00	0.44	−0.37	BK	−0.91	0.13	0.94	0.00	0.59	−0.08
DE	−0.73	0.06	0.71	0.00	0.41	−0.16							
IT	−0.37	0.03	0.30	0.00	0.18	−0.42							

(continued)

Table 3.1 (continued)

	Min	Median	Max	Mean	St Dev	Skewness	Min	Median	Max	Mean	St Dev	Skewness
OE	−0.98	0.00	0.92	0.00	0.49	−0.04						
PL	−0.57	0.05	0.62	0.00	0.32	−0.16						
HU	−1.10	0.06	1.05	0.00	0.58	−0.01						
OC	−0.51	−0.04	0.76	0.00	0.35	0.58						
BK	−1.00	0.00	0.88	0.00	0.53	−0.19						
Maize												
FR	−1.71	0.07	1.10	0.00	0.65	−0.97						
DE	−1.27	0.23	1.04	0.00	0.62	−0.37						
IT	−1.61	0.06	0.88	0.00	0.65	−0.75						
OE	−0.88	0.03	0.80	0.00	0.53	−0.15						
PL	−1.77	−0.03	1.15	0.00	0.67	−0.52						
HU	−2.30	0.25	2.22	0.00	1.24	−0.18						
OC	−1.35	0.04	1.19	0.00	0.86	−0.10						
BK	−1.82	−0.08	1.31	0.00	0.79	−0.51						
Rye												
DE	−1.44	0.05	1.29	0.00	0.70	−0.24						
OE	−0.47	−0.03	0.59	0.00	0.26	0.25						
PL	−0.46	0.01	0.40	0.00	0.22	−0.56						
HU	−0.73	0.03	0.55	0.00	0.33	−0.37						
OC	−0.54	−0.13	0.73	0.00	0.40	0.73						
BK	−0.57	−0.02	0.62	0.00	0.31	0.09						

Country code: *EN* England, *WA* Wales, *SC* Scotland, *NI* Northern Ireland, *FR* France, *DE* Germany, *IT* Italy, *OE* EU−15, excluding UK, FR, DE and IT, *PL* Poland, *HU* Hungary, OC −EU−12, excluding Poland, Hungary and Balkan countries, *BK* Balkan Countries, i.e. Romania and Bulgaria outcomes is defined by stretching the minimum and the maximum by 5 % whenever such a range needs to be imposed. Sensitivity analysis will be carried out when it appears that there is a need to widen the range further.

After estimation is completed, goodness-of-fit tests, namely the Kolmogorov–Smirnov test (KS test) and the Anderson–Darling test (AD test) are employed to test whether given sample data are drawn from the assumed distribution.

- Parametric distributions

For parametric distributions, there is no consensus in the literature as to which one is preferred. (Sherrick et al. 2004) examines the effect of crop yield distribution choices on crop insurance payout (Sherrick et al. 2004). It ranks the beta and Weibull distributions as the top two candidates as they are able to accommodate negative skewness, which is commonly found in crop yield distributions. As the majority of the crop yields in the EU are also found to be negatively skewed,[1] the estimation starts with the beta distribution. The probability density function of a beta distribution for a random variable x is shown in Eq. (3.2).

$$f(x) = \frac{\Gamma(\alpha+\beta)}{\Gamma(\alpha)\Gamma(\beta)} x^{\alpha-1}(1-x)^{\beta-1}, 0<x<1 \tag{3.2}$$

where α and β are the positive parameters to be estimated with the sample data. The maximum likelihood estimation (MLE) is used and therefore a range needs to be imposed so that all the absolute deviates can be converted into proportions to fall into the 0 and 1 support of the density function.[2] With the estimated values under the MLE method, the hypothesis that the observed crop yield deviates are drawn from a beta distribution cannot be rejected for all the crops apart from sunflower in Italy. Weibull distributions are also estimated using the historical deviate data. The probability density function of a Weibull distribution for a random variable x is shown in Eq. (3.3).

$$f(x) = \begin{cases} \frac{k}{\lambda}\left(\frac{x}{\lambda}\right)^{k-1} e^{-\left(\frac{x}{\lambda}\right)^k}, & x > 0 \\ 0, & x<0 \end{cases} \tag{3.3}$$

[1] See Table 3.1. We acknowledge that the distribution assumption is an area that requires further investigation.

[2] The method of moment, which estimates parameter values by equating population moments and sample moments, has also been applied to a subset of crops, i.e. crops in the four countries in the UK. The advantage of the method of moment is that it estimates not only the parameters α and β but the lower bound and the range as well. However, there is no guarantee that the resultant estimates will be valid (i.e. positive α and β). Unrealistic estimated values may also occur (e.g. a lower bound that will cause the crop yield to be negative). The goodness-of-fit results are not as good as those of the MLE. Therefore, this method is not widely applied and no stochastic draws are made based on estimations resulting from this method.

where k and λ are the positive parameters to be estimated. The MLE method is used with an imposed range, which defines 5 % below the minimum of each sample as its lower bound. But the goodness-of-fit test results indicate that the Weibull distribution is not a good candidate.

- Empirical (nonparametric) distributions

Without a parametric distribution assumption, an empirical distribution can be derived based on the sample data. To illustrate the idea, the basic method, which is adopted in Richardson (2000), is repeated here (Richardson et al. 2000).

Let D_{in} denote the deviates, where i represents the crops and n represents the annual observation. N denotes the total size of each sample. Define

$$D_{i,\text{low\_bound}} = 1.05 * \min_n D_{\text{in}} \tag{3.4}$$

$$D_{i,\text{upper\_bound}} = 1.05 * \max_n D_{\text{in}}. \tag{3.5}$$

Then sort D_{in} from the smallest to the $P\big(D_{i(N)}\big) = \big(\frac{1}{N}\big) + P\big(D_{i(N-1)}\big)$ largest and denote the sorted D_{in} as $D_{i(j)}$, where j represents it as the jth smallest observation of $\{D_{\text{in}}\}$. Assign the probabilities to $D_{i(j)}$ as follows:

$$P\big(D_{i,\text{low\_bound}}\big) = 0 \tag{3.6}$$

$$P\big(D_{i(1)}\big) = \left(\frac{1}{N}\right) * 0.5 \tag{3.7}$$

$$P\big(D_{i(2)}\big) = \left(\frac{1}{N}\right) + P\big(D_{i(1)}\big) \tag{3.8}$$

$$P\big(D_{i(N)}\big) = \left(\frac{1}{N}\right) + P\big(D_{i(N-1)}\big) \tag{3.9}$$

$$P\big(D_{i,\text{upper\_bound}}\big) = 1 \tag{3.10}$$

The resultant distribution is referred to as a simple empirical distribution in this chapter. The cumulative density function of the simple empirical distribution of the wheat yield deviates in England is shown on the left panel of Fig. 3.2. It is essentially a collection of linear functions, resulting in a non-smooth appearance. Non-smoothness can be avoided using more advanced methods, for example the kernel method. The kernel method assigns probabilities to all possible outcomes within the range based on a weighted average of observations in their neighbourhoods (the so-called bandwidth). Kernel, which gives the name of the method, is essentially the function determining the weights. More details of the method are provided in Li and Racine (2006). The cumulative density functions of the kernel empirical distributions for wheat deviates in England are plotted against the beta

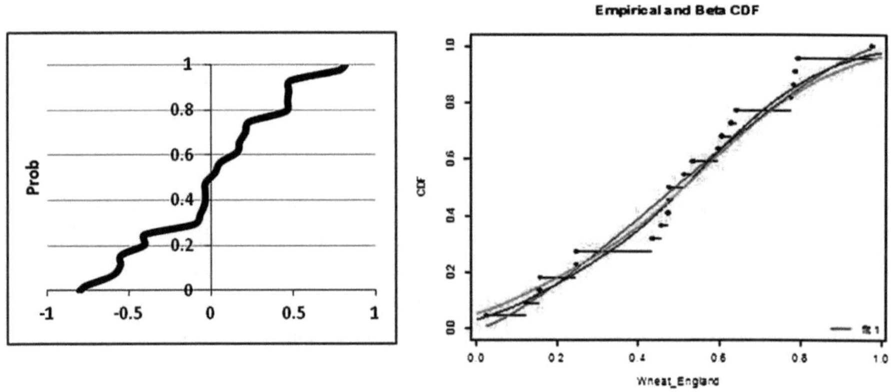

Fig. 3.2 Estimated cumulative density functions of wheat deviates in England

MLE distributions on the right panel of Fig. 3.2. The beta MLE distribution and the kernel empirical distribution are close for this crop, but this is not always the case.

- Distribution Choices

After estimation utilising different assumptions and methods, a choice on what to use for further simulations must be made. This is not an easy task, particularly given the limited size of the sample. Take wheat in England as an example. Its largest deviate is 0.776 in 1996 in absolute terms and the second largest is 0.483. With a sample size of 22, this implies the probability that a deviate less than 0.483 is 0.932 when the simple method is used; however, in the kernel method, smoothing reduces this probability to around 0.85, shifting more densities to deviates greater than 0.483 (on the right panel of Fig. 3.2, the black dots are above the cumulative density function lines). This difference is essentially caused by the limited number of observations. Provided that there are enough data, the smoothing in the kernel method is local and the resultant estimation should not deviate far away from the simple one.

It should be noted that there is no champion for all in distribution choice as it depends on not only the data fit, but also the question under investigation and other factors as well. For example, in the case of insurance analysis, the right segment of the cumulative distribution function is less important as no payment is needed for a positive shock whatever its size is. In contrast, in the analysis of biofuel, which diverts crop production, the right segment of the cumulative distribution function is crucial in determining the likelihood of tight supply demand balances.

Distribution choice is another area for further investigation in our study. Simulations within this chapter are based on the estimated beta throughout.

3.2.2.2 Correlation Matrix

Correlation is concerned with whether a shock to one random variable implies a shock to another random variable and measures the direction and magnitude of this relationship. Positive (negative) correlation between a pair of random variables means the realisation of shocks of the two are in the same (opposite) direction. The correlation matrix of sample deviates of some selected crops is shown in Table 3.2.[3] Correlations between soft wheat and barley in the same country are very strong. Correlations among neighbouring countries of the same crop also tend be strong. For example, correlations between wheat in France and that in England, Germany and Italy are all greater than 0.5. In contrast, correlations between wheat in England and that in Germany and Italy are smaller, and there is hardly any correlation between England and Poland. Negative correlations of sample data are also possible and probably reflect weather patterns across the EU. However, within our data set, their values (in absolute terms) are generally smaller and negativity occurs only between more distant countries. In all, the positive correlations dominate.

3.2.2.3 Generation of Stochastic Inputs

The procedure to generate stochastic inputs to the modelling systems is based on (Richardson et al. 2000) and is described below.

First, generate a matrix of independent standard normal random variables, R, of dimension I by T, where I is the total number of stochastic inputs and T is the total of projection periods. Secondly, if correlations among the stochastic variables are taken into account, then apply Cholesky factorisation to the correlation matrix denoted by C. That is to find a lower triangle matrix L such that $LL' = C$, where $c_{ij} = \text{correlation}(e_i, e_j)$ for every element c_{ij} in C. Left multiply R by L and denote the resultant matrix as B, i.e. $B = LR$. Then the correlation between row i and row j of elements in B will be approximately equal to c_{ij} for all i and j. Next, retrieve the cumulative probability for every b_{it}, i.e. obtain P_{it} such that $P_{it} = P(x \leq b_{it})$. Finally, calculate the deviates d_{it} from the estimated distributions whose cumulative probability equals P_{it}. The matrix D, consisting of $\{d_{it}\}$, is the stochastic input to be fed into the modelling system. To obtain 500 draws of the stochastic inputs, the above procedure is repeated 500 times.

3.3 Simulation Results

This section presents the simulation results.

---

[3] There is insufficient room for the full matrix. This is available from the author upon request.

Table 3.2 Selected correlation of crop yield deviates in the EU

	WS_EN	BA_EN	WS_SC	WS_FR	BA_FR	WS_DE	BA_DE	WD_IT	WS_PL	BA_PL	WS_HU	WS_BK	WS_IT
WS_EN	1.00												
BA_EN	**0.76**	1.00											
WS_SC	*0.24*	*0.34*	1.00										
WS_FR	**0.56**	*0.35*	0.07	1.00									
BA_FR	**0.50**	**0.42**	0.16	**0.90**	1.00								
WS_DE	*0.35*	*0.23*	0.05	**0.63**	**0.59**	1.00							
BA_DE	0.15	0.12	0.13	**0.60**	**0.65**	**0.83**	1.00						
WD_IT	*0.28*	*0.23*	0.07	*0.31*	*0.29*	0.13	0.05	1.00					
WS_PL	0.02	−0.22	0.02	**0.45**	*0.40*	*0.34*	*0.42*	*0.21*	1.00				
BA_PL	−0.19	−0.33	−0.20	*0.26*	*0.27*	*0.31*	**0.42**	−0.06	**0.84**	1.00			
WS_HU	−0.13	−0.12	0.17	*0.27*	*0.29*	**0.48**	**0.53**	*0.29*	**0.52**	**0.45**	1.00		
WS_BK	−0.13	−0.30	0.13	*0.20*	0.17	*0.29*	**0.44**	*0.24*	**0.50**	*0.36*	**0.76**	1.00	
WS_IT	**0.45**	*0.28*	0.05	**0.61**	**0.52**	0.15	0.06	**0.48**	−0.07	−0.33	−0.09	−0.05	1.00

Note (1) WS and BA stand for soft wheat and barley respectively; (2) EN, SC, FR, DE, IT, PL, HU and BK stand for England, Scotland, France, Germany, Italy, Poland, Hungary and Balkan countries; (3) correlations no less than 0.4 are bolded and those between 0.2 and 0.4 are *single underlined*

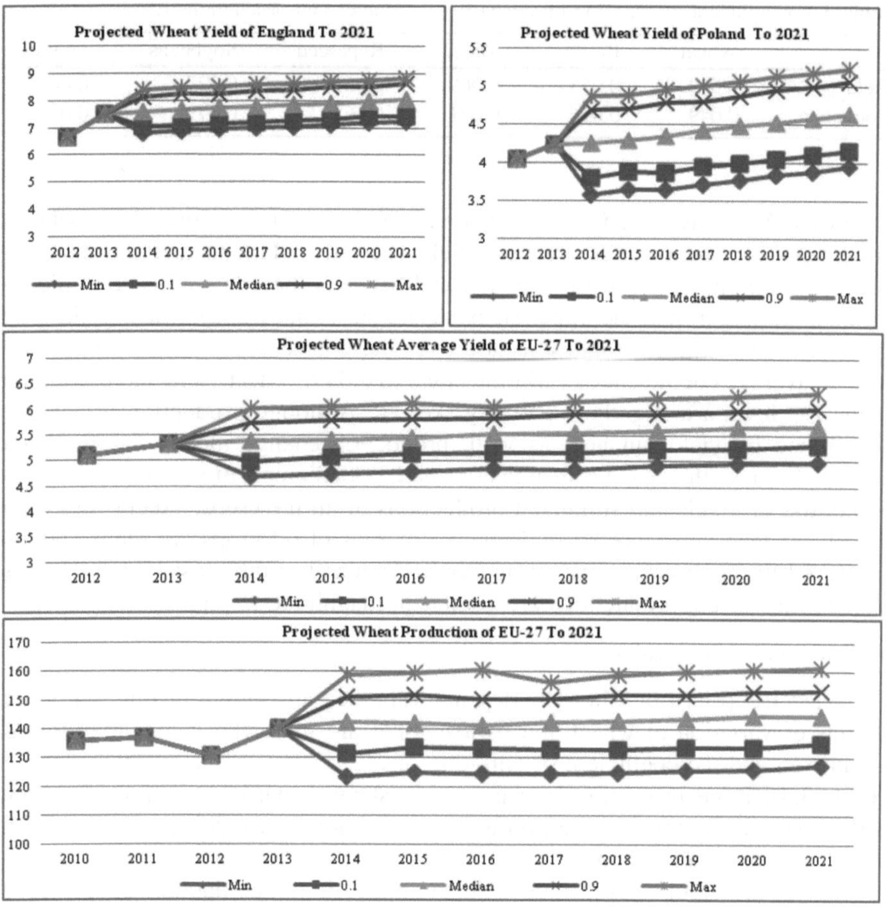

Fig. 3.3 Projected stochastic wheat yield (tonne/hectare) and production (million tonnes) to 2021

3.3.1 Stochastic projection of baseline 2012–2021

- Yield and production

Examples of projected stochastic wheat yields (England, Poland and the EU-27) are shown in Fig. 3.3. Historical wheat yields in England have never been below 7 tonnes/hectare since 1993 except for 2012. In the stochastic projection, wheat yields in England vary between just below 7 to above 8.4 tonnes/hectare. In 2014, the proportion of simulations with a yield level below 7 tonnes/hectare is about 10 % and falls to zero at the end of the projection period. Nevertheless, wheat yields in England are higher and exhibit lower variability than those in Poland (median of 7.6–8 compared to 4.2–4.6 tonnes/hectare; coefficient of variation

Table 3.3 Coefficient of variation of stochastic crop yields and productions EU-27

	Wheat	Barley	Maize	Rapeseed	Soybeans	Sunflower
Yield	0.049	0.055	0.069	0.067	0.077	0.067
Production	0.049	0.056	0.071	0.067	0.079	0.069

across the 500 draws[4] of 0.052 compared to 0.074). Furthermore, the stochastic projections of wheat yields are negatively skewed in Poland and the distances between the minimum and the median, and between the 10 % percentile and the median, are larger than the upper sides. Averages of the EU-27 wheat yield grow from below to over 5.5 tonnes/hectare over the course of the projection period, with an average coefficient of variation equal to 0.049. Wheat yield and production at the EU level are slightly negatively skewed. Although production is not only determined by yield but area as well, production change mostly follows yield changes because the changes in area are small.[5] Coefficients of variation across the 500 draws of yields and simulated productions of all the crops, except for rye, at the EU-27 level are reported in Table 3.3. Yields of oil crops are more variable compared to the grains crops; among grains, maize has the most variable yields. Variabilities of the productions are similar in magnitude to yields.

- Price and output value

Projected EU prices of wheat, barley and rapeseed are shown in Fig. 3.4. Coefficients of variation of wheat and barley prices are somewhat larger than that of production and the coefficient of variation of rapeseed is only marginally smaller (Table 3.4). Price projections are positively skewed, particularly for wheat and barley. This is contrary to the production projections since price and production move in opposite directions.

In terms of output value, which is the multiplication of production and price, the coefficients of variation at the EU-27 level are small, indicating the effects of 'natural hedge', i.e. the offsetting impact of price. However, this mechanism is the most effective at the EU-27 level, the level at which commodity prices are determined. At more disaggregated levels, e.g. at the individual country level, the coefficient of variation of output value is bigger. The coefficient of variation of output value for wheat in England is the same as that for the wheat price; for barley and rapeseed, the coefficients of variation of their output values are even greater than their prices. This arises due to the mismatch of timing of price and yield at the local level. For an individual region, it is possible that local adverse events result in loss in production without the loss being compensated by higher prices, which

[4] Simulating the models stochastically generates one coefficient of variation each projection year. Values reported here are the averages over 2014–2021.

[5] Area is a function of price and thus the equilibrium areas are different among the draws in the simulation. However, the differences are limited due to the small elasticities of acreage to price change.

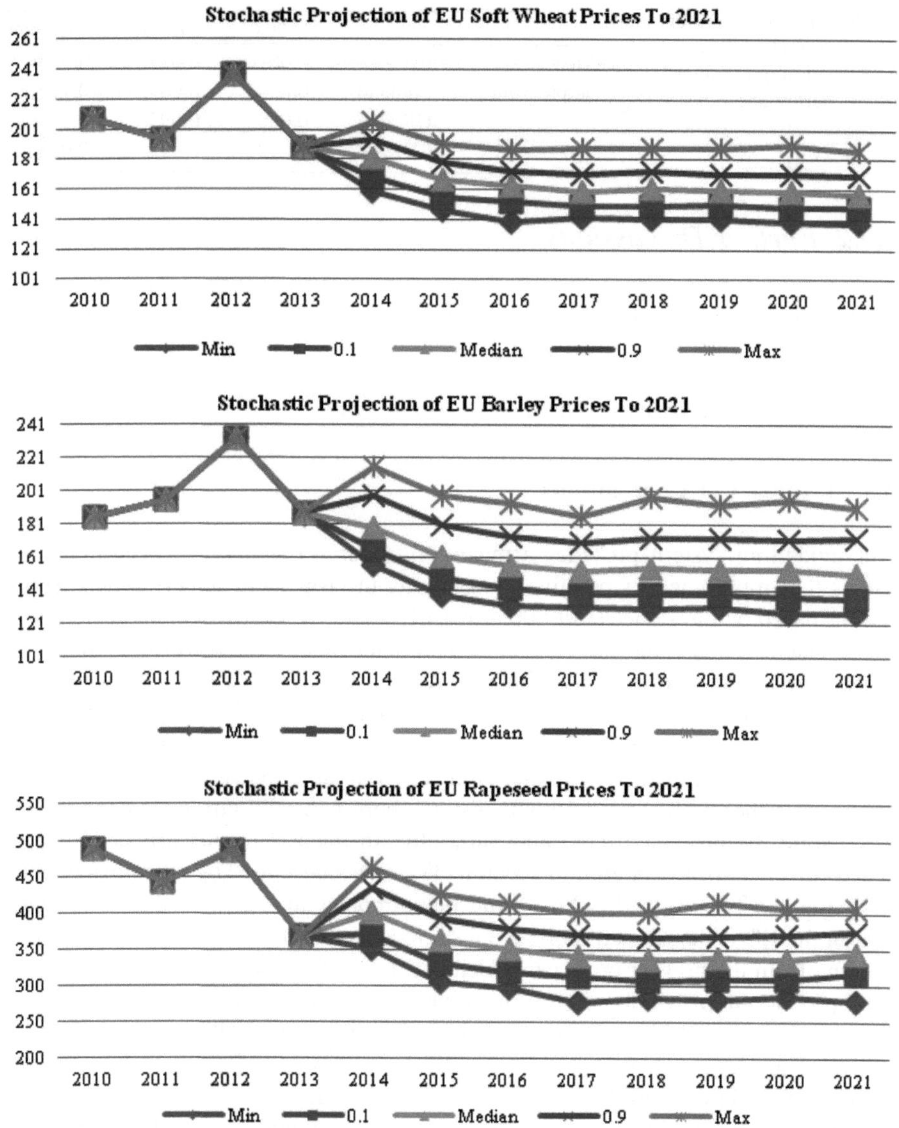

Fig. 3.4 Projected stochastic price of wheat, barley and rapeseed (Euro/tonne) to 2021

are determined at the EU-27 level; and vice versa. Variability of output value is substantial even when there is only one uncertainty source, as revealed here. It could potentially become larger when more uncertainty sources are introduced. The results in Table 3.4 highlight the importance of addressing the issue of risk at a disaggregated level.

Table 3.4 Coefficient of variation of stochastic crop prices and output values: wheat, barley and rapeseed

		Wheat	Barley	Rapeseed
Price	EU-27	0.051	0.079	0.065
Output value	England	0.060	0.084	0.112
	UK	0.059	0.095	0.108
	EU-27	0.012	0.034	0.018

3.3.2 *Further Discussions*

This section is divided into two parts. First, how the correlations, most of which are positive, contribute to the crop price/production variabilities are highlighted. Next, the simulated crop yield/price variabilities over time are presented and compared to those based on the data of recent years.

3.3.2.1 Impact of Correlations

Our modelling system provides a means to undertake a 'natural experiment' to assess the importance of positive correlations in crop yield uncertainties. Simulations based on the same set of 500 draws of stochastic yields without correlations being taken into account are run for comparison. Figure 3.5 shows the simulated wheat production projections of the UK and the EU-27, both of which are aggregated values. That is, the UK is the aggregate of England, Wales, Scotland and Northern Ireland and the EU-27 is the aggregate of the UK, France, Germany, Italy, Ireland, other EU-15, Poland, Hungary, Balkan countries and other EU-12. Comparing the 'with' and 'without' results, there is hardly any difference for the UK, while the projected band and variability in the EU results are much larger when correlations are taken into account. This is because the UK wheat production is dominated by England, which contributes more than 92 % of the total, while that of the EU is distributed more evenly across its member states with about 27 % coming from the biggest producer which is France. The coefficients of variation of total production in which correlation is taken into account are consistently higher for all the crops (Table 3.5), among which the difference is the most marked for wheat. The large variability for the with correlation results arises because the positive correlations amplify rather than diversify the uncertainties.

The larger variabilities when correlations are accounted for are also mirrored in the price results (Fig. 3.6). Furthermore, price variabilities in wheat more than double (Table 3.6).

3.3.2.2 Variability in Crop Yield Versus Price Volatility Over Time

The 500 random draws result in 500 possible paths of evolvement of all the variables. When variability/volatility are calculated along the time path, the results roughly correspond to what are observed in reality, which can be regarded as a

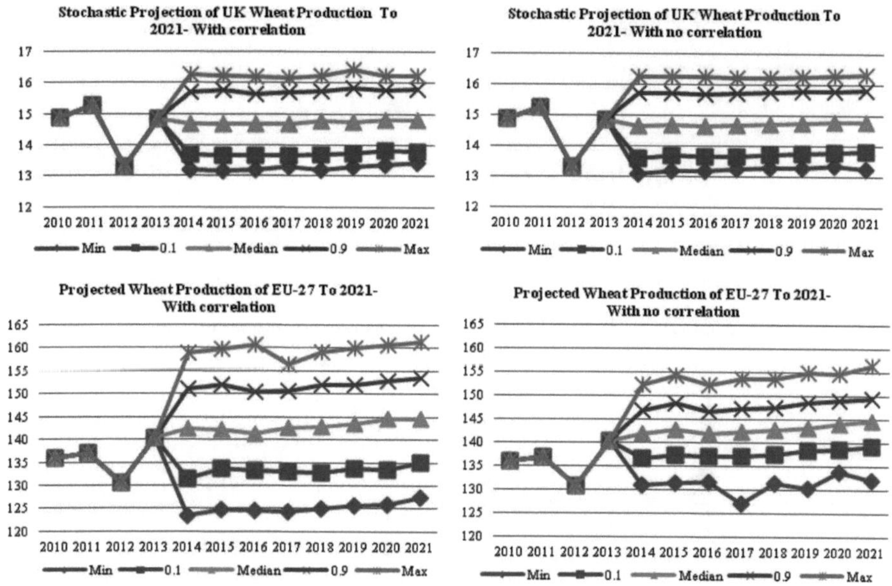

Fig. 3.5 Projection bands of UK and EU wheat production (million tonnes) to 2021: with versus without correlation

Table 3.5 Coefficient of variation of crop production EU-27 (average of 2014–2021): with versus without correlation

	Wheat	Barley	Maize	Rapeseed	Soybeans	Sunflower
With correlation	0.049	0.056	0.071	0.067	0.079	0.069
Without correlation	0.027	0.047	0.054	0.042	0.062	0.068

bench mark for evaluation of model simulations. It should be noted that the comparisons are at best rough for at least two reasons: (1) models cannot and should not be expected to replicate reality. When stochastic modelling is concerned, it is developed to understand the impacts of the key uncertainties and therefore not all sources of uncertainties will be included. In terms of the analysis in this chapter, only one source of uncertainty, i.e. internal yield variability, is incorporated, and simulated price volatility will be only partial compared to the observed ones; (2) in the case of crop yield, correlations across time are not accounted for in our modelling.

The minimum, median and maximum of simulated yield variability and price volatility of selected crops and their observed counterparts are presented in Table 3.7.[6] In the table, the simulated values are variability/volatility of the last

[6] Yield variability and price volatility are defined as the standard deviation of annual changes, which is the same as the OECD study that examines the driving forces of price volatility (Taya 2010).

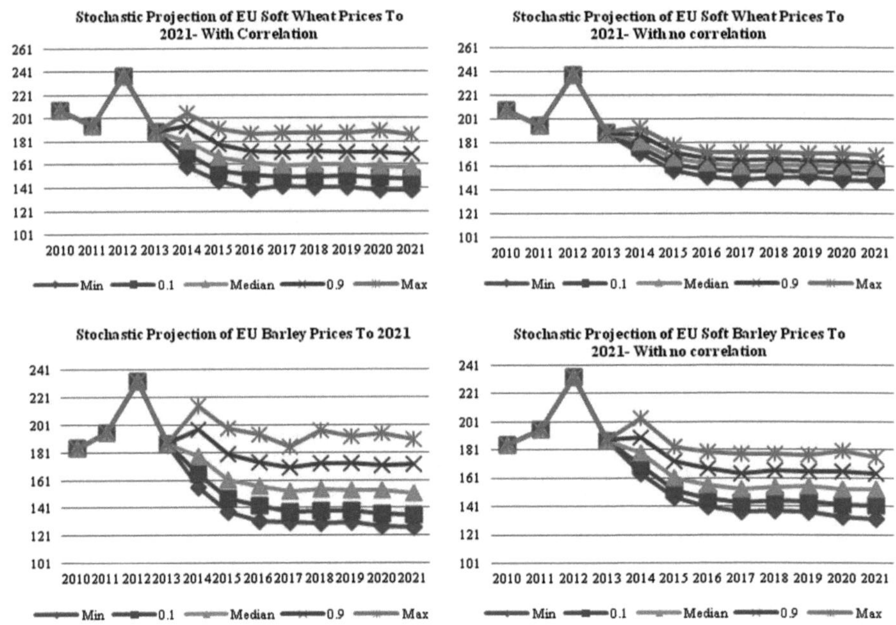

Fig. 3.6 Projection bands of EU wheat and barley prices (Euro/tonne) to 2021: with versus without correlation

Table 3.6 Coefficient of variation of crop price EU-27 (average of 2014–2021)

	Wheat	Barley	Maize	Rapeseed	Soybeans	Sunflower
With correlation	0.051	0.079	0.093	0.065	0.008	0.025
Without correlation	0.023	0.051	0.060	0.040	0.004	0.023

five projection years. For all the crops, the medians of the variabilities of the 500 simulated paths are close to the observed values that are calculated using the 1990–2011 data. Barley is the only crop that has a smaller variability for the period 2007–2011 than that of the full sample period.

Again, simulated price volatilities are slightly higher than the yield variabilities. Simulated wheat price volatility is about 31 % of the observed one for the period 1990–2011 and 17 % of the observed one for the volatile 2007–2011 period. For barley, these percentages reach 50 and 30 %. Those of rapeseed are slightly smaller than barley. The comparison suggests that uncertainties other than internal yield variability are more important in the wheat market than in the barley and rapeseed markets, which corresponds with the fact that wheat is more of an international commodity than the other two crops. Furthermore, in view of the fact that yield variability is the only uncertainty source in the study and that not all the

Table 3.7 Simulated yield variability and price volatility versus observed

		Wheat				Barley		Rapeseed
		EU-27	England	France	Poland	EU-27	England	EU-27
Yield variability	Observed between 1990 and 2011	0.034	0.032	0.039	0.051	0.033	0.036	0.040
	Observed between 2007 and 2011	0.038	0.041	0.038	0.039	0.020	0.025	0.039
	Simulated min	0.005	0.003	0.005	0.009	0.005	0.006	0.009
	Simulated median	0.030	0.029	0.039	0.046	0.033	0.035	0.042
	Simulated max	0.071	0.073	0.084	0.116	0.077	0.078	0.097
Price volatility	Observed between 1990 and 2011	0.102				0.093		0.106
	Observed between 2007 and 2011	0.189				0.183		0.187
	Simulated min	0.007				0.011		0.009
	Simulated median	0.032				0.054		0.048
	Simulated max	0.073				0.131		0.127

major producers are covered in the model,[7] these figures indicate that the impact of yield variability on price volatility is significant. In the future, additional sources of uncertainty will be taken into account in the analysis.

3.4 Conclusions

In the light of increased volatility on both internal and external agricultural commodity markets, the FAPRI-UK and EU-GOLD partial equilibrium modelling system is being developed to produce stochastic projections. By varying assumptions about certain exogenous variables, stochastic modelling provides a means to capture some of the uncertainty associated with agricultural production systems. The aim of the stochastic modelling is to assess the implications of these uncertainties on the sector and in the future examine the impact of policies that will be triggered under certain circumstances, including existing and potential risk management policies.

This chapter documents the first attempt to develop the stochastic extension to the FAPRI-UK and EU-GOLD modelling system. Stochastic modelling involves estimating the distributions of the deviates and calculating correlations of the historical deviates of the variables under investigation. Subsequently, stochastic inputs are generated and fed into the partial equilibrium model for simulation.

[7] Major producers outside the EU-27 are modelled in the "deterministic" way, in which normal weather and macro-economic conditions are assumed.

Only one source of uncertainty, i.e. crop yield variability, is examined in this study. Crop yield variability is an important source of uncertainty in the agricultural sector. The underlying reason is that positive correlations among crop yields in neighbouring production regions amplify rather than mitigate uncertainties. Our simulation experiment shows the level of variation in EU production and price is significantly greater, in some cases double, when correlation is taken into account compared to when it is not.

Variabilities in crop yield obtained from the simulations are similar to observed values, thereby validating the methodology for capturing uncertainties from this source. Area, production and price are simultaneously determined in the model. Since changes in area are limited, variability of production is determined predominantly by changes in yield. Variability of price is higher than production and yield. Simulated volatility of price evolution over time is smaller than observed, which is unsurprising as only one source of uncertainty has been incorporated within this analysis. However, the magnitude of the simulated price variability indicates that the impact of yield variability within the EU is substantial.

The policy relevance of the stochastic modelling approach adopted in this analysis is illustrated in the results of the output value variable (price multiplied by production). At the EU-27 level, variability of output value is smaller than either that of production or price due to the offsetting effect of price. However, variability of output value is significantly greater at the individual country level compared to the aggregate EU-27 level. This will probably become even larger once other uncertainty sources are incorporated within the modelling framework. Capturing the variability of output value at the individual country level will be vital once the modelling system is used to assess the implications of policies designed to mitigate price volatility, e.g. income stability mechanism.

The analysis described in this chapter represents the initial attempts of producing stochastic projections. It is important to recognise that there are issues that need to be furthered examined within the methodology, such as the estimation method for the distribution of the deviates. Also, other key sources of uncertainty will be incorporated in the future.

References

Li, Qi, & Racine, J. S. (2006). *Nonparametric econometrics: Theory and practice*. Princeton, New Jersey, USA: Princeton University Press.

Moss, J., Zhang, L., Patton, M., Kim, I. S., Binfield, J., & Westhoff, P. (2010). *A Stochastic Analysis of the Impact of Volatile World Agricultural Prices on European and UK Agriculture". FAPRI-UK Technical Report, Belfast: Agri-Food & Biosciences Institute, Queen's University Belfast*. Columbia, Missouri: University of Missouri.

Richardson, J. W., Klose, S. L., & Gray, A. W. (2000). An applied procedure for estimating and simulating multivariate empirical (MVE) probability distributions in farm-level risk assessment and policy analysis. *Journal of Agricultural and Applied Economics*, *32*(2), 299–315.

Sherrick, B. J., Zanini, F. C., Schnitkey, G. D., & Irwin, S. H. (2004). Crop insurance valuation under alternative yield distributions. *American Journal of Agricultural Economics, 86*(2), 406–419.

Taya, S. (2010). Stochastic Model Development and Price Volatility Analysis, OECD FOOD, Agriculture and Fisheries Working Papers, No. 57, OECD Publishing. http://dx.doi.org/10.1787/5k95tmlz3522-en

Part II
Estimating Income and Performance Levels

Chapter 4
Alternative Specifications of Reference Income Levels in the Income Stabilization Tool

Robert Finger and Nadja El Benni

Abstract In this chapter, different approaches for the specification of reference income levels in the income stabilization tool (IST) are analyzed. The current proposal of the European Commission suggests a 3-year average or a 5-year Olympic average to specify the farm-level reference income that is used to identify if and to what extent a farmer is indemnified in a specific year. Using Monte Carlo simulations, we investigate the impact of income trends on indemnification if these average-based methods are used in the IST. In addition, we propose and investigate a regression-based approach that considers observed income trends to specify reference income levels. Furthermore, we apply these three different approaches to farm-level panel data from Swiss agriculture for the period 2003–2009. We find that average-based approaches cause lower than expected indemnification levels for farmers with increasing incomes, and higher indemnifications if farm incomes are decreasing over time. Small income trends are sufficient to cause substantial biases between expected (fair) and realized indemnification payments at the farm level. In the presence of income trends, average-based specifications of reference income levels will thus cause two major problems for the IST. First, differences between expected and realized indemnification levels can lead to significant mismatches between expected and real costs of the IST. Second, indemnity levels that do not reflect farm-level income losses do not allow achieving the actual purpose of the

R. Finger (✉)
Wageningen University, Wageningen, The Netherlands
e-mail: r.finger@ilr.uni-bonn.de

R. Finger
Bonn University, Bonn, Germany

N. El Benni
University of Applied Sciences HTW Chur, Chur, Switzerland

N. El Benni
ETH Zurich, Zurich, Switzerland

C. Zopounidis et al. (eds.), *Agricultural Cooperative Management and Policy*, Cooperative Management, DOI: 10.1007/978-3-319-06635-6_4,

IST of securing farm incomes. Our analysis shows that a regression-based approach to specify reference income levels can contribute to bound potential biases in cases of decreasing or increasing income levels.

4.1 Introduction

Supporting farmers in managing their income risks has become one of the focal points on the agenda of agricultural policy makers. This is underlined, for instance, by the dissemination of all kinds of risk management tools and agricultural insurances in Europe and North America (e.g., Bielza Diaz-Caneja et al. 2008; Turvey 2012). Currently, the introduction of the Income Stabilization Tool (IST) is discussed in Europe (e.g., EC 2011; Meuwissen et al. 2011).[1] In evaluating the IST from a policy-maker perspective, particular emphasis is usually laid on actuarial evaluations, governmental costs, impacts optimal farm programs, as well as the identification of potential beneficiary groups among farmers (e.g., EC 2009; dell' Aquila and Cimino 2012; Pigeon et al. 2012; Liesivaara et al. 2012; Mary et al. 2013).

Past research on insurance schemes in agriculture in general has revealed that technical aspects of the implementation, such as the choice of trigger values for indemnification or the data used to calculate premiums (e.g. aggregated or farm-level data), can have significant impacts on the performance and effects of insurances (see, e.g., Atwood et al. 2003; Hennessy 2009; Just and Weninger 1999; Ker and Goodwin 2000; Ramirez et al. 2003). These aspects will also be of great importance for the outcome and applicability of the IST, but have not been addressed so far. In this study, we specifically focus on the specification of reference income levels. In the IST, the reference income level serves as an evaluation point (i.e., baseline) to identify if and how much a farmer is indemnified by the insurance in a specific year. An important aspect is that this reference income has to be specified in advance, i.e., it has to be estimated. Depending on data availability at the moment of specifying insurance contracts, a 1- or 2-year ahead estimation may be required.[2] The commonly applied strategies to calculate these baselines at the farm-level are based on averages of income observations from previous years. More specifically, an average of the last 3 years of available (e.g., income) observations or a 5-year Olympic average (removing the highest and lowest observation from the sample before calculating the average) are frequently used in agricultural insurance schemes (e.g. Barnett and Coble 2012; Cooper 2010; Glauber 2013). These are also the proposed measures to be used in the IST (EC 2011).

---

[1] See El Benni et al. (2013) for discussions on the Swiss case.

[2] For instance, the ACRE program in the US uses a 2-year ahead specification (Zulauf et al. 2008).

In general, averages provide highly efficient estimators for the calculation of expected values and by using the Olympic average, robustness against observations that deviate from the pattern described by the majority of observations (i.e., outliers) can also be achieved.[3] However, these estimators will cause biased estimates of reference income levels in the presence of trends as illustrated by the forecasts of income levels in Fig. 4.1. It shows that both the 3-year average and the 5-year Olympic average will under- or overestimate future incomes if there is a decreasing or increasing trend underlying the income observations, respectively. This has implications for potential indemnifications within the IST: If a farmer faces a deterministic increase of his income over time (e.g., due to rising output price levels), he is less likely to get indemnification under the currently proposed reference income specifications—even though he faces the same 'risk,' i.e., non-deterministic fluctuations of his income, as if no trend would be underlying his income. In contrast, farmers with decreasing income are more likely to be indemnified in the subsequent year. Along these lines, it is not only relevant whether there is a trend in farm incomes, but also its magnitude determines the effect on indemnity levels. This has two major potential implications for the IST: First, income trends may cause significant differences between expected and realized indemnification levels and thus may constitute a threat for the budget of such an instrument. Second, not considering trends in income levels may cause indemnity levels that do not reflect farm-level income losses, which should be the actual purpose of the IST.

Even though it has been recognized that underlying trends, for instance in income, will limit the applicability of insurance mechanisms (e.g., Meuwissen et al. 2003), neither a critical evaluation of the reference income specification measures has been conducted, nor have alternative measures been suggested for any insurance scheme that uses average-based indicators.

We contribute to filling this gap by analyzing and quantifying potential implications of misspecification of the reference income and by investigating a potential alternative reference income specification approach based on regression models. To this end, a two-fold approach is chosen. First, we use Monte Carlo simulations to illustrate the effect of average-based (i.e., 3-year average and 5-year Olympic average) and regression-based reference income specification methods for the IST, assuming (a) different periods of data availability, (b) different time trends, and (c) different forecast periods. Second, we illustrate our findings using FADN data from Swiss agriculture as an example. The results presented in this chapter can be used by policy makers and researchers to gain a deeper understanding of how different reference income specification methods can affect governmental costs and the distribution of governmental support across different sectors and regions. The results are also applicable beyond the scope of the IST for all insurance schemes using average-based reference values.

---

[3] See, e.g., Huber (1972) for details on trade-offs with respect to robustness and efficiency for trimmed means.

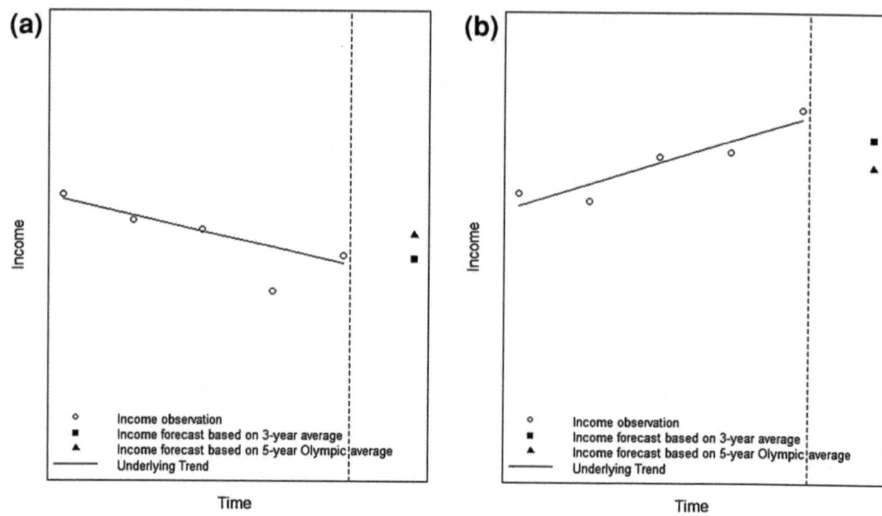

Fig. 4.1 Specification of reference income levels in the presence of trends

4.2 Background and Methodology

In the IST (as proposed by the European Commission,[4] EC 2011), a farmer is indemnified if he experiences an income loss of more than 30 % compared to the farm-specific reference income level I_R. Thus, historical income observations at the specific farm are used to calculate the expected income in period t, namely E_t. This expected income level is used to derive the reference income level which triggers indemnification, i.e., to determine if and to what extent indemnification takes place. More specifically, the reference income $I_R = 0.7E_t$ triggers indemnification if the income observation is below this level. In the IST, the level of indemnification is 70 % of the lack between actual income I_i and expected income E_t: $\text{Indemnity} = \left\{ \begin{array}{c} 0 \text{ if } I_i \geq I_R \\ 0.7(E_t - I_i) \text{ if } I_i < I_R \end{array} \right\}$

4.2.1 *Approaches to Specify Reference Income Levels*

In the current proposal of the EC (2011) and other agricultural insurances (e.g. Barnett and Coble 2012; Cooper 2010; Glauber 2013; Zulauf et al. 2008), the 3-year average and the 5-year Olympic average are employed to specify expected,

[4] Though Switzerland is not member of the European Union, we assume that an IST would be similar to specifications used in other European countries (El Benni 2012).

and thus, reference income levels. Besides considering these in our analysis, we additionally investigate a regression-based approach. The three measures to derive E_t are defined as follows:

(1) 3-year average: the average income of the previous 3 years is used as expected income in the future: $E_t = \frac{1}{3}\sum_{t-4}^{t-1} I_i$
(2) 5-year Olympic average: the average income of the previous 5 years discarding the highest and lowest observation from the analysis. With $I_{(i)}$ denoting the order statistic, this can be written as $E_t = \frac{1}{3}\sum_{2}^{4} I_{(i)}$
(3) Regression approach: We estimate a linear regression model, describing the relationship between income observations I and time Y as $I = \alpha_0 + \alpha_1 Y + \varepsilon$ with α_0 and α_1 being the parameters to be estimated and ε representing the error term. The expected income level in time t is calculated as the extrapolation of the relationship described by the estimated coefficients $E_t = \alpha_0 + \alpha_1 t$.

In order to fit the regression coefficients in step (3), we use the M-estimator. The idea underlying this estimator is to replace the equal residual weighting used for Ordinary Least Squares regression by a function that reduces the influence of outlying observations. For the M-estimator, the regression coefficients are thus chosen to minimize the sum of weighted residuals as follows: $\min_{\widehat{\beta}} \sum_{i=1}^{n} \rho(r_i)$, where $\widehat{\beta}$ are the coefficient estimates, r_i the regression residuals, and ρ a loss function representing the weighting scheme. We employ the Huber-type loss function that down-weights outliers, but does not assign zero weights to observations (see Hampel et al. 1986, for details). The estimates are derived using iteratively re-weighted least squares (e.g., Maronna et al. 2006). Finger (2013) investigated the performance of the M-estimator in comparison to other outlier-resistant regression techniques and showed that this estimator offers a good compromise between efficiency and robustness properties. Furthermore, the M-estimator is frequently used in agricultural economic applications (e.g., Harri et al. 2009; Ker and Coble 2003).

4.2.2 Monte Carlo Simulation

We use Monte Carlo simulations to simulate repeatedly (9,999 times) time series of income observations, with each time series representing income observations from a single farm. Based on a farm-specific set of income observations, we specify reference income levels using the 3 different methods introduced above, and finally analyze if and to what extent the farm is indemnified under the IST scheme. In our simulations, we account for two different periods of data availability (5 and 10 years) and two different forecast periods (1- and 2-year ahead forecasts). Furthermore, we account for cases of (i) decreasing, (ii) constant, and (iii) increasing trends in income data. The simulation setups are summarized in Table 4.1.

The Monte Carlo simulations are conducted using the following steps (an example is given below for the 2-year-ahead forecast based on a 5-year data history): First, a set of seven income observations is simulated. Second, the first five years are used to specify reference income levels for period 7 using the 3-year average (over income observations from year 3–5), the Olympic average and the regression-based approach, respectively. Third, we use the simulated income observation for period 7 to derive the level of indemnification in this period (for each of the three different reference income levels). Fourth, this process is repeated 9,999 times. Fifth, the 9,999 indemnification levels are used to derive (a) the percentage of indemnification events and (b) the average level of indemnification within the farm population.

Different trends in incomes are considered in our simulations to reflect that not a single unique trend in the income across all farmers is expected. Even though the development of farm income over time is driven by changes in boundary conditions such as agricultural policy and market conditions that are faced by all farmers; income realizations and developments are also very sector and even farm specific. For instance, sectors (e.g., crop vs. livestock production) may face different market developments and some farmers may be able to take advantage of changes over time, while others do not.

We consider two different lengths of time series of income data (5 and 10 years) that are available to specify reference income levels to illustrate their effect on indemnification. More specifically, we expect a superior performance of the regression-based approach under a longer time series because loss of efficiency compared to average-based procedures caused by a lower degree of freedom is of decreasing relative importance for larger sample sizes.

In all simulations, the identical distribution of income levels is assumed (Table 4.1). Thus, the expected frequency and level of indemnification are identical in all simulations because the "income risk" expressed as the standard deviation is the same for all simulations. More specifically, the normal distribution of incomes assumed in our Monte Carlo analyzes suggests that the expected probability of indemnification is 4.95 % and the expected average indemnification would be 724 CHF per year. These are the benchmark values used in our evaluation. Furthermore, we use these sets of Monte Carlo simulations to derive 83 % confidence intervals of the frequency and average level of indemnification.[5] These confidence intervals are used to test for significant differences across scenarios at the 5 % level, which are indicated by nonoverlapping confidence intervals (Payton et al. 2003).

[5] We use non-parametric bootstrap based on 9,999 replicates.

Table 4.1 Summary of assumption and approaches used in Monte Carlo simulations

Parameter	Assumptions and scenarios
Data availability	(i) 5 Years (ii) 10 Years
Income trends[a]	(i) –3000 CHF/year (ii) 0 (iii) +3000 CHF/year
Forecast period	(i) 1 Year (ii) 2 Years
Income distribution[a]	*N(55.000, 10.000)*
Specification of reference income levels	(i) 3 year average (ii) 5-year Olympic average (iii) Regression approach
Number of replications	9,999
Theoretical outcomes (identical for all scenarios and approaches)	–Expected probability of indemnification: 4.95 % –Expected level of indemnification: 724 CHF/year

[a] The trends and income distribution used for simulations are within the characteristics of the farm incomes we observed in the Swiss FADN data used in this research (see below)

4.2.3 Application to Farm-Level Income Data

Next, we apply the different reference income specification approaches to farm-level income observations from Switzerland. Thus, the 3-year average, 5-year Olympic average, and the regression approach are used to specify reference income levels. We also consider both 1- and 2-year ahead forecasts. Due to the lack of available panel data (see below), we restrict our analysis to a period of 7 years of data: Years 1–5 are used to specify reference income levels and years 6–7 are used to identify the level of indemnification.

By applying the different reference income specification methods to observed income data, we aim to answer the following points. First, we test if the regression-based approach is applicable in practice. Even though it can perform well on simulated data, real world income data neither is usually homogeneously distributed nor does it follow similar trends. Second, we test if the application of the regression-based approach makes, in practice, a difference compared to the average-based estimations of the reference income levels. Third, we investigate if differences in indemnification levels across methods differ across farms, and whether farm characteristics can be used to explain these differences.

To this end, differences in indemnification levels between different reference income specification approaches are calculated for each farm. We then present frequencies and average levels of indemnification across all farms and test for significant differences across indemnification levels resulting from the three methods to specify reference income levels using (paired) Wilcoxon tests. In particular, we investigate the difference between (a) the 3-year average and the regression approach, and (b) the 5-year Olympic average and the regression approach.

Next, paired and unpaired Wilcoxon tests are used to test for (a) differences between the methods used to specify reference income levels dependent on farm characteristics (comparison of methods between groups) and (b) differences between groups of farms with different characteristics dependent on the method used (comparison of groups between methods).

To this end, we summarize the results of indemnification levels of the different methods for groups of farms with respect to production regions and farm types. More specifically, we distinguish valley, hill, and mountain regions,[6] as well as five different farm types following the categorization of the Swiss Agriculture Research Station: Type 1—noncattle roughage animals (horses/sheep/goats, and others), Type 2—dairy production and combined dairy and crop production, Type 3—suckler cow and cattle farms, Type 4—poultry and pork production, Type 5—arable crops and horticulture (for more details, see Meier 2000).

The analysis and comparison of indemnity levels based on different reference income specification methods would be biased if absolute values are used. If income levels differ, for instance, across regions, this will also imply larger differences in indemnities across the methods to calculate reference incomes. Thus, we construct relative values by expressing the difference in indemnity levels relative to the average income level of the farm. We remove farms with negative incomes at this stage of the analysis, because negative signs would bias the inference of differences in relative indemnification levels across methods. Average differences in indemnification levels across all farms in a specific region and with a specific farm type are presented. To limit the influence of outliers on inference, we present mean values that use a 1 % trimming of observations.

4.2.4 Data

We use farm-level data on net farm incomes, region, and farm type of the Swiss Farm Accountancy Network (FADN) over the time period 2003–2009 to evaluate the effect of different reference income specification methods on the probability and level of indemnification through an income insurance tool as proposed by the EC (2011).

The different reference income specification methods are applied to net farm income, which is defined as the sum of farm revenues minus fixed and variable input costs, depreciation, wages as well as rents and interests and aims to compensate family labor and equity (Hausheer Schnider 2011). To avoid potential biases caused by differences in farm characteristics in an unbalanced panel approach, we select a balanced panel data set that includes all farms with data

[6] The definition of regions (valley, hill, mountain) is based on climatic and topographic conditions (LZV 2008) and is provided with the FADN data.

Table 4.2 Sample data used for the analyses in comparison to the full data set

	Unweighted (weighted) selected panel data set		**Unweighted (weighted) full data set**	
	No. of observations	% over all panel data farms	No. of observations	% over all sample farms
Region				
Valley	552 (7,661)	43% (43%)	1,444 (22,141)	43% (46%)
Hill	369 (4,709)	29% (27%)	1,057 (13,374)	31% (28%)
Mountain	353 (5,193)	28% (30%)	871 (12,893)	26% (27%)
Farm Type				
Type 1: non-cattle roughage	154 (2,912)	12% (17%)	469 (8,865)	14% (18%)
Type 2: dairy	638 (7,701)	50% (44%)	1,597 (18,572)	47% (38%)
Type 3: cattle	174 (3,073)	14% (17%)	439 (7,929)	13% (16%)
Type 4: non-ruminant	249 (2529)	20% (14%)	604 (6,114)	18% (13%)
Type 5: crops	59 (1,348)	5% (8%)	263 (6,928)	8% (14%)
All observations	1,274 (17,563)	52% (36%) compared to the full data set	3,372 (48,408)	100%

entries for each year over the time period 2003–2009.[7] The final data set comprises 1,742 farm observations that are used to calculate indemnification through the income insurance. Based on the farm weights (that are provided by the Swiss Agriculture Research Station for the representativeness of single farm observations) for the year 2009; the resulting selection represents 17,563 farms, which makes up about 36 % of the Swiss farm population (i.e., of the weighted full data set) at large (see also Table 4.2).

Table 4.2 shows the number of farms (of the selected panel data set) that are located in the valley, hill, and mountain region as well as the number of farms distinguished by different farm types. In the first two columns, the number of observations and the percentages of selected (panel data) farms are presented, once for the unweighted and once for the weighted data set. The last two columns show the number and percentages of observations for the unweighted and weighted full data set. With regard to the unweighted data set, it shows that the percentage of farms within different regions and with different farm types is similar in the panel and the full data set. Thus, the analysis is not biased with regard to sample selection. Comparing the weighted data sets in the selected panel and the full data set shows that farms of farm type 2 are slightly overrepresented in the selected panel, while farms of farm type 5 are underrepresented. Furthermore, mountain farmers are

[7] We did not increase the time period considered as the number of observations would have decreased considerably. For instance, increasing the time period to 10 years (2000–2009) would have reduced the available farm observations by more than the 50 %.

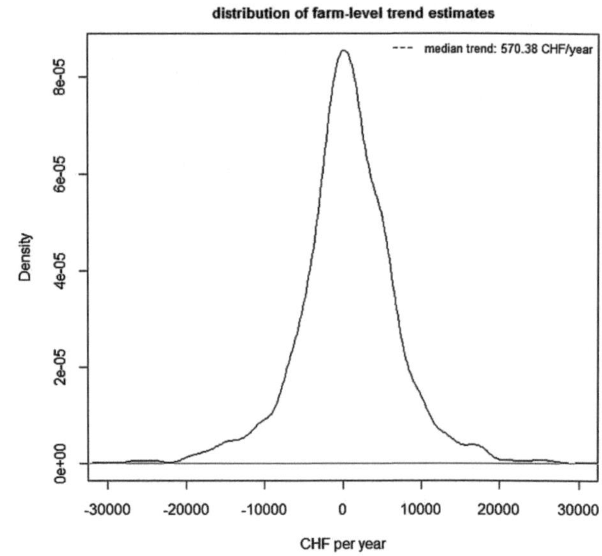

Fig. 4.2 Distribution of farm-level income trends 2003–2007

slightly over-represented and valley farmers are slightly underrepresented. However, the differences in the weighted panel and full data set are not very big.

Figure 4.2 shows the distribution of the estimated trends for all sample farms over the period 2003–2007, which is used in our analysis to specify reference income levels. On average, over all farms, a positive income trend of 570 CHF/year can be observed. Compared to the expected income levels in Swiss agriculture,[8] this income trend is, however, marginal. Figure 4.2 also shows that farms differ widely with respect to their income trends. In total, for 707 farms, a positive income trend, and for 567 farms, a negative income trend is observed.

4.3 Results

In a first subsection the results of the Monte Carlo simulations are presented providing information on indemnification through the IST using different methods to specify the reference income level in the case of no, decreasing and increasing income trends. In a second sub-section farm-level FADN data are used to investigate if and to what extent differences between the methods and between farms with different characteristics might occur in practice.

[8] For instance, the average income in our sample was 71,569 for the year 2008 and 62,298 for the year 2009.

4.3.1 Results of the Monte Carlo Simulations

The results from the Monte Carlo simulations are summarized in Table 4.3. For the case of no trends in income (Case 1, upper panel of Table 4.3), we find that the 3-year average and the 5-year Olympic average lead, on average, to an indemnification for 6.49 and 5.89 % of all farms, respectively. The resulting average indemnification levels are 1,029 and 909 CHF/year and farm. These values are very close to the theoretically expected values that have been derived based on the underlying income distribution.[9] In contrast, the regression approach leads to an indemnification of 11.68 % of all farms (with an average indemnification of 2,321 CHF/year and farm) if 5 years of observations are available. Considering a 2-year instead of a 1-year ahead specification of the reference income levels does not lead to significant changes for the average-based approaches—but leads to even more inaccurate results using the regression approach. Thus, the regression approach is inferior to the average-based approaches if no trend is present in farmer's incomes. This is due to the fact that even though both average- and regression-based approaches are assumed to deliver unbiased estimates of reference income levels, the regression approach lacks efficiency compared to the average-based approaches (e.g., Finger 2013). Using the regression approach, a trend is indicated too easy even though no trend may be present and robust estimation techniques are used. Note that this result also stems from the fact that we have not distinguished between significant and insignificant trends in our analysis.[10] However, the results based on the regression approach come closer to the expected values if more data is available (10 instead of 5 years[11]).

Next, the suitability of the different approaches is addressed in situations where income trends are present. The middle panel of Table 4.3 (Case 2) shows the simulation results for the case of decreasing income levels. As expected, this leads to a massive overshoot of the percentage of indemnifications and the indemnification levels, resulting from the use of average based specifications of reference income levels. This effect is even more emphasized if a 2-year ahead specification is required because income expectation and realization are drifting further apart. For instance, the frequency of indemnification using the Olympic average approach for 2-year ahead specification is 42.74 %, with an average level of indemnification of 6,563 CHF/year. To put this result into perspective: in this case the actual indemnification exceeds its expected value (724 CHF) by more than

[9] The expected probability of indemnification was 4.95 % and the expected average indemnification was 724 CHF/year.

[10] Alternative robust regression approaches such as the MM-estimator may perform better than the M-estimator if trends approach or are equal to zero (You 1999). Considering alternative estimation techniques may thus contribute to overcome disadvantages of the regression approach in this situation.

[11] Note that higher data availability is not relevant for the average-based approaches because only the previous 3 or 5 years of observations matter.

Table 4.3 Results from monte carlo simulations

	1-year ahead specification				2-year ahead specification			
	3-year average	5-year Olympic average	Regression trend – based on 5 years	Regression trend – based on 10 years	3-year average	5-year Olympic average	Regression trend – based on 5 years	Regression trend – based on 10 years
Case 1: constant income levels								
Indemnifications (in %)	6.49 [6.16-6.83]	5.89 [5.57-6.21]	11.68 [11.24-12.12]	9.26 [8.86-9.66]	6.27 [5.94-6.60]	5.82 [5.50-6.15]	14.54 [14.06-15.03]	10.04 [9.63-10.45]
Average indemnification (in CHF)	1,029 [974-1083]	909 [858-959]	2,321 [2229-2414]	1,687 [1612-1761]	990 [937-1043]	894 [844-944]	3,207 [3095-3321]	1,905 [1824-1986]
Case 2: decreasing income levels								
Indemnifications in %	26.44 [25.84-27.04]	32.25 [31.61-32.88]	20.71 [20.16-21.26]	18.91 [18.37-19.43]	36.17 [35.51-36.82]	42.74 [42.07-43.42]	25.70 [25.11-26.30]	21.51 [20.95-22.08]
Average indemnification (in CHF)	3,763 [3674-3854]	4,801 [4701-4902]	3,352 [3255-3450]	2,697 [2615-2779]	5,309 [5205-5412]	6,563 [6451-6675]	4,486 [4369-4604]	3,112 [3024-3203]
Case 3: increasing income levels								
Indemnifications in %	0.86 [0.73-0.98]	0.39 [0.31-0.49]	5.76 [5.43-6.08]	3.61 [3.36-3.87]	0.41 [0.32-0.50]	0.14 [0.09-0.19]	7.51 [7.15-7.89]	3.86 [3.60-4.13]
Average indemnification (in CHF)	153 [131-176]	58 [53-83]	1,378 [1299-1458]	791 [734-849]	72.42 [57.06-88.38]	23.40 [14.95-32.44]	1,999 [1900-2100]	913 [850-977]

Results are based on the Monte Carlo simulations with N=9,999. Ranges in brackets are 83% confidence intervals derived with non-parametric bootstrap (N=9999)

factor 9. Thus, trends underlying the farm incomes may constitute a high risk for the financial resources available in the IST scheme. The 3-year average performs slightly better than the 5-year Olympic average because it accounts for the most recent years (of higher incomes), while high income situations are not considered in the Olympic average. Our results show furthermore that the regression approach performs superior to the average-based approaches in the presence of decreasing income levels. But, the regression approach can only reduce and bound the biases made (e.g., it reduces frequency of indemnification by about 40 % compared to the Olympic average approach[12]), but not solve the problem entirely because the frequency and level of indemnification are still far away from the theoretically expected levels. The regression approach may tend to overestimate trends that are underlying the income observations. Therefore, too high levels of indemnification are observed if negative income trends are present. The errors made are, however, still substantially smaller than those made with average-based approaches. Furthermore, we find the relative gains from using the regression approach instead of average-based approaches to increase if a 2-year instead of a 1-year-ahead forecast is considered. In addition, a substantial and significant improvement in the performance of the regression-based approach is found if 10 instead of 5 years of observations are available to estimate reference income levels.

Similar results are found if an increasing trend in income data is assumed (bottom panel of Table 4.3, Case 3): the average-based specification approaches lead to almost no indemnification (less than 1 % of farmers are indemnified). In contrast, the regression approach almost perfectly reflects the theoretically expected frequency and levels of indemnification, in particular if the analysis is

[12] (Almost) all reported differences are significant. This is, however, also due to the large number of (simulated) observations underlying every point estimate (N = 9,999).

based on a 10-year data history. The fact that the regression approach performs substantially better for increasing than for decreasing income levels is due to the fact that a falsely identified trend under decreasing income levels leads to more indemnification. In contrast, a falsely identified (too) positive trend in incomes does not lead to higher indemnification.

In summary, we find that average-based specifications of reference income levels cause excess indemnification under decreasing income levels, but lead to too low frequencies and levels of indemnification if income levels are increasing over time. The associated overshoots in costs in cases where farm incomes have a decreasing trend are enormous. In contrast, there is almost no chance for indemnification (despite falling below 'expected' values) if income tends to increase over time. Our results show that identifying reference income levels using a regression approach bounds these errors. This result is even more emphasized if more than a 5-year income data history is available to identify reference income levels. Furthermore, longer time series reduce the inefficiency of the regression-based approach if no trend is underlying the income data.

4.3.2 Results for the Farm-Level Data

Next, we employ the three different reference income specification techniques to farm-level income observations from Switzerland over the period 2003–2009. Based on the first five years of observations (2003–2007), reference income levels for both 1-year and 2-year-ahead specifications are calculated and used to elaborate if and to what extent a farm would have been indemnified in the years 2008 and 2009, respectively.

Table 4.4 shows the percentage and average level of indemnification across all farms. It shows that both the relative number of indemnified farms as well as average indemnification per farm differ between reference income specification method and year (i.e., the years 2008 and 2009 represent the 1-year and 2-year ahead forecast of indemnification, respectively). For instance, in 2008 the percentage of farms indemnified ranges between 10.20 and 11.62 % if the reference income is specified based on the 3-year average- and regression-based approach respectively. As shown in Table 4.4, the average indemnification in 2008 ranges between 1,128 and 1,524 CHF per farm. In 2009, however, twice as many farmers would have received indemnity payments which also increase the average indemnification to between 2,477 and 3,581 CHF per farm.[13] In general, these levels of indemnification are in line with other research on IST in Swiss agriculture (El Benni 2012; El Benni et al. 2013; Finger and El Benni 2014).

[13] Higher income losses in 2009 are caused by an increase in cost levels, but in particular by decreasing output price levels, which was especially distinct for the milk market where the abandonment of the quota regime led to sharp decreases in milk prices (Chavaz 2010; Mann and Gairing 2011; El Benni and Finger 2013).

Table 4.4 Frequency and average level of indemnification

	2008: 1 -year ahead			2009: 2-years ahead		
	3-year average	5-year Olympic average	Regression trend	3 -year average	5-year Olympic average	Regression trend
Percentage of indemnification	10.20	10.91	11.62	20.09	20.25	24.73
Average indemnification (in CHF)	1,128	1,216	1,524	2,477	2,613	3,581

For both years, indemnification levels based on both average-based approaches differ significantly (1 % level) from those derived using reference incomes with the regression-based approach

The numerical results underline that the IST is not aiming at frequent indemnification, but focusses on (rare) catastrophic events (dell' Aquila and Cimino 2012). While we find only small differences in indemnifications across reference income specification methods, our results highlight that the financial means for the IST can differ widely between years.

Indemnification levels based on different baseline specification methods are significantly positively (but not perfectly) correlated with each other (details are presented in the Appendix). For the years 2008 and 2009 (paired) Wilcoxon tests show significant differences in indemnification between the average-based approaches (i.e., the 3-year average and the 5-year Olympic average) and the regression-based approach. Indemnifications based on the regression approach are significantly higher than under the average-based approaches (i.e., the 3-year average and 5-year Olympic average). In contrast, no significant differences can be observed between indemnification levels based on the 3-year average and the 5-year Olympic average.

This is in line with the fact that the average trend in farm-level incomes is positive. Thus, this result supports the previous findings of the Monte Carlo Simulations in that indemnification levels will be downward-biased in the presence of positive trends in income data if average-based approaches are used.

Next, we present differences in relative indemnification levels between the three methods to specify the reference income levels by production region and farm type. More specifically, we consider the production region (valley, hill, and mountain) as well as five different farm types: Type 1—noncattle roughage animals (horses/sheep/goats, and others), Type 2—dairy production and combined dairy and crop production, Type 3—suckler cow and cattle farms, Type 4—poultry and pork production, Type 5—arable crops and horticulture. Differences between the regression and the 3-year average approach, as well as the regression and the

Table 4.5 Differences in indemnification relative to income levels between the average-based approaches and the regression-based approach by region and farm type

	2008: 1-year ahead		2009: 2-year ahead	
Relative indemnification	3-year average versus regression	5-year Olympic average versus regression	3 -year average versus regression	5- year Olympic average versus regression
Region				
Valley	0.96[a]	0.37	3.77[c]	2.75[c]
Hill	1.76[b]	1.65	5.52[c]	4.98[c]
Mountain	3.08[c]	2.50[b]	6.37[c]	6.39[c]
Farm type				
Type 1: non-cattle roughage	2.39	0.61	7.77[b]	2.12[a]
Type 2: dairy	1.96[c]	1.79[c]	5.51[c]	5.40[c]
Type 3: cattle	2.14	1.53	2.29	2.78
Type 4: non-ruminant	1.05	−0.38	4.42[c]	1.75
Type 5: crops	2.98	6.18[b]	5.96[b]	10.64[c]

[a], [b], [c] denote significant differences (from 0) at the 10, 5, and 1 %. Presented values are 1 % trimmed means

Olympic average approach, are presented as trimmed (1 %) means across all farms in the respective group. Differences are given in relative terms as a percentage of the income level at the farm.[14] Wilcoxon tests are used to investigate if the hypotheses of zero differences can be rejected. The results are presented in Table 4.5.

The values presented in Table 4.5 are differences between relative indemnities under the regression and the average-based approaches. A positive value thus shows that the indemnification under the regression-based approach was higher than, for instance, by using the 3-year average to specify the reference income level. The difference is positive in virtually all cases depicted in Table 4.5, which reflects the—on average—tendency of incomes to increase within the considered period (cp. Fig. 4.1). The stronger the positive income trend, the higher the indemnification using the regression- compared to the average-based approaches. Moreover, we find that differences between the regression approach and the 3-year average approach tend to be (though not significant) larger than those between the regression and the Olympic average approach. From the simulation results (Table 4.3), we would have expected the opposite. This may indicate problems with outlying observations rather than income trends because—in contrast to the 3-year average approach—both the Olympic average and the regression approach (using the M-estimator) provide robustness against outliers. The performance of

[14] At this step, we removed farms with negative income observations. This step reduced the sample size from initially 1,274–1,242 for 2008 and to 1,227 for 2009.

IST reference income specification methods in the presence of outliers is beyond the scope of this analysis, but should be addressed in future research.

Comparing the results for the 1-year and 2-year ahead forecast, we find that the differences between average-based and regression-based approaches become much more relevant if a longer time horizon (2 instead of 1 year) is considered. For instance, comparing the regression with the 3-year average approach for the hill region, the difference is 1.76 % (of the farm's income level) for the 1-year-ahead specification and 5.52 % for the 2-year ahead specification. In line with the results from the simulation study (Table 4.3), the effects of underlying income trends are more emphasized if reference income levels have to be specified for periods further in the future.

Though differences are (on average) positive in all regions, we find the differences to be largest for the mountain and smallest for the valley region.[15] This pattern is the same for all cases considered in Table 4.5. This indicates that income trends in the mountain region (for the period 2003–2007) tend to be higher than in the valley region (note that this could indicate more farms with positive trends, farms with stronger positive income trends, or both). Thus, this result indicates that farms in the mountain region would—on average—suffer most in terms of underproportional indemnification levels compared to their objective income risks.

When comparing differences in indemnity payments across farm types, a less distinct pattern is found than for the farm's region. However, when comparing all four cases presented in Table 4.5, we find that (on average) the differences between the reference income specification methods are smallest for farm types 3 and 4, and most distinct for farm types 2 and 5. Thus, dairy and crop producers in our sample seem to face (slightly) stronger income increases over time (for the period 2003–2007) than other farm types. Thus, dairy and crop producers would have benefit more compared to other farm types in terms of indemnification levels. No significant differences are found across farm types.

4.4 Discussion and Conclusion

Our analysis revealed that not considering income trends when specifying reference income levels in the (IST) may cause substantial biases. Thus, the currently suggested approaches based on the 3-year average and the Olympic average may provide misleading results. More specifically, these approaches will cause a lower indemnification for the farmer in the case of increasing incomes, and a higher

[15] This difference across zones is, however, only significant (at the 10 % level) if comparing the mountain and the valley region based on the difference between the regression and the 3-year average approach for the 1-year-ahead specification (not shown).

indemnification if income levels are decreasing over time. Our Monte Carlo simulations underline that small income trends can be sufficient to cause substantial biases. In the setting of the IST, particularly the case of decreasing incomes can be a threat for the system. The IST is not aiming at frequent indemnification but focusses on (rare) catastrophic events, and indemnification levels are thus not intended to (at least on average) be large. If income levels are increasing and farmers receive lower indemnification levels than they should, based on objective income risks, because the reference income was specified with an average based procedure, the gap between expected and realized indemnity payments is thus usually small and bounded (because the indemnity payment can be zero in minimum). In contrast, farmers receive too high indemnification using average-based approaches if incomes decrease over time. This overshooting is not bounded because higher income losses always imply higher indemnification. In the presence of income trends, average-based specifications of reference income levels will thus cause two major problems for the IST. First, it may cause large differences between expected and realized indemnification levels, potentially causing budgetary problems. Second, it may cause indemnity levels that do not reflect farm-level income losses, and it may thus not allow achieving the actual purpose of the IST of securing farm incomes.

Our results show that the specification of reference income levels using income trends estimated in a regression approach could partly overcome this problem and bound the levels of indemnity payments in cases of trends in farm incomes. This approach, however, performs inferior to the average-based approaches if no trend is present in the income data in that the probability of indemnification increases. But the differences for this situation were indicated to be small by our simulation study, compared with the potential biases if trends are present in farm incomes.

The case of trends in farm incomes is expected to be highly relevant. Even if farm incomes may not reveal substantial income trends at the aggregated level, the incomes of the individual farmers may still reveal trends. Trends differ across farms, either because of differences in farmers' managerial abilities or due to market developments (e.g., price developments of specific products). Developments of increasing and decreasing income trends may outweigh each other in the farm population at large (leading to no or small trends at aggregated levels), but these farm-level trends are critically important for the IST. This is due to the fact that the IST uses farm-level data to specify farm individual indemnification levels. Furthermore, the effects of increasing incomes for some farmers may not necessarily outweigh the effects of decreasing incomes for other farmers because implications are not symmetric (see above). Thus, the relevance of alternative specifications of reference income level, e.g., using regression-based approaches, is also very relevant if no income trends are present at aggregated levels. In other words, neglecting income trend effects due to the observation of small or zero trends in aggregated levels can cause biases in indemnification levels. These biases may cause cost increases in the IST and/or indemnity levels that do not reflect farm-level income losses.

Besides threatening the budget of the insurer or the causing of nonindemnification of the farmer, even though income losses occurred, the nonconsideration of trends may cause redistribution of incomes within the farm population that are unintended by an income stabilization tool. All farms are expected to contribute to the costs of the IST via a kind of mutual fund (EC 2011). Our results show that farms with decreasing incomes will profit more (in terms of indemnification) than farms with positive income trends. This may cause redistribution from farms that are able to improve their productivity over time towards farms with decreasing performance. The average-based calculation of reference income levels in IST may thus slow down structural change. Income trends may also differ across sectors (i.e., farm types), and redistribution may also take place between them. Differences in indemnity levels across sectors may also be caused by different trends underlying farm incomes in these sectors if average-based approaches are used to specify reference income levels. In order to avoid these asymmetries, sector-level mutual funds may be used.

We considered income trend estimation based on the M-estimator, a robust regression technique in our analysis. This estimator was chosen because it allows some robustness against outliers, which are especially expected if working with farm-level data, but still provides a high efficiency compared to other robust estimators (e.g., Finger 2013). However, the choice of the regression technique used to estimate income trends and reference income levels can influence the results, i.e., the indemnification in the IST. Thus, future research should consider a larger set of regression techniques ranging from highly efficient but nonrobust (OLS) to very robust but inefficient regression techniques.[16] Further research should also address dynamic effects of the choice of the reference income specification technique. This is expected to be important because indemnifications also influence reference income levels in future periods.

Acknowledgments This work was supported by the Swiss Federal Office for Agriculture and the Swiss National Science Foundation. We would like to thank the Agroscope Reckenholz-Tanikon Research Station for providing the FADN data.

[16] Clearly, no technique will dominate all others in all situations. However, the choice of the regression technique may be based on the characteristics underlying the situation at hand, e.g., with respect to income variability. For instance, following You (1999), the MM-estimator may allow to overcome problems of the regression approach if income trends approach zero.

A.1 4.5 Appendix

Table 4.6 (Appendix)

Table A1 Correlation coefficients (Pearson/Spearman) between farm-level indemnities based on different methods of specification of the reference income level

	2008: 1-year ahead		2009: 2-years ahead	
	Olympic average approach	Regression approach	Olympic average approach	Regression approach
3-year average approach	0.90/0.84	0.78/0.70	0.92/0.87	0.66/0.64
Olympic average approach	–	0.66/0.59	–	0.53/0.52

A/B denote Pearson and Spearman correlation coefficients, respectively. The null hypothesis of zero correlation is rejected in all cases

References

Atwood, J., Shaik, S., & Watts, M. (2003). Are crop yields normally distributed? A reexamination. *American Journal of Agricultural Economics, 85*(4), 888–901.

Barnett, B. J., & Coble, K. H. (2012). Understanding regional differences in farm policy preferences. *American Journal of Agricultural Economics, 94*(2), 528–534.

Bielza Diaz-Caneja, M., Conte, C. G., Dittmann, C., Gallego Pinilla, J., & Stroblmair, J. (2008). *Agricultural insurance schemes*. ispra, Italy: European Commission, Joint Research Center Institute for the Protection and Security of Citizens JRC.

Chavaz, J. (2010). *Summary: The swiss milk market. panel of the high level experts group on milk*. Brussels, Belgium: Deputy Director General, Federal Office for Agriculture, Bern, Switzerland.

Cooper, J. C. (2010). Average crop revenue election: A revenue-based alternative to price-based commodity payment programs. *American Journal of Agricultural Economics, 92*(4), 1214–1228.

dell'Aquila C., & Cimino O. (2012). Stabilization of farm income in the new risk management policy of the EU: A preliminary assessment for Italy through FADN data, Paper presented at the 126th EAAE Seminar: *New challenges for EU agricultural sector and rural areas. Which role for public policy*?, Italy: Capri.

EC. (2009). *Income variability and potential cost of income insurance for EU*. Brussels, Belgium: Directorate-General for Agriculture and Rural Development, European Commission.

EC. (2011). *Proposal for a regulation of the european parliament and of the council on support for rural development by the European agricultural fund for rural development (EAFRD)*. Brussels, Belgium: European Commission.

El Benni, N. (2012). *Einkommensrisiko in der Schweizer Landwirtschaft und Risikomanagementinstrumente*. Bern, Switzerland: Report on behalf of the Swiss Federal Office for Agriculture.

El Benni, N., & Finger, R. (2013). Gross revenue risk in swiss dairy farming. *Journal of Dairy Science, 96*(2), 936–948.

El Benni, N., Finger, R., & Meuwissen, M. (2013). Potential effects of the income stabilization tool in swiss agriculture, Paper presented at the 133rd EAAE Seminar. *Developing integrated and reliable modeling tools for agricultural and environmental policy analysis*, Chania, Greece, 15–16 June, 2013.

Finger, R. (2013). Investigating the performance of different estimation techniques for crop yield data analysis in crop insurance applications. *Agricultural Economics, 44*(2), 217–230.

Finger, R., & El Benni, N. (2014). A note on the effects of the Income Stabilisation Tool on income inequality in agriculture. *Journal of Agricultural Economics*. (in Press).

Glauber, J. W. (2013). The growth of the federal crop insurance program, 1990–2011. *American Journal of Agricultural Economics, 95*(2), 482–488.

Hampel, F. R., Ronchetti, E. M., Rousseeuw, P. J., & Stahel, W. A. (1986). *Robust statistics*. New York, USA: Wiley.

Harri, A., Coble, K. H., Erdem, C., & Knight, T. O. (2009). Crop yield normality: A reconciliation of previous research. *Review of Agricultural Economics, 31*(1), 163–182.

Hausheer Schnider, J. (2011). *Glossar der zentralen auswertung von buchhaltungsdaten, research station*. Ettenhausen, Switzerland: Agroscope Reckenholz-Tänikon ART.

Hennessy, D. A. (2009). Crop yield skewness and the normal distribution. *Journal of Agricultural and Resource Economics, 34*(1), 34–52.

Huber, P. J. (1972). The 1972 Wald lecture robust statistics: A review. *The Annals of Mathematical Statistics, 43*(4), 1041–1067.

Just, R. E., & Weninger, Q. (1999). Are crop yields normally distributed? *American Journal of Agricultural Economics, 81*(2), 287–304.

Ker, A. P., & Goodwin, B. K. (2000). Nonparametric estimation of crop insurance rates revisited. *American Journal of Agricultural Economics, 82*(2), 463–478.

Ker, A. P., & Coble, K. H. (2003). Modeling conditional yield distributions. *American Journal of Agricultural Economics, 85*(2), 291–304.

Liesivaara P., Myyrae, S., & Jaakkola, A. (2012). *Feasibility of the income stabilisation tool in finland*, Paper presented at the 123rd EAAE Seminar: Price volatility and farm income stabilisation-modelling outcomes and assessing market and policy based responses, Dublin, Ireland.

LZV (2008). Verordnung über den landwirtschaftlichen Produktionskataster und die Ausscheidung von Zonen, Stand 1. Januar 2008. http://www.admin.ch/ch/d/sr/912_1/index.html

Mann, S., & Gairing, M. (2011). Post milk quota experiences in Switzerland. *EuroChoices, 10*, 16–21.

Maronna, R. A., Martin, R. D., & Yohai, V. J. (2006). *Robust statistics—theory and methods*. New York, USA: Wiley.

Mary, S., Santini, F., & Boulanger, P. (2013). An ex-ante assessment of CAP income stabilisation payments using a farm house hold model, paper presented at the 87th annual conference of the agricultural economics society, UK: University of Warwick 8–10 April, 2013.

Meier, B. (2000). *neue methodik fuer die zentrale auswertung von buchhaltungsdaten an der FAT*. Ettenhausen, Switzerland: Research Station Agroscope Reckenholz-Taenikon ART.

Meuwissen, M. P. M., Huirne, R. B. M., & Skees, J. R. (2003). Income insurance in European agriculture. *EuroChoices, 2*, 12–17.

Meuwissen M. P. M., van Asseldonk, M., Pietola, K., Hardaker, B., & Huirne, R. (2011). *Income insurance as a risk management tool after 2013 CAP reforms*?, Paper presented at the EAAE 2011 Congress, Zurich, Switzerland: Change and Uncertainty August 30–September 2, 2011.

Payton, M. E., Greenstone, M. H., & Schenker, N. (2003). Overlapping confidence intervals or standard error intervals: What do they mean in terms of statistical significance? *Journal of Insect Science, 3*(34), 1–6.

Pigeon, M., Henry de Frahan, B., Denuit, M. (2012). *Actuarial evaluation of the EU proposed farm income stabilisation tool*, Paper presented at the 123rd EAAE Seminar: Price Volatility and Farm Income Stabilisation-Modelling Outcomes and Assessing Market and Policy Based Responses, Dublin, Ireland, 2012.

Ramirez, O. A., Misra, S., & Field, J. (2003). Crop-yield distributions revisited. *American Journal of Agricultural Economics, 85*(1), 108–120.

Turvey, C. G. (2012). Whole farm income insurance. *Journal of Risk and Insurance, 79*(2), 515–540.

You, J. (1999). A monte carlo comparison of several high breakdown and efficient estimators. *Computational Statistics and Data Analysis, 30*(2), 205–219.
Zulauf, C. R., Dicks, M. R., & Vitale, J. D. (2008). Average Crop Revenue Election (ACRE) farm program: Provisions, policy background, and farm decision analysis. *Choices, 23*, 29–35.

Chapter 5
Developing of Modelling Tool for Policy and Economic Rent in Agriculture

A. Bezat-Jarzębowska, W. Rembisz and A. Sielska

Abstract In this chapter, a completely original microeconomic model of the producer's choice is presented. To capture the impact of agricultural policy on the agricultural producer's income two main sources of income growth were included in the model, namely efficiency of production (economic rent) and funds obtained from solutions under the CAP agricultural policy (policy rent). In this chapter, the micro-level (plant, animal, and mix production types) and macro-level agricultural data were used. The agricultural producer optimises his choices, i.e. reaches equilibrium when it comes to these two sources of income for the objective function (income maximising). Therefore, the purpose of the article was to show a certain range of substitution between these sources of producer's revenue. We have observed that the rate of substitution of these two sources of income growth is not equal to one, which means that replacing one with the other is not without any effect on the level of income. We find it important to assess not only the substitution between the two sources of income which are in our case production efficiency and political rent, but also to study to what extent investment decisions (which create a basis for future income) depend on political rent.

5.1 Introduction

The purpose of the chapter is to show the problem of substitution between the income effects of policy and economic rent and to provide a certain illustration in this regard without making any specific hypothesis. In our reasoning, we use certain mathematical representations for showing a model as the basis of reasoning about the issue of substitution that has been raised. This is in line with the microeconomic model of the producer's choice, but the approach is completely original.

A. Bezat-Jarzębowska (✉) · W. Rembisz · A. Sielska
Institute of Agricultural and Food Economics-National Research Institute (IAFE-NRI), Warsaw, Poland
e-mail: bezat.jarzebowska@gmail.com

C. Zopounidis et al. (eds.), *Agricultural Cooperative Management and Policy*, Cooperative Management, DOI: 10.1007/978-3-319-06635-6_5,

We assume income to be a basis of the farm producer's objective function. Next, we show two fundamental sources of this income. The first one is efficiency of production. The second source is agricultural policy income effects. In the analysis, we leave out relations of prices received and paid as external sources (independent from the producer's), using them as data on the basis of the *ceteris paribus* principle. The first source associated with efficiency, we call economic rent. The second, related to the agricultural policy, is policy rent.[1] We make this simplification, using the principle of reductionism to extract the essence of the issue raised in the chapter. Of course, the producer chooses both of these sources of shaping and growth of income. What is more, in their rational behaviour, which is a somewhat classic premise in microeconomics, producers choose the source which is more favourable[2] or more useful, i.e. it produces greater effects in relation to the costs (efforts) associated with it.

The purpose of the chapter is to show a certain range of substitution between these sources of revenue of the producer. We do not put forward a claim on substitution whereby increasing the scope of policy support leads to a reduction of efforts to improve efficiency or vice versa. We can then experience a synergy effect.

The data being used in the study is mainly the Farm Accountancy Data Network (FADN). The database was used to present the extent of policy and economic rent and substitution between these types of rent. The relationship between the policy rent and the producer's investment was shown on the basis of the FADN dataset as well. The Polish example was used. Poland was selected as an important EU member relevant to the agricultural sector. The background for the study (e.g. the rate of growth of agricultural production) was presented as an example of selected European countries by use of FAO and EUROSTAT data. The OECD Stats data was used to present the value of agricultural production and support of producers in the EU.

5.2 Income as an Objective Fuction of the Farm Producer

From the literature, it follows that increasing agricultural production "meets inelastic demand, causing a fall in real prices of agricultural products. As a result, farmers' incomes are not growing in proportion to the rate of growth of

[1] These concepts known in economics were introduced to agricultural economics by Wilkin (2005). In the article policy rent is understood as income effect of support for agricultural producers resulting from implementation of various programmes and mechanisms of the CAP. Due to the fact that our main focus is producer's income, some impacts of agricultural policy (e.g. animal welfare policies and regulations) are not taken into consideration separately.

[2] We may refer this to the theory of rational expectations by Lucas and Sargent from the 1970s, who assumed that economic operators (producers and consumers) adapt flexibly, for their own economic benefit or maximisation of their own aim or function, to anticipated changes in regulations and economic policy.

production" (Woś 2004). In fact, this also depends on the growth rate of production efficiency. However, agricultural producers cannot rely "on the increase in prices of products as the source of increase in their income" (Rembisz 2007). Due to the increasing role of the processors and the recessive nature of the market, agricultural producers who seek to improve their income are forced to use the possibilities in the field of productivity of production factors. This also applies to the labour factor.[3]

Let us assume that—having regard to the uncertainty associated with the environmental and climate aspects of the process of management in agriculture—we can express the objective function of the agricultural producer as follows:

$$\max_{R} E(D_t) \tag{5.1}$$

We define the income of the agricultural producer[4]:

$$D_t = \{C_R \cdot R - N \cdot C_N(R)\} \tag{5.2}$$

where:

$R \cdot C_R$	the revenue (production value) of the agricultural producer (agriculture sector) as the product of the volume of production (supply) and the prices of products,
$N \cdot C_N(R) \Rightarrow C_K \cdot K + C_L \cdot L$	the cost of using manufacturing factors, i.e. the factor of capital and the labour factor for a given level of agricultural production (on a producer or sector scale),
C_K, K	remuneration (price) of the capital factor and the involvement of capital factor,
C_L, L	remuneration (price) of labour factor and employment of the labour factor,
E	expected value

Next we can assume, in a simplistic manner, in order to extract the essence of the issue raised, that the income of agricultural producers is only a product of the remuneration of the labour factor and its employment. So we have a simple relation:

$$C_R \cdot R - K \cdot C_K \approx L \cdot C_L \tag{5.3}$$

[3] The models used in agriculture include, in particular, the indicators of labour productivity and productivity of the land. This is also the essence of the developed model in relation to agricultural production and agricultural producers. For a detailed explanation see, e.g. (Bezat-Jarzębowska et al. 2012).

[4] In the same way, we define the function of producer in mathematical economics.

As we know, the income of agricultural producers (income in agriculture) is currently being increased as a result of the effects of existing agricultural policy solutions (CAP). We will denote this with the symbol: T_B. It is, i.e. revenue, also being reduced, although to a small extent by imposing tax and other burdens, which is denoted as: P_T. Therefore, we can express the income of agricultural producers as

$$L \cdot C_L + (T_B - P_T) = D_R \tag{5.4}$$

where:

T_B value of different forms of transfers, subsidies and support for agriculture producing the income effect (direct payments, maintaining prices, quotas on prices, quotas on import and other regulations, production and intervention activities),

P_T value of different tax burdens and other payments imposed on the agricultural holding,

D_R income of agricultural producers (agriculture)

We further assume that income, such defined, is a maximised objective function of the agricultural producer. This is a certain simplification, because of the multi-criterion objective function of the producer[5]; however, it is needed for extracting the essence of the discussed problem of substitution of the agricultural producer's two income sources.

5.3 Economic and Policy Rent in Realising the Objective Function of the Farm Producer

To capture the impact of agricultural policy on the farm producer's objective function expressed in such a way, and, to be more precise, on the paths of its maximisation, let us identify, in the context of this formula, two main sources of income growth. First, this source is improved efficiency of production (with a given relation of prices obtained for products to prices paid for inputs). Second, this source is also funds obtained from solutions under the CAP agricultural policy. We can also write it as follows:

$$D_t = \max_R f\{(\mathrm{EP}) + g(B)\} \tag{5.5}$$

[5] Cf. Sielska (2012, p. 28), where the decision problem of the agricultural producer is shown, using the multi-criterion approach, as a space of assessing decision options.

where:

EP production efficiency in its technical basis is: $\frac{y}{K+L}$,

$g(B)$ income effect of support for agricultural producers related with implementation of the various programmes and mechanisms of the CAP with an assumption facilitating further reasoning that: $P_T \approx \text{const} \approx \Delta P_T \approx 0$

Using $\{(\text{EP}) + g(B)\}$ as a component of this objective function, we can place the following dilemma facing a reasonably acting and progressive agricultural producer[6]: Will they be more focussed on income benefit of the CAP agricultural policy or on the benefit of improving the efficiency of production?

These former advantages related with agricultural policy are referred to as policy rent. The latter, related to the improvement of efficiency are called economic rent. Admittedly, the former seem easier to obtain than the latter. Whether this view is true or not, there is a different mechanism of achieving both types of income advantages. This is an interesting question in itself, which we leave for another occasion. At this point, we are interested in the question of the possible substitutability between these choices made by the producer (in the agricultural sector as a set of agricultural producers).

Let us note that production efficiency,[7] as a source of income growth, which depends on the producer, is an endogenous determinant. However, the benefits of agricultural policy, as well as changes in the relations of prices obtained to those paid, which is here assumed on the basis of the principle *ceteris paribus* in a short run,[8] is a condition independent from the producer (an exogenous factor).

5.3.1 Economic Rent

Producers operating on a competitive market seek possibilities for maximisation of expected profit by increasing production, especially by non-decreasing returns to scale (we assume that in a competitive market equilibrium, the price is given (fixed) for producer and processor in the agri-food sector.[9]) But on most of the markets in countries with a high GDP per capita (e.g. countries of Western Europe or North America) the rate of production growth in a sector is determined by low demand increment. Research confirms that a given growth of demand for agri-food products, occurring at a specific time, determines also the output growth in the agri-food sector (Figiel and Rembisz 2009). The low growth rate of demand for

---

[6] Also in line with the aforementioned assumptions of the theory of rational expectations.

[7] Production efficiency is determined by a given production function for the agricultural producer (production technique): $R_t = f(K_t, L_t)$.

[8] However, price relations are a surface source rather than a fundamental source of change in profitability—with given efficiency—and thus revenue.

[9] The agri-food sector is understood as a farm sector and processor sector.

agricultural products can limit the growth in the agri-food sector, and consequently, the agricultural production growth inducing technical change. Since the agri-food sector's revenue growth is primarily caused by the increase in demand for agricultural products, hence, the gross income of farm producers does not increase significantly and satisfactorily. Therefore, the low growth rate of demand for agri-food products must determine the change of efficiency-based relations treated as a main growth factor in the sector (Bezat-Jarzębowska and Rembisz 2013).

The further considerations are related only to the farm sector. The first element of Eq. (5.5)—economic rent—is production efficiency determined, for the convenience of this reasoning, in value terms rather than technically, as follows:

$$\mathrm{EP} = (C_R \cdot R - N \cdot C_N)_R \tag{5.6}$$

Of course, prices are fixed here.[10] In the TFP approach (*Total Factor Productivity*), this efficiency can also be expressed as

$$\mathrm{EP} = \frac{R \cdot C_R}{N \cdot C_N} = \frac{R \cdot C_R}{K \cdot C_K + L \cdot C_L} \tag{5.7}$$

And assuming unchanging price relations, i.e. prices obtained to those paid (price scissors), in dynamic terms appropriate for TFP, we can express this as

$$\frac{\Delta \mathrm{EP}}{\mathrm{EP}} = \frac{\Delta R}{R} - \frac{\Delta N}{N} \approx \frac{\Delta R}{R} - \left(\frac{\Delta K}{K} + \frac{\Delta L}{L}\right) \tag{5.8}$$

and:

$$\frac{\Delta \mathrm{EP}}{\mathrm{EP}} = 0 \Rightarrow \mathrm{TFP} \uparrow \ \text{when} \ \frac{\Delta R}{R} > \left(\frac{\Delta K}{K} + \frac{\Delta L}{L}\right) \tag{5.9}$$

The rate of growth of agricultural production in selected countries of the European Union is shown in Fig. 5.1. The value of the rate of growth of agricultural production $\left(\frac{\Delta R}{R}\right)$ has fluctuated over the years 2000–2009. The biggest changes (over 10 %) have been observed in France, Germany, Poland and Spain. The EUROSTAT and FAO dataset was used for calculation.

The rate of growth of remuneration of the labour factor in selected countries of the European Union is shown in Fig. 5.2. The value of the rate of growth of the labour factor $\left(\frac{\Delta L}{L}\right)$ was close to zero for most countries, which indicates that the number of employees in agriculture did not change much in the period 2000–2010.

---

[10] When we assume price volatility $c = C_R/C_N$ (price scissors), this representation is expressed by the profitability index: $\mathrm{OP} = (C_R^t \cdot R_t - N_t \cdot C_N^t)_R$.

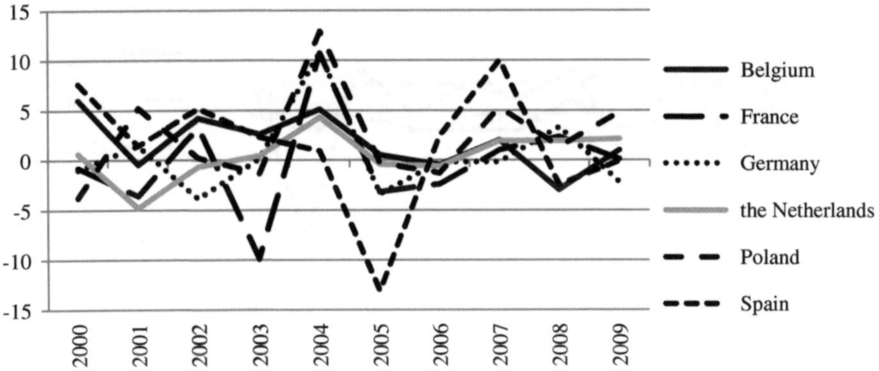

Fig. 5.1 The rate of growth of agricultural production $\left(\frac{\Delta R}{R}\right)$ in selected countries of the European Union in the years 2000–2009. *Source* own calculations based on the FAO and EUROSTAT data

The exceptions are countries like Poland (year 2002) and Spain (year 2009), which indicates a decrease of 0, 1 in the rate of growth of the labour factor.

We will not develop the issue of sources and measurement of production efficiency improvement,[11] and we will limit ourselves to the above characteristics of the process of improving efficiency in the sense of changes in TFP, which at the same time includes many of the production factors[12] .

Let us note that efficiency improvement is a source of income growth, whose triggering concerns a longer period when technical changes are possible (manufacturing techniques in the above formula for changes in relation: $\left(\frac{\Delta K}{K} / \frac{\Delta L}{L}\right)$, as a result of the investment). This is a source which is invisible on the surface of phenomena, in contrast to changes in relations of product prices and manufacturing factors. It should be noted that change in efficiency relations based on including new technical solutions in the production process involves investment.

5.3.2 Policy Rent

Equally important is the question of the income effects of agricultural policy. The function of these effects, $g(B)$ expressed in formula (5.5), can be written as follows:

$$g(B) = \bar{T}_R + T_B \cdot Z_t \qquad (5.10)$$

where:

[11] This area is the topic of the work by Bezat and Rembisz (2011).

[12] Measuring TFP based on the grain industry in Bezat (2008).

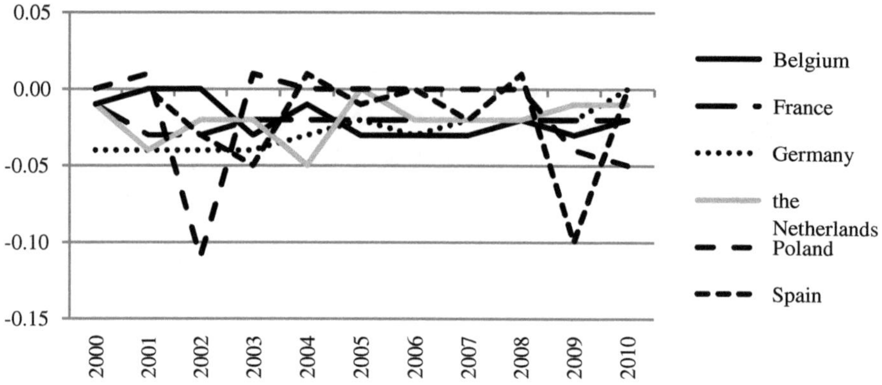

Fig. 5.2 The rate of growth of the labour factor $\left(\frac{\Delta L}{L}\right)$ in selected countries of the European Union in the years 2000–2010. *Source* own calculations based on the FAO and EUROSTAT data

$\bar{T}_R$ the income effects related to market intervention in the organisation of common markets (CMO) expressed as the average level of financial support per agricultural holding;

$T_B \cdot Z_t$ direct area payments per hectare of arable land and the area of this arable land in the agricultural holding at a given time, having a direct impact on the income of agricultural producers

We can use income effect (payments) expectations of agricultural policy according to the following function:

$$E[g(B)] = p(t) \cdot (T_B \cdot Z_t) \tag{5.11}$$

where[13]:

$p(t) = p(f(R_{t-1}) = p(f(K_{t-1}, L_{t-1})$ payments linked to production achieved from the previous base period

Figure 5.3 shows the costs and value of production and the proportion of support in the framework of the agricultural policy in production value. As we can see the proportion of support in production value has increased in the period 2004–2009.

In the considerations as part of the analysis, Producer Support Estimate (PSE)[14] was provided, which presents how income counted in producer prices is higher as a

---

[13] Cf. the concept of Hennessy (1998) and Ghobin and Guyomard (1999).

[14] Other indicators are: Market Price Support (MPS), which specifies the impact of price adjustment on the amount of retransfers to agricultural holding, Consumer Support Estimate (CSE), which characterises the costs incurred by consumers as a result of the support system used and consumer NPC presenting the relation between domestic price and international price, without the support system, paid by the consumer. The total size of the transfers is represented by Total Subside Estimate (TSE)—describing retransfers from consumers and producers adjusted by transfers of producers to the budget (including from taxes paid) (Czyżewski and Kułyk 2009).

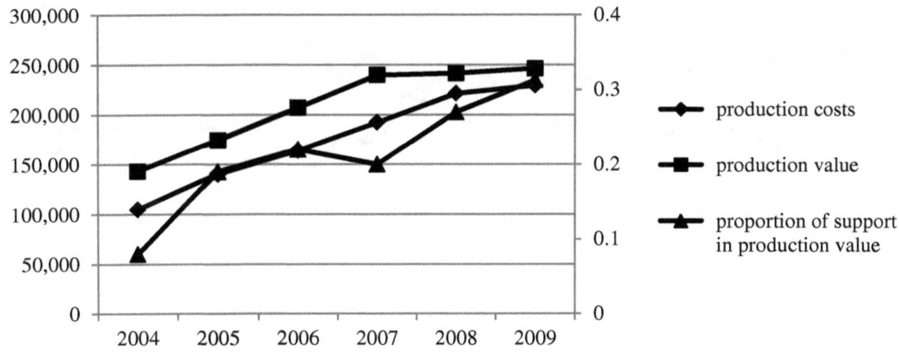

Fig. 5.3 The costs and value of production and the proportion of support in the framework of the agricultural policy in production value in the years 2004–2009. *Source* own calculations based on FADN data

result of the obtained support in comparison to results without the support system. This indicator covers: price support, payments for production, payments for acreage and livestock, payments for indirect consumption, payments limiting the involvement of current means of production, supporting income and other re-transfers (Fig. 5.4).

The income effect of agricultural policy included in formula (5.11) is the essence of policy rent. It is associated with expecting support which is receivable, almost as per its definition. This expectation is important here. This is in a sense a reference to the theory of rational expectations and adaptive behaviour resulting from this. Could it not reflect on the ambition to maximise the objective function by improving efficiency of production? Let us examine this further.

5.4 Substitution Between Economic and Policy Rent

As demonstrated above, the producer maximises his objective function—income—based on two arguments: (a) production efficiency and (b) support and transfers resulting from the agricultural policy. Both these sources of income require cost (effort). Under the terms of rational choice (also rational expectations as mentioned before), producers seek balance by appropriately substituting the more expensive and demanding source with a relatively cheaper one, which does not require much effort to obtain. Improving economic efficiency (and profitability), in particular improving the use of efficiency of the factors of production at a given price relation, is always difficult. Using transfers is not costless, but it seems cheaper. Therefore, we upheld the above argument that the producer behaving rationally will always be willing to adopt cheaper (more effective) solutions[15] .

[15] A more effective solution means achieving a certain level of income with lower effort.

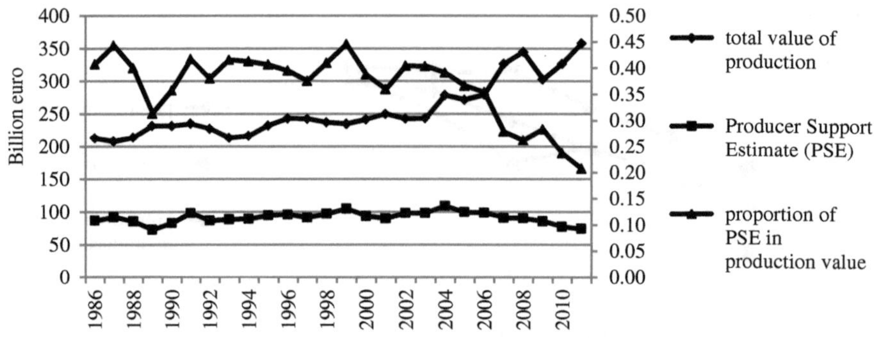

Fig. 5.4 The value of agricultural production and support of producers (PSE) in the EU in the years 1986–2011. *Source* own calculations based on OECD stats

To prove this observation, let us assume full and continuous substitution of these two factors (sources) of changes in the income of an agricultural producer, which we derived above. In addition, we assume that we consider this phenomenon for a given level of income. Increasing the use of a single source (factor), without changing the income, must therefore be at the expense of another factor. As a result, the differential of income equation:

$$D = f(\mathrm{EP}, B) \Rightarrow \max \tag{5.12}$$

is equal to zero, so we have:

$$dU_R = \Delta\mathrm{EP}\frac{\partial U_R}{\partial \mathrm{EP}} + \Delta B\frac{\partial U_R}{\partial B} = 0 \tag{5.13}$$

where:

$\Delta\mathrm{EP}\frac{\partial U_R}{\partial \mathrm{EP}}$ the income effect of improving the efficiency of production,

$\frac{\partial U_R}{\partial B}$ marginal utility efficiency of an agricultural producer, that is, from the point of view of realising his objective function,

$\Delta B\frac{\partial U_R}{\partial B}$ the income effect of increasing the scope of support of the agricultural producer under the CAP,

$\frac{\partial U_R}{\partial B}$ income marginal utility of support under the CAP for realisation of the objective function of an agricultural producer

Therefore, the agricultural producer optimises his choice, i.e. reaches equilibrium when it comes to these two sources of income for the objective function (income maximising), when we have:

$$\pm\Delta\mathrm{EP}\frac{\partial U_R}{\partial \mathrm{EP}} = \mp\Delta B\frac{\partial U_R}{\partial B} = 0 \tag{5.14}$$

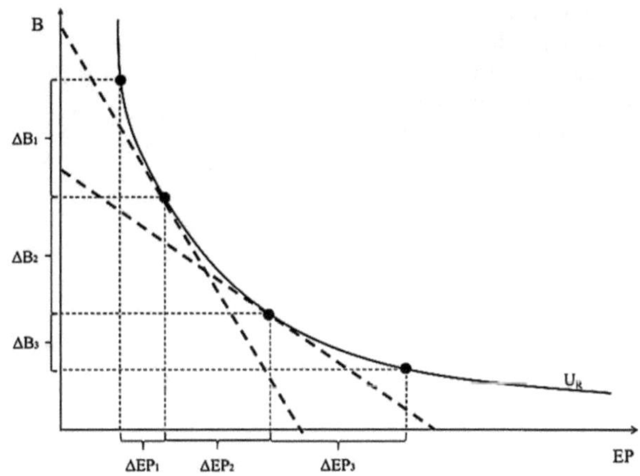

Fig. 5.5 The relationship between the level of efficiency (EP) and the level of support (B). *Source* own work

that is, when equalising the benefits of measures to improve production efficiency and measures to use the benefits of any intervention and support. In fact, by acting reasonably, they compensate for marginal utility of these two sources to improve their objective function. We omitted the minus sign here, so as not to suggest the direction of substitution between these two sources to improve income of an agricultural producer.

This condition means that the agricultural producer has reached a balance, i.e. they maximise their objective function—income, when the income effect of a policy equates to the loss of income effect as a result of deterioration in the efficiency of production. This decrease in production efficiency stems from the fact that support resulted in a decreased pressure to improve efficiency that would exist if it were not the support[16] . But we need to remember that these are relative and unit values because they refer to a given level of production (on a given isoquant), as shown in Fig. 5.5.

Acting rationally, the agricultural producer will choose an easier solution (the policy rent), although they can be discouraged from this by ever-increasing bureaucratic and cumbersome procedures (that generate increasing transaction costs associated with obtaining a transfer under agricultural policy instruments). In addition, on the basis of rational expectations, they can always provide for adaptation of the level of support to deteriorating economic climate in agriculture to lower profitability, etc. They have considerable political reference material and scientific support.

Formally, the condition of substitution between these two sources of realisation of the objective function can be written as follows:

[16] The direction of this substitution discussed on the basis of the above formula may go the other way round, i.e. growing income effects of improved efficiency replace the need for support of the agricultural policy. It seems, however, less likely.

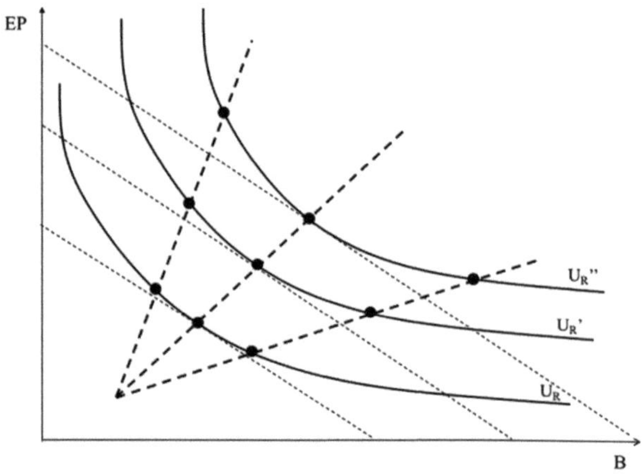

Fig. 5.6 Sample income expansion path depending on the selection of the combination of economic and policy rent. *Source* own work

$$s_{\mathrm{EP}/B} = \frac{\Delta EP}{\Delta B} = \frac{\frac{\partial U_R^{\mathrm{EP}}}{\partial \mathrm{EP}}}{\frac{\partial U_R^B}{\partial B}} \tag{5.15}$$

The (marginal) rate of the substitution factor, which we defined as economic rent with a factor that we adopted as policy rent, is determined by the relation of their impact on the (objective) utility function of the agricultural producer. This rate of substitution is determined by the relation of the utility of these two sources of maximising income for an agricultural producer. This approach can be called an attempt to describe the mechanism of behaviour or choice of the agricultural producer. This mechanism is being referred at this point to a relation between economic and policy rent and generally the condition of agricultural policy as an exogenous condition. Economic rent, i.e. striving to improve efficiency, is an endogenous condition in this context.

Seeking to maximise utility, and thus the income existing in the objective function, the producer selects the more favourable combinations of available sources of their growth, i.e. a combination of economic and policy rent. The producer's behaviour refers here to consumer behaviour, maximising the utility of their basket of goods. The system of the combination of economic and policy rent can be defined as a growth (expansion) path of income (Fig. 5.6). The curve of the expansion of income is conditioned by the substitution rate between economic rent and policy rent.

Thus, we can assume that the condition described with the above formula (5.16) highlights substitutability between these two sources of the improving income objective function of the agricultural producer. It shows the essence of the problem. There is some substitutability of support effects of agricultural policy in relation to the agricultural producer's efforts aimed at improving efficiency, as a primary source of income growth. This is a potential threat, because it may hinder those efforts which rely on structural change and processes of concentration. Of

Table 5.1 Changes in economic and policy rent and the rate of substitution between these types of rent among agricultural holdings in Poland in the years 2005–2009 (year t-1 = 100, delta EP and delta B in PLN)

Description	2005	2006	2007	2008	2009
Farms with predominant animal production (a)					
ΔEP	−2445	22651	−3425	−26560	−5981
ΔB	11063	9798	−6670	10880	2322
Rate of substitution	−0.22	2.31	0.51	−2.44	−2.58
Farms with predominant plant production (b)					
ΔEP	−10534	−2833	21642	−36165	−7532
ΔB	15375	14446	−7783	21540	16122
Rate of substitution	−0.69	−0.20	−2.78	−1.68	−0.47
Farms with any specified type of production (c)					
ΔEP	−5215	7902	6703	−29295	−6777
ΔB	12592	11206	−5757	16751	8332
Rate of substitution	−0.41	0.71	−1.16	−1.75	−0.81

Source own calculations based on FADN data

course, potentially this does not mean that it is a real threat. There may be a completely different synergy process, when the income effects of support under the agricultural policy have an impact on investment and the associated modernization of manufacturing techniques and technology, and, as a result, on improved efficiency of production. This requires separate empirical research and an additional analytical framework. The one in this analysis refers to static rather than dynamic conditions without taking account of investments. The investments are shown in the next part of the chapter.

The changes in economic and policy rent have been shown in Table 5.1. The rate of substitution is negative in most analysed years, which confirms analytical relationships shown in Eq. (5.15).

The assessment is carried out on the basis of data collected from farms across Poland within the framework of Farm Accountancy Data Network (FADN). The dataset included all the farms collecting the data within the framework of the FADN.[17] The sample covers farms within three groups: (a) with the predominant animal production (more than 50 % of revenue coming from the animal production) (b) with the predominant plant production (more than 50 % of revenue coming from the plant production) (c) with any specified type of production. The selection of a specific group is made because of different production technologies

[17] One needs to remark that data collected within the framework of the FADN includes only the farms with an economic size bigger than 2 ESU. European size unit, abbreviated as ESU, is a standard gross margin of EUR 1,200 (Eurostat 2012). Economic size thresholds applied by the Commission (in ESU) from year 2008 for Poland amounts to 2 ESU. European size unit, abbreviated as ESU, is a standard gross margin of EUR 1,200 (FADN 2012).

in farms focussed on plant, animal or mixed types of production. The production data is reported as revenue/expenditure denominated in PLN in constant prices.

The rate of substitution (Eq. 5.15), as shown above, with the relation of marginal utility of improved production efficiency (economic rent) and agricultural policy (policy rent) for income, should be weighed against the costs to obtain these utilities. This is not, however, easy because it would be difficult to assume any limit on these costs as a condition for the objective function of the producer due to the two factors discussed. It would be easier to determine the costs of achieving marginal utility from improved efficiency than the costs of achieving this thanks to a policy (participation in specific programmes or mechanisms).[18] This requires additional analysis and research. The one in this analysis refers to static rather than dynamic conditions without taking account of investments.

5.5 Policy Rent and Agricultural Producer's Investment

As we previously stated, a rational agricultural producer should substitute the source of income, which requires more effort to achieve a certain level of income with a relatively cheaper one. We find it important to assess not only the substitution between the two sources of income which are in our case production efficiency and political rent, but also to study to what extent investment decisions (which create the basis for future income[19]) depend on political rent.

The main aim of this section is to provide insight into the relationship between income, investment and subsidies on the basis of publicly available datasets. Due to the fact that "farmers are by far the largest source of investment in agriculture" (FAO 2012, p. XIII) we concentrate on the micro (producer's) level.

Among agricultural producers, there are different motivations for increasing savings and/or investments (Odoemenem et al. 2013). Therefore, complete analysis of the ways, in which they are created and how their values change in different groups of agricultural households requires exact microeconomic data including not only economic characteristics of individual households, but also demographic variables. The reason for that is the fact that decisions on whether to invest are likely to depend on the individual characteristics of the decision maker, i.e. his

[18] We can probably assume that: $kd_{Ep} > kd_B$, where: kd_{EP} —the cost of achieving income effects owing to economic rent, kd_B—the cost of achieving income effects owing to policy rent. This, as we can assume, determines the direction of substitution in the scope of both types of rent analysed here. Policy rent somewhat supersedes economic rent, as it were. A wider analysis will be done on a separate occasion.

[19] In the long run producer's income depends not only on labour input, but on capital input as well. Because of this fact in this part of the paper, we lift simplifying assumptions presented in formula (5.3).

attitude towards risk.[20] Unfortunately, at the micro level, such data can be obtained only by the survey.[21]

Despite the fact that the data on precise determinants of investment are usually unavailable, we can assume that the level of investment undertaken by an individual producer can result from its income from previous periods, according to the following formula:

$$D_{t-1} \Rightarrow S_{t-1} \Rightarrow I_t \tag{5.16}$$

where:

D_{t-1} the income achieved in t-1 period,
S_{t-1} savings in t-1 period,
I_t investments

Changes in net investment and farm income data in Poland for the period 2004–2009 are shown in Fig. 5.7. Variables are highly correlated (Pearson's correlation coefficient between D_{t-1} and I_t takes the value of −0.68). As we can see in Fig. 5.8, higher income results in higher investments only for the farms characterised by the greatest economic size. Therefore, in spite of the lack of data on savings (essential for investment), economic size can be used as one of the potential explanatory variables in an investment forecast problem.

As we indicated before, growing efficiency of production subsidies increase agricultural producer's income; therefore, they may have a positive impact on producer's investment possibilities. It can be denoted as follows:

$$\mathrm{EP} + B \Rightarrow D \tag{5.17}$$

and:

$$\Delta B \Rightarrow \Delta S \Rightarrow \Delta I \tag{5.18}$$

Relationships between net investment and different types of subsidies are presented in Fig. 5.9. Despite high correlations between those variables (values of Pearson's correlation coefficients are shown in Table 5.2), an increase in subsidies does not result in growth of investment undertaken by agricultural producers. Only in groups of farms characterised by an economic size greater than 100 ESU, is there a positive correlation between investment and all subsidies considered by us.

---

[20] In the case of some decision-makers (households) "savings serve in part as a buffer against stochastic decreases in income" (Birdsall et al. 1996). When these precautionary motives for savings are concerned, decision-makers are less prone to undertake investments because of their irreversibility (Pender and Fafchamps 1997).

[21] FADN database contains variables related to the demographical characteristics of an agricultural household that may help explain investment behaviour, but in our opinion they should not be treated like a complete set of information explaining investment decisions.

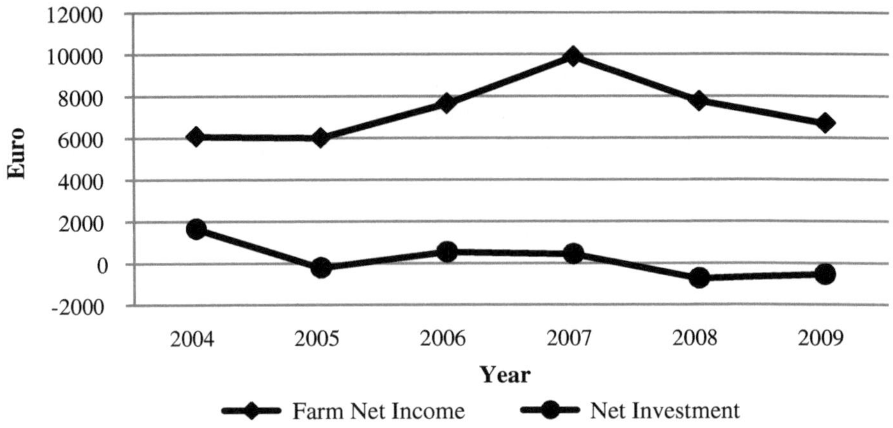

Fig. 5.7 Net investment and farm net income in Poland in the years 2004–2009. *Source* FADN public database

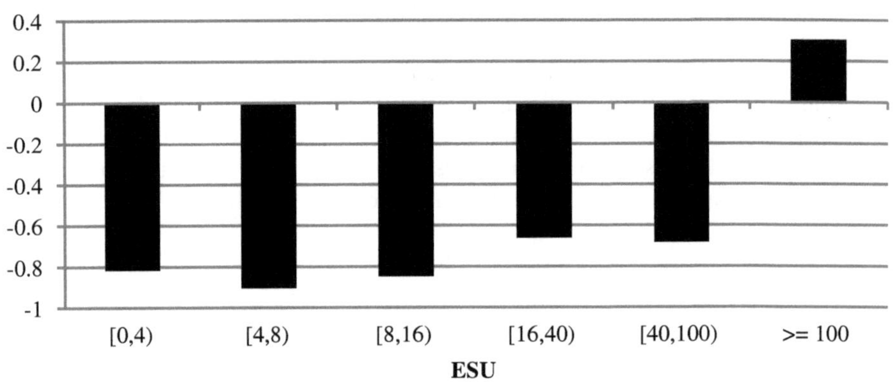

Fig. 5.8 Values of Pearson's correlation coefficient between income D_{t-1} and investment I_t for Polish farms in the period 2004–2009 by the economic size of farm. *Source* own calculations based on FADN public database

Together with our earlier observations, it may lead to the conclusion that subsidies are treated as an incentive for investments only in that group of agricultural producers.

It should be noted, however, that increasing subsidies (and growing income) are not the only factors influencing investment decisions.

When we concentrate on the dynamics of the process, i.e. the relationship between growth rates of subsidies (denoted as: ΔB) and growth rates of investments (denoted as: ΔI), it is clear that in order to speak about positive effects of agricultural policy on development in the agricultural sector, the relation between these two values has to be greater than 1 (formula 5.19). That means, the increase

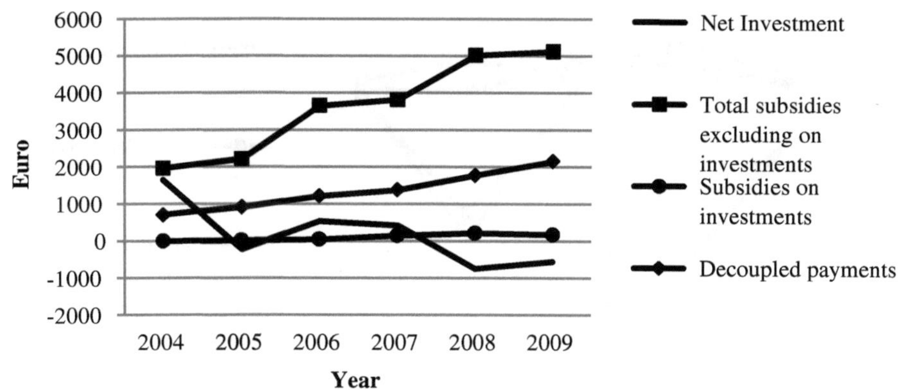

Fig. 5.9 Net investment, total subsidies excluding subsidies on investment, subsidies on investment and decoupled payments for Polish farms in the period 2004–2009. *Source* FADN public database

Table 5.2 Pearson's correlation coefficients between investment and subsidies B_{t-1} in Poland in the years 2005–2009 by type of subsidy and economic size of the farm

Economic size	Total subsidies excluding investments	Subsidies on investments	Decoupled payments
0–<4 ESU	−0.439	−0.433	−0.607
4–<8 ESU	−0.708	−0.703	−0.731
8–<16 ESU	−0.732	−0.788	−0.925
16–<40 ESU	−0.414	−0.440	−0.696
40–<100 ESU	−0.342	−0.392	−0.599
>=100 ESU	0.221	0.300	0.145
Total	−0.501	−0.515	−0.717

Source own calculations based on FADN data

in investments is faster than the increase in subsidies, because subsidies (support) multiply investments as the stable basis for growth.

$$\frac{\Delta I}{\Delta B} > 1 \tag{5.19}$$

The growth rates for farm net income, subsidies (including subsidies on investment), net investment and the factor $\left(\frac{\Delta I}{\Delta B}\right)$ are presented in Fig. 5.10. Growth rates for the first two variables have similar values close to 1, while changes in net investment were more rapid in the period 2005–2009. That resulted in dynamic changes of $\frac{\Delta I}{\Delta B}$, which in all years were below 1.

If we assess the above-mentioned relation within groups of agricultural producers specified on the basis of economic size, it is clear that in the majority of groups the value of $\frac{\Delta I}{\Delta B}$ is positive (Table 5.3).

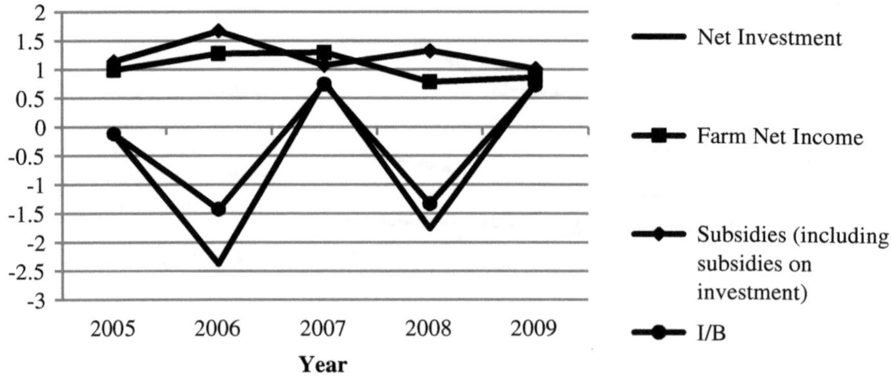

Fig. 5.10 The growth rate of net investment, farm net income subsidies (including subsidies on investment) and $\frac{\Delta I}{\Delta B}$ in Poland in the years 2005–2009 (year t-1 = 1). *Source* FADN public database

Table 5.3 Relation between growth rate of investments and subsidies (including subsidies on investment) in Poland in the years 2005–2009 by economic size of the farm

	0–<4 ESU	4–<8 ESU	8–<16 ESU	16–<40 ESU	40–<100 ESU	>=100 ESU	Total
2005	1.1459	1.4851	0.0249	1.1372	0.9684	−0.1034	−0.2459
2006	0.5458	0.5751	1.8665	1.1322	0.8961	−13.5830	−1.1259
2007	0.7960	1.1422	0.2608	0.8408	1.2254	0.7261	0.7887
2008	1.2679	1.3748	−5.0611	0.2849	0.2794	1.9261	−1.7493
2009	0.8419	0.7230	0.7448	1.3645	1.2556	0.3046	0.6902

Source own calculations based on FADN data

What is more, we are able to distinguish periods in which it is greater than 1, i.e. we may assume that in the future producers will benefit from greater investments undertaken in this period. It is important to point out that this is not only the case of farms characterised by the greatest economic size, but also small ones. It means that in the long term, all specified groups are able to benefit from greater investment opportunities resulting from subsidies.

5.6 Conclusions

The chapter discusses the topic of the sources of income shaping and its growth, as a basis of the objective function of the agricultural producer. By maximising this function and striving to increase the level of utility, the producer chooses between two major sources of this growth, namely production efficiency and transfers resulting from agricultural policy. Those two elements—referred to in the chapter in the same way as by other researchers, and for simplification purposes as economic and policy rent—determine the producer's income effects. Under the assumption that the producer's decisions are rational, it is clear that the dominant source of shaping income and its growth will be the type of rent which is more useful—it generates a given income level at lower cost.

Results obtained from FADN data show that the rate of substitution of these two sources of income growth is not equal to one, which means that replacing one with the other is not without any effect on the level of income. This is due to the fact that, first, transaction costs of achieving each type of rent, and second, changes in efficiency terms are related to investment, which do not otherwise exist in the case of transfers from agricultural policy. In most cases the rate of substitution is negative which positively verifies our assumptions and reasoning.

There is a high negative correlation between subsidies and investment. An increase in subsidies does not result in growth of investment undertaken by the agricultural producer. Only in the case of a group of farms characterised by an economic size greater than 100 ESU is a correlation between investment and all types of subsidies considered by us positive. Subsidies can be treated as an incentive for investments only in that group of agricultural producers. On the other hand in the majority of groups of farms, investments increased faster than subsidies, which means that in the long term the majority of groups are able to benefit from greater investment opportunities resulting from the income effect of agricultural policy.

The chapter is based on microeconomic analytical formulas describing the choices of the agricultural producers. The data collected from national and international statistics are mainly to illustrate regularities or conclusions derived from the model and the selected analytical formulas.

Acknowledgments This publication was prepared in the Independent Laboratory of Mathematics Application in Agricultural Economics (Institute of Agricultural and Food Economics, National Research Institute) as a contribution to the research conducted within the Multi-Annual Programme 2011–2014 "competitiveness of the Polish food economy in conditions of globalization and European integration".

References

Bezat, A. (2008). DEA-based Malmquist TFPC index as a toll for measuring of the productivity change over time. In Z. Binderman (Eds.), *Metody ilościowe w badaniach ekonomicznych: wielowymiarowa analiza danych.* Quantitative methods in economic research: multidimensional data analysis (vol. 9, pp. 19–28). Warsaw: SGGW.

Bezat, A., & Rembisz, W. (2011). Zastosowanie funkcji typu Cobba-Douglasa w ocenie relacji czynnik-produkt w produkcji rolniczej. Using the function of Cobb-Douglas type in assessing the relation factor-product in agricultural production. Komunikaty. Raporty. Ekspertyzy, no. 557, Warsaw: IAFE-NRI.

Bezat-Jarzębowska, A., Rembisz, W., & Sielska, A. (2012). *Growth model of agri-food production*. Warsaw: IAFE-NRI.

Bezat-Jarzębowska, A., & Rembisz, W. (2013). Efficiency-focused economic modeling of competitiveness in the agri-food sector. *Procedia—Social and Behavioral Sciences, 81*(s), 359–365.

Birdsall, N., Pinckney T. C., & Sabot R. H. (1996). Why low inequality spurs growth: savings and investment by the poor. *Inter-American Development Bank Office of the Chief Economist*, Working Paper 327. http://citeseerx.ist.psu.edu/viewdoc/download?doi=10.1.1.195.8650&rep=rep1&type=pdf

Czyżewski, A., & Kułyk, P. (2009). Relacje między otoczeniem makroekonomicznym a rolnictwem w warunkach zmiennej koniunktury gospodarczej w UE-15 i Polsce w latach 1990–2008. Relationships between macroeconomic environment and agriculture in the context of volatile economic climate in the EU-15 and Poland in the years 1990–2008. In: *Makroekonomiczne uwarunkowania rozwoju gospodarki żywnościowej The macro-economic conditions in the development of food economy*, (pp. 7–19) A. Borowska & A. Daniłowska (Eds.), Warsaw: SGGW.

EUROSTAT. (2012). *Glossary: European size unit* (ESU). September 2012. http://epp.eurostat.ec.europa.eu/statistics_explained/index.php/Glossary:European_size_unit_%28ESU%29

FADN. (2012). *Field of survey*. September 2012. http://ec.europa.eu/agriculture/rica/methodology1_en.cfm

FAO. (2012). *The state of food and agriculture 2012*. May 2013. http://www.fao.org/docrep/017/i3028e/i3028e.pdf

Figiel, S. Z., & Rembisz, W. (2009). Przesłanki wzrostu produkcji w sektorze rolno-spożywczym—ujęcie analityczne i empiryczne [The conditions of production growth in the agri-food sector—analytical and empirical approach]. Multi-annual Programme 2005–2009, nr 169, Warsaw: IAFE-NRI.

Ghobin, A., & Guyomard, C. (1999). *Measuring the degree of decoupling of alternative internal support policy instruments*. Warsaw: EAAE Conference.

Hennessy, D. (1998). The production effects of agricultural income support policies under uncertainty. *American Journal of Agricultural Economics,80*.

OECD Stats. (2012) November 2012. http://stats.oecd.org/

Odoemenem, I., Ezihe, J., & Akerele, S. (2013). Saving and investment pattern of small-scale farmers of benue state, nigeria. *Global Journal of Human Social Science (CD), XIII*(WI) Version I.

Pender, J., & Fafchamps, M. (1997). Precautionary saving, credit constraints, and irreversible investment: Theory and evidence from semi-arid India. *Journal of Business and Economic Statistics, 15*(2), 180–194.

Poczta, W. (2002). *Ocena polskiego rolnictwa pod kątem jego konkurencyjności na rynku unijnym [Assessment of Polish agriculture for its competitiveness in the European market]*. Warsaw: RSSG.

Rembisz, W. (2007). *Mikroekonomiczne podstawy wzrostu dochodów producentów rolnych [Microeconomic foundations of the income of agricultural producers]*. Warsaw: Vizja Press and It.

Sielska, A. (2012). *Decyzje producentów rolnych w ujęciu wielokryterialnym—zarys problemu [Decisions of agricultural producers based on the multi-criterion approach—outline of the problem]*. Warsaw: IAFE-NRI.

Wilkin, J. (2005). Pogoń za rentą przy pomocy mechanizmów politycznych [Pursuit of rent using political mechanisms.] In: Teoria wyboru publicznego. Wstęp do ekonomicznej analizy polityki i funkcjonowania sfery publicznej [Theory of public selection. Introduction to economic analysis of policy and the functioning of public sphere]. Warsaw: Scholar, chapter 10: 204–219.

Woś, A. (2004). *W poszukiwaniu modelu rozwoju polskiego rolnictwa [In search of a model of development of Polish agriculture]*. Warsaw: IAFE-NRI.

Zawadzki, H. (2009). *Problemy optymalizacyjne w ekonomii matematycznej [Optimization problems in mathematical economics]*. Katowice: Wyd. AE Katowice.

Zegar, S. J. (2009). *Dochody rolników po akcesji do unii Europejskiej [Income made by farmers after accession to the European Union]*. http://www.jard.edu.pl/pub/15_2_2009.pdf

Author Biographies

Agnieszka Bezat-Jarzębowska Adjunct Professor at National Research Institute of Agricultural and Food Economics and Warsaw University of Life Sciences (Faculty of Economic Sciences). The main research and publishing issues in microeconomics, agricultural economics and efficiency.

Wlodzimierz Rembisz Professor at National Research Institute of Agricultural and Food Economics and Warsaw University of Finance and Management, Poland. Expert in EU and UNDP projects. The main research and publishing issues in microeconomics and agricultural economics—specifically: Producers incomes, price risk management, farm market and intervention.

Agata Sielska Assistant at National Research Institute of Agricultural and Food Economics and Ph.D.-candidate at Warsaw School of Economics.

Chapter 6
Performance Evaluation of Rural Governance Using an Integrated AHP-VIKOR Methodology

Giuseppa Romeo and Claudio Marcianò

Abstract The objective of this study is to provide an assessment of the performance of the development process of the Leader Approach in the Rural Development Programme 2007–2013 in Calabria. In particular, on one hand, it measured and evaluated the application of the 'Good Governance' criteria by the identification and selection of an appropriate set of process indicators, procedures and actions practiced by Local Action Groups (LAG) in the planning phase of the Local Development Plans. On the other hand, those indicators have been employed to construct an evaluation tree through which the performance of each LAG is tested and compared to an ideal model of 'Good Governance'. To this end, it used an integrated multi-criteria model, which combines two techniques: the Analytic Hierarchy Process (AHP) and VlseKriterijumska Optimizacija the Kompromisno Resenje (VIKOR). The results deal with the definition of a possible model for assessing the quality of the integrated planning process of rural governance. The results can be useful to policy makers at the regional level and also to the LAGs in highlighting eventual elements of criticism and possible virtuous behaviours of their own planning process that can be considered in the future EU programming period of 2014–2020.

6.1 Introduction

In the last twenty years the European Community has identified in LEADER the most suitable instrument to promote the integrated development of the rural areas so that, in the 2007–2013 EU planning period, it lost its experimental modality of 'laboratory' to assume that of 'approach'. Such a passage has strengthened and consolidated the typical features of the participation and of the bottom-up ability

---

G. Romeo · C. Marcianò (✉)
Agraria Department, Mediterranean University of Reggio Calabria, Reggio Calabria, Italy
e-mail: claudio.marciano@unirc.it

C. Zopounidis et al. (eds.), *Agricultural Cooperative Management and Policy*, Cooperative Management, DOI: 10.1007/978-3-319-06635-6_6,

related to the strategies of territory development. The purpose of the Leader Approach is to promote the local development through a process of participated and bottom-up concerted actions, to understand better the real needs of local communities (Storti et al. 2010). More in detail, through integrated planning, it is possible to strengthen the capacity of a territory to set systematic relations among subjects of a different nature, able to carry out an active role in the realization of a local development process. The same Council Regulation (EC 2005) No. 1698/2005 states the necessity of building a *joint strategy of local development*, through a public-private partnership able to work and carry out multi-sector strategies, important as a *start-up* in improving synergic actions among actors and projects belonging to the different sectors of rural economy.

The Local Action Groups (LAGs), as an expression of the Partnership, are deputed to favour the conservative meeting among the public, private and social components of the territory. The instrument of the LAGs making such a synergy possible is the Local Development Plan (LDP). Therefore, the Partnership becomes the organizational form of the institutional model of *governance* which, by involving the local actors of development and civil society, represents a modality potentiating the quality of public policies and their probabilities of success, widening the consensus and the responsibility of each partner (Aa 2007). However, the will of enforcing the dimension of the participatory democracy, felt at a European level, through a greater involvement of the civil society in the decision-making process, has contributed to developing the normative concept of 'Good Governance' (GG) (EC 2001). Playing a very important role within the implementation of rural development policies (Birolo et al. 2012), such a term was introduced for the first time in 1989 by the Word Bank (Chowdhury and Skarstedt 2005). In spite of its growing use, an agreement has still not been decided on the essence of its concept (Santiso 2001; Börzel et al. 2008; McCall and Dunn 2012). In the literature, there are several definitions affected by the research lines of the followed program; in particular, regarding studies about development, the concept of GG is usually singled out with that of *Governance* (Börzel et al. 2008).

In 2001 the European Commission published the 'White Paper' on the GG which contributed to modifying radically the mechanisms of functioning, government and production of European policies (Zurla 2010). The GG is focused on five principles: opening, participation, responsibility, effectiveness, coherence, the application of which also interests policies at a local level. Such principles hold a growing importance in different sectors, among which that of rural development where they have reached a key role. On one hand, it is evidence of the insertion of the GG among the main four dimensions to improve the quality of life in rural areas; on the other, it has been important to recognize the purpose of the Leader Approach to potentiate governance from the aspects of the cooperation, planning and participation of local actors (Secco et al. 2011). In this view, an appropriate assessment of the quality of governance can represent a useful instrument of support to decisions both in the steps of formulation, realization, revision of social policies and in the development projects of rural areas (Franceschetti et al. 2012a).

However, the literature offers us various studies aimed mainly at evaluating the results and the impact of the policies worked out by the partnerships; on the contrary, a lack of studies is noticed analyzing the activities and the relations characterizing the planning phase of local development strategies. Therefore, the choice of focusing on the planning phase has as a purpose the analysis and the assessment of the different aspects with which the partnerships have operated, in order to determine and formulate development actions. This is motivated by the assumption according to which the result of a process depends largely on the quality of the same planning process (Reeb 2004). It comes out that the quality of the organizational and planning capacity of the partnership influences both the management of the financial resources and the achievement of excellent performance levels in the implementation phase, with positive or negative effects on the impact of the policies on the territory.

On the basis of such preconditions, the analysis of the elaboration process of the LDP represents a first useful step to understand how the LAGs have been thinking and organizing a development process in their own territory. It is important to investigate the mechanisms of the rural governance related both to the activities carried out for the decision-making and the existent relationships among the institutional actors and stakeholders (partners or not) of the LAG's. Such a concept relates to the models of interaction, in which coherence and effectiveness of the management of the territorial processes depend mainly on the horizontal and vertical coordination among different institutional and social actors, and on their ability to attain a sharing of purposes, a negotiation of agreements and a cooperation in order to achieve them (Governa 2004).

In this way, the present study has as a purpose an evaluating analysis of the performance of the development process of the Leader Approach in the Rural Development Programme 2007–2013 in Calabria and the empirical part of the study assesses how the LAGs have been using the existing planning instruments. In particular, on one hand, we want to measure and evaluate the application of the 'GG' criteria, studying them in depth, by the identification and selection of an appropriate set of process indicators (Franceschetti et al. 2012a, b), procedures and actions practiced by LAGs in Calabria to reach the purposes of the program. On the other hand, those indicators will be employed to construct the tree evaluation through which the performance of each LAG will be tested compared to an ideal model of '*GG*'. The object of the assessment is not the decision itself, which represents the result to achieve, but the process adopted to reach it (Kørnøv and Thissen 2000). The focus of the analysis of the evaluation process of the governance at a local level is not on *what* has been carried out, but *how* the development process has been structured. So, the activity of assessment is aimed at investigating the process with a more exploratory and developmental character than a confirmatory one (Brinkerhoff 2002). In the first step they have singled out the conceptual framework of the criteria (key-dimensions) and sub-criteria (sub-dimensions) of the GG. For each sub-criterion, a quantitative and qualitative indicator has been identified, the origin of which came out from an initial analysis of the literature, the transformation of the information gathered through interviews with experts and from the definition of new indicators (Franceschetti et al. 2012a).

Appropriate scales of measurement have been associated with the indicators of a qualitative nature. In the second step, after the indicators' measurement, the conceptual frame designed before has been used to build the evaluation tree of the performance of rural governance within the fields of the Leader Approach. For this purpose, we carried on with the implementation of the AHP and the VlseKriterijumska Optimizacija I Kompromisno Resenje (VIKOR) multi-criteria integrated model. With the first technique the weighing of both the GG criteria and sub-criteria has been carried out, while the second one has got the ranking of the performance of the rural governances under examination. With the AHP-VIKOR integrated method, unlike what is found in the literature, where the assessment of the quality of governance is confined to one or some dimensions of the GG (Secco et al. 2011; Birolo et al. 2012; Franceschetti et al. 2012b), it was possible to take into consideration all the dimensions at the same time. The structure of the paper is as follows: Sect. 6.2 presents the criteria and sub-criteria, while Sect. 6.3 describes briefly the methodology that has been applied. In Sect. 6.4, the AHP and VIKOR methods are applied for the ranking of the 11 case studies and the results are presented. The final section gives some concluding remarks and suggests potential future directions.

6.2 Features of the Good Governance Dimension

In the planning and implementation phases of integrated rural development processes, an essential component is represented by the public-private local actors and by the representatives of civil society. Such subjects, through the coordination of the governance, are asked to manage in a cooperative way the planning of the territory and the financial resources addressed to it. In this view, it is important to succeed in translating into action the principles of the governance, to be able to start up qualified processes of development (Di Iacovo and Scarpelli 2006) and the role of the various organizational and interactive modalities adopted within the partnership process becomes fundamental. Because of the fact that the principles affecting the GG have effects also on the practices of the planning processes (Annunzi 2006) of the local policies, they have been used as a reading key to evaluate the quality of the operational process of the rural governances. For this reason, the concept of GG has been subdivided into seven dimensions to which 39 sub-dimensions have been associated, identified to facilitate the individualisation of the process indicators, listed in Table 6.1. Social Capital and the Sustainability have been added to the traditional GG dimensions. The dimensions identified are listed as follows:

Table 6.1 Conceptual frame for good governance

Sub-criteria	Indicator	Indicator specification
Criterion 1: social capital		
1.1 Heterogeneity partnership	Network heterogeneity index	$NTh = -\sum_{i=1}^{N} f_i * \ln f_i$ N = maximum number of categories (I) potentially present in a LAG, f_i is the proportion of the number of actors belonging to the I category on the total actors in the LAG
1.2 Consensus partnership	Which percentage of decisions approved unanimously in the partnership assemblies rather than the ones agreed by the majority? (Abrams et al. 2003)	0–100 %
1.3 Construction mode partnership	Which channel has been mainly used to point out the actors of the partnership?	• Direct research (1) • Open meetings (3) • Expression of interest (5) • Other (7)
1.4 Collaborative capability	In which measure do the partnership actors recognize that the cooperation is able to produce qualitatively different results from the possible ones obtained working alone? (Kelly et al. 2012)	• No clear or marked benefit in working in partnership (1) • Some, but not all the participants recognize an added value in working in partnership (3) • All the participants believe that working in partnership leads to meaningfully better results than those possible to obtain working alone (5)
1.5 Agreements with stakeholders	Which is the number of collaborations through protocols of understanding/ letters of intent/agreements undersigned with the stakeholders present the LAG's territory (research boards/university, public boards etc.)?	Number of agreements
Criterion 2: efficiency		
2.1 Listening to the territory	Which is the percentage of time dedicated to the animation step for the specific activities of listening and individualisation of the needs of the territory?	0–100 %
2.2 Duration of partnership's assembly	Which is the average duration of a partnership assembly?	Duration expressed in minutes
2.3 Number of animators	Rating of density of the animation	Number of operators/Population density of the Leader's area

(continued)

Table 6.1 (continued)

Sub-criteria	Indicator	Indicator specification
2.4 Number of partnership's assemblies	Which was the number of partnership assemblies in the LDP's working-out step?	Number of assemblies
2.5 Period of consultation in partnership's assembly	How much of the joint activity of the interventions/actions to insert in the LDP has been carried out during the partnership assemblies?	0–100 %
2.6 Proactive skills in writing of the partnership	Which is the total number of protocols of understanding/proposals/written ideas were acquired from the partnership actors?	Number of written proposals
Criterion 3: effectiveness		
3.1 Conflict	Were there any non-constructive conflicts during the concertation of the assembly phase?	• No (0) • Yes, for the allocation of the resources to the single interventions (1) • Yes, for a not well-defined division of the activities and the responsibility among the actors (3) • Yes, for a poor transparency of the decision-making process (5) • Yes, for the dominance of some partners in the decision-making process (7) • Yes, other (9)
3.2 Presence of active actors	Which is the percentage of active actors within the partnership?	0–100 %
3.3 Contribution knowledge stakeholders	How much does it affect the fund information acquired by the stakeholder (partner) in the pointing-out of the strategy?	0–100 %
3.4 Meetings of pre-consultation	Which is the frequency of the pre-concertation meetings between and among the partnership's actors? (Abrams et al. 2003)	Void (1) Low (2) Medium (3) High (4)
3.5 Informal meetings	In order to acquire knowledge and exchange information, were the informal meetings or the formal ones more effective?	Informal (1) Formal (2)
Criterion 4: participation		
4.1 Meetings with stakeholders	Number of consultation meetings/territory understanding which have involved the stakeholders (partners or not) before the LDP's drawing up	Number of meetings

(continued)

Table 6.1 (continued)

Sub-criteria	Indicator	Indicator specification
4.2 Gathering info for the SWOT	Through which modalities was it possible to single out the necessary information for the SWOT analysis build-up?	Analysis desk (1) Consultations *ad personam*/ opinion leader (3) Interview with a questionnaire (5) Public meetings (7) Workshop (9) Thematic tables (11) Other (13)
4.3 Degree of representativeness of the territory	Rating of representativeness of the territory	Categories present in the LAGs / total categories of the LAGs
4.4 Use of participatory techniques	Was any participative modality or technique employed?	No (0) Yes, Focus Group (1) Yes, GOOP (3) Yes, PGIS (5) Yes, EASW (7) Yes, WEB (forum on-line) (9) Yes, METAPLAN (11) Yes, other (13)
4.5 Participation in initiatives of animation	Which is the percentage of the actors invited to the public meetings promoted by the LAG who accepted to participate?	0–100 %
4.6 Stakeholders' involvement in strategy	Which is the incidence of the stakeholder partners on the total number of the partnership's actors?	0–100 %
4.7 Expression of the views of the partnership	Which is the level of expression of ideas of the partnership's actors in the dialogue and confrontation phrase?	Void (1) Satisfactory (2) High (3)
4.8 Publication of collective animation activities	Which is the level with which the activities of collective animation are promoted to the stakeholders (not partners)?	Void (1) Low (2) Medium (3) High (4)
Criterion 5: transparency		
5.1 Verbal publication on the WEB	Are the minutes of the Board of Director, the members and the partnership published in the website?	No (0) Yes (1)
5.2 Decision support models	Are models of support to the decisions for the allocation of the resources to the single interventions employed?	No (0) Yes (1)

(continued)

Table 6.1 (continued)

Sub-criteria	Indicator	Indicator specification
5.3 Accessibility of information to stakeholders	Do the stakeholders (partners or not) have access to all the necessary information to understand the situation and participate in an effective way? (GFI 2009)	No (0) Yes, in part (1) Yes (2)
5.4 Openness of partnership assemblies	Are the partnership assemblies open to everybody?	No (0) Yes (1)
5.5 Communication animators - partnership	Do the animators synthesize and present to the partnership assembly all the information acquired in the territorial analysis and animation steps?	No (0) Yes (1)
Criterion 6: accountability		
6.1 Influence of stakeholders in decision-making	Which is the intensity of the level of participation of the partners in the partnership's assembly? (McCall 2003)	Void (0) Informative (1) Consultation (2) Decisional (3)
6.2 Presence of leader	In which measure do key individuals (leaders or participants) share, motivate or dominate the process and inspire the others in the participation? (Kelly et al. 2012)	• No clear Leader or individual actor appears as "champion" (1) • An individual holds a more relevant but not dominant role (2) • An individual holds a strong leadership role and "champion" partnership (3)
6.3 Decision support: technical group	Does the technical group recover a key role in the decision-making phase of the concertation?	No (0) A little considerable (1) Quite considerable (2) Very considerable (3)
6.4 Skills: private actors	In which measure have the private actors been more proactive in the planning stage?	0–100 %
Criterion 7 sustainability		
7.1 Continuity of planning	Which is the percentage of the call investments which could have a prosecution after the current planning?	Programmed amount of the actions of measures 121, 123, 311, 312, 323/Total amount measure 410 (0–100 %)
7.2 Involvement of private actors	Which is the incidence of the number of private actors on the total of the partnership's actors?	0–100 %
7.3 Involvement of private capital	Rating of mobilization of private resources	Shares of private co-financing/total expenditure planned in the LDP

(continued)

Table 6.1 (continued)

Sub-criteria	Indicator	Indicator specification
7.4 Stakeholder analysis	Has it been carried out in the preliminary step an analysis of the stakeholders to involve? (Reeb 2004)	No (0) Yes (1)
7.5 Integration with other plans	How many nets or plans of territorial development is the LAG partner or promoter of?	Number of nets or plans
7.6 Presentation of the plan to stakeholders	Which is the number of public meetings organized to present to the stakeholders (not partners) the plan before its last approval?	Number of meetings

Source authors' own elaboration

- *Social capital.* Considering that the partnership presents itself as a system of relations among actors, the social capital has become an essential element of the dynamics of local development (Aa 2007; Franceschetti 2009; Nardone et al. 2010; Lopolito et al. 2011). Such a dimension is particularly important both to understand how the collective actions have been carried out (Cundill and Fabricius 2010) and to verify if the strategies have been integrated and shared on the basis of the principles of co-partnership and cooperation which, through the synergic action, strengthens the production capacity of the same territory. With the Social Capital dimension it is possible to assess some intangible aspects of the planning process as the development of the trust and the cooperative capacity among the actors, and the ability to reach joint choices (Marcianò and Palladino 2013).
- *Efficiency.* This refers to the advantages of the information acquired by the actors involved in the planning phase, because they are crucial to define a development strategy based on territorial needs. Since the acquisition of information reduces the margin of uncertainty on the choices, the research of information is intended as one of the main factors of the GG's organizational factors (Augustyn and Knowles 2000).
- *Effectiveness.* The partnership's effectiveness is generally measured through its own capacity to reach a certain purpose. In the phase of policy planning the economic indicators are not enough to determine the partnership's effectiveness (Augustyn and Knowles 2000). So, such a dimension refers to the partnership's actors' ability to reduce the less constructive conflicts and, at the same time, to optimize the exchange of knowledge and information necessary to produce and carry out effective and joint decisions.
- *Participation.* The principle of participation derives from the acceptance of the fact that the community is the heart of the development (Srivastava 2009). It emerges that participation implies both the stakeholders' and the local communities' involvement in the development process, and that each of them play a role within the decision-making process. So, the participation gets a role of

functional utility, because it consents to using the learning capacity of the stakeholders and gives them a voice in the identification and planning of development interventions fitting the effective needs of the territory.

- *Transparency*. Transparency refers to the decision process, assuring that the information is freely available for all those who are involved or interested in the decisions taken. So, transparency means free access to the information (Căluşer and Sălăgean 2007).
- *Accountability*. Accountability reflects the values of democracy (Callahan 2007) and is linked to the responsibilities each actor holds in the activities he is involved in (McCall and Dunn 2012). In particular, such a dimension focuses on the role and degree of influence the actors take on in the decision process.
- *Sustainability*. sustainability concerns the partnership's ability to build up policies projected in a medium- and long-term period of vision without exhausting their utility at the end of the planning cycle. Within this strategic view it is important to be able to mobilize both private capitals and, even more so, actors of the local entrepreneurial tissue to let the LDP policies create a synergy with those from other plans.

As previously hinted, in the following conceptual frame the related sub-criteria are reported for each criterion with the corresponding process indicators used for the evaluation of the LAGs' performance in relation to an ideal model of GG (Table 6.1).

6.3 Methodology

6.3.1 Analytic Hierarchy Process

The Analytic Hierarchy Process (AHP) (Saaty 1988) is a Multiple Criteria Decision Making (MCDM) method which allows the decision-makers to represent the interaction of a variety of factors in complex situations (Abdi et al. 2011). The decision process is modelled on the building of a hierarchical tree in more levels, the number of which grows according to the level of data disaggregation. In respect of other multi-criteria methods, the AHP presents the advantage, above all in those situations where the subjective opinions are a fundamental part of the decision process, of being able to integrate tangible and non-tangible criteria (Abdi et al. 2011). It has been shown that the AHP helps to make those decisions more consistent, the criterion of which is expressed through subjective measures based on the experience (Schniederjans and Garvin 1997; Baykasoglu and Kaplanoglu 2008). So, the AHP capacity of assigning appropriate weights to the elements affecting the different levels of the decision tree lets its application be suitable for the performance assessments (Chia-Chi 2011).

Table 6.2 Judgment scale assessment

Value w_{ij}	Linguistic judgment	Interpretation
1	Equal importance	*i* and *j* are equally important
2	Moderate importance	*i* is moderately more important than *j*
3	Strong importance	*i* is highly more important than *j*
4	Very strong importance	*i* is decidedly more important than *j*
5	Extreme importance	*i* is extremely more important than *j*

The AHP is set out in three steps:

Step 1: Individualisation of the goal and the factors to evaluate, to which follows the decomposition or the hierarchical structuring of the decision problem of levels and sub-levels, each one characterized by certain components.
Step 2: Formulation of the comparative judgments through the pair-wise comparison among the components pointed out for each level. Formulation of the comparative judgments through pair comparison among the components pointed out for each level. The AHP's crucial point is the determination of the weights, specified on the basis of the subjective opinions expressed by experts through the delivery of the questionnaire. Considering *n* elements for each level of the tree, the procedure establishes the construction of a square matrix of the pair-wise comparisons 'A', in which w_{ij} denotes the importance of the element *I* in respect of the element *J*, through the attribution of a numeric score related to a semantic evaluation scale (numeric/linguistic), which varies from 1 to 5 (Saaty 1988) (Table 6.2).

As for mutual property, when w_{ij} = k, it automatically follows that w_{ji} = 1/k, while for the property of symmetry, all the elements on the diagonal are equal to 1, as in the following matrix:

$$A = \begin{bmatrix} 1 & w_{12} & \cdots & w_{1n} \\ w_{21} & 1 & \cdots & w_{2n} \\ \vdots & \vdots & \ddots & \vdots \\ w_{n1} & w_{n2} & \cdots & 1 \end{bmatrix} = \begin{bmatrix} 1 & w_{12} & \cdots & w_{1n} \\ 1/w_{12} & 1 & \cdots & w_{2n} \\ \vdots & \vdots & \ddots & \vdots \\ 1/w_{1n} & 1/w_{2n} & \cdots & 1 \end{bmatrix} \quad (6.1)$$

Step 3: Numerical determination of the weights through the normalization of the weight vector pointed out with matrix A. Analytically, the works go on with the calculation of the geometric mean (GM) of each ith line, giving the following equation:

$$\mathrm{GM}_i = \left(\prod_{j=1}^{n} w_{ij}\right)^{1/n} \quad (6.2)$$

where:
i, *j* = 1,2, …., *n*

$w_{ij} =$ Matrix A valuation judgment value
$n =$ Matrix A dimension number

Successively the normalization of the weight vector is worked out, obtained from the ratio between each single element of the column with the values of the GM and the total of the column, as expressed in the following equation:

$$W_i = \frac{\mathrm{GM}_i}{\sum_{j=1}^{n} \mathrm{GM}_i} \tag{6.3}$$

To verify the consistency of the results obtained, the self-value associated to the self-vector of the maximum module of the matrix related to the pair-wise comparison matrix A is determined, called $\lambda_{\max}$:

$$\lambda_{\max} = W_i * \sum_{j=1}^{n} w_{ij} \tag{6.4}$$

which enables to determine the Index of Consistency (IC), given by the following equation, in which n represents the number of criteria considered:

$$\mathrm{IC} = \frac{\lambda_{\max} - n}{n - 1} \tag{6.5}$$

Dividing the IC for the Random Index (RI), whose value derives from a pre-set table and is associated to the number of n considered criteria, the Consistency Ratio is obtained (CR):

$$\mathrm{CR} = \frac{\mathrm{IC}}{\mathrm{RI}} \tag{6.6}$$

The higher the value undertaken, the lower the consistency of the subjective judgments expressed by the decision-maker. In general, the CR's tolerance borderline is 10 %, but values of 20 % can also be tolerated (Fanizzi and Misceo 2010).

6.3.2 VIKOR

The VIKOR method was developed by Opricovic and Tzeng (2002, 2004, 2007) and refers to the theory of the displaced ideal of Zeleny (1975, 1982). VIKOR is a compromise ranking method which has been introduced as a useful technique to implement within MCDM (Opricovic and Tzeng 2003) and opens itself well to solving discrete decision problems with non-commensurable (different units) and conflicting criteria. This method has the advantage of letting the decision-maker to obtain a ranking of the alternatives analyzed, whose performance score takes into account simultaneously the set of criteria that are considered (performance matrix) (Tavakkoli-Moghaddam et al. 2011; Ou Yang et al. 2013). Assuming that each alternative is evaluated according to each criterion function, the compromise ranking is performed by comparing the measure of closeness to the ideal solution (Opricovic and Tzeng 2004; Kahraman and Kaya 2010), introducing the multi-criteria ranking index (Vinodh et al. 2013). The VIKOR method uses the aggregative function *L_p-metric* (Ou Yang et al. 2013), to formulate the ranking measures:

$$L_{P,K} = \left\{ \sum_{J=1}^{n} [W_J (|f_J^* - f_{K,J}|)/(|f_J^* - f_J^-|)]^P \right\}^{1/P} \tag{6.7}$$

where:

$K = 1, 2, \ldots\ldots, m$ represents the alternatives $A_1, A_2, \ldots, A_m$;
$J = 1, 2, \ldots\ldots, n$ represents the criteria $C_1, C_2, \ldots, C_n$;
$f_{K,J}$ = performance score for alternatives A_K with respect to the criteria C_J;
W_J = weight on the Jth criteria

The positive f_J^* and negative f_J^- ideal point respect to the Jth criteria among all alternatives is defined empirically. $L_{1,K}$ (as S_K Eq. 6.13) and $L_{\infty,K}$ (as R_K Eq. 6.14) are used to formulate ranking measures (Yucenur and Demirel 2012).

The compromise ranking algorithm VIKOR is summarized as follows:

Step 1: Decision matrix construction

The decisional problem criteria and alternatives are indentified. Qualitative or quantitative values are assigned to the criteria employed for each level of the constructed decisional tree, which can be concisely expressed in a matrix format (Bazzazi et al. 2011):

$$D = \begin{array}{c} \\ A_1 \\ A_2 \\ A_3 \\ \vdots \\ A_m \end{array} \begin{array}{c} \begin{array}{ccccc} C_1 & C_2 & C_3 & \cdots & C_n \end{array} \\ \begin{bmatrix} x_{11} & x_{12} & x_{13} & \cdots & x_{1n} \\ x_{21} & x_{22} & x_{23} & \cdots & x_{2n} \\ x_{31} & x_{32} & x_{33} & \cdots & x_{3n} \\ \vdots & \vdots & \vdots & \ddots & \vdots \\ x_{m1} & x_{m2} & x_{m3} & \cdots & x_{mn} \end{bmatrix} \end{array} \tag{6.8}$$

where, $x_{m,n}$ is the original rating of alternative A_K with respect to criterion C_J.

Step 2: Normalized decision matrix

The linear normalization is used to eliminate criteria function measurement units. Thus, D decisional matrix values are normalized through the following equation:

$$f_{K,J} = \frac{x_{m,n}}{\sqrt{\sum_{K=1}^{m} x_{m,n}^2}} \tag{6.9}$$

The subsequent matrix D^* is then obtained:

$$D* = \begin{array}{c} \\ A_1 \\ A_2 \\ A_3 \\ \vdots \\ A_m \end{array} \begin{array}{c} \begin{array}{cccccc} C_1 & C_2 & C_3 & \cdots & C_n \end{array} \\ \begin{bmatrix} f_{11} & f_{12} & f_{13} & \cdots & f_{1n} \\ f_{21} & f_{22} & f_{23} & \cdots & f_{2n} \\ f_{31} & f_{32} & f_{33} & \cdots & f_{3n} \\ \vdots & \vdots & \vdots & \ddots & \vdots \\ f_{m1} & f_{m2} & f_{m3} & \cdots & f_{mn} \end{bmatrix} \end{array} \tag{6.10}$$

$f_{m,n}$ is the normalized rating of alternative A_K with respect to criterion C_J

Step 3: Determination of best and worst values

For all criteria functions, the best value is f_J^* and the worst value is f_J^-; that is, for criterion $j = 1, 2, \ldots, n$, we have formulas (6.11) and (6.12), illustrated below:

$$f_J^* = \left\{ \begin{array}{ll} \text{Max}_K f_{K,J} & \text{if Jth function represents a benefit} \\ \text{Min}_K f_{K,J} & \text{if Jth function represents a cost} \end{array} \right\} K = 1, 2, \ldots\ldots, m \tag{6.11}$$

$$f_J^- = \left\{ \begin{array}{ll} \text{Max}_K f_{K,J} & \text{if Jth function represents a benefit} \\ \text{Min}_K f_{K,J} & \text{if Jth function represents a cost} \end{array} \right\} K = 1, 2, \ldots\ldots, m \tag{6.12}$$

Step 4: Determination of the weights

The weights represent the relative importance of the *J*th criterion and can be determined using the AHP or any other method. In this study, the AHP method will be used.

Step 5: Distance calculation

In this step, S_k and R_K are calculated. The former represents the distance of the *K*th alternative according to the positive ideal solution; however the latter implies maximal regret of each alternative (Liuo and Chuang 2010).

For $K = 1, 2, \ldots, m$ S_k and R_k are computed with the relation:

$$S_K = L_K^{P=1} = \sum_{J=1}^{n} W_J \left(f_J^* - f_{K,J}\right) / \left(f_J^* - f_J^-\right) \tag{6.13}$$

$$R_K = L_K^{P=\infty} = \max_J \left[W_J \left(f_J^* - f_{K,J}\right) / \left(f_J^* - f_J^-\right) | J = 1, 2, \ldots, n \right] \tag{6.14}$$

where W_j are the weights of the criteria, expressing their relative importance.

Step 6: P_K value calculation

$$P_K = v(S_K - S^*)/(S^- - S^*) + (1 - v)(R_K - R^*)/(R^- - R^*) \quad (6.15)$$

where:

$$S^* = \text{Min}\,[(S_K)|K = 1, 2, \ldots., m] \quad S^- = \text{Max}\,[(S_K)|\,K = 1, 2, \ldots., m] \quad (6.16)$$

$$R^* = \text{Min}\,[(R_K)|K = 1, 2, \ldots., m] \quad R^- = \text{Max}\,[(R_K)|\,K - 1, 2, \ldots., m] \quad (6.17)$$

Coefficient v is called the 'strategy coefficient' (Vucijak et al. 2013) and is introduced as a weight for the strategy of maximum group utility, whereas $1 - v$ is the weight of the individual regret (Opricovic and Tzeng 2007). Coefficient v always belongs to the interval [0, 1]. Normally, the value of v is taken as 0.5. A compromise ranking list for a given v can be obtained by ranking with P_k measure. The best alternative is the one having the minimum $\mathrm{P_k}$ value (Chatterjee et al. 2009; Caliskan et al. 2013), because its value is closest to the ideal level.

6.4 The Implementation of Proposed Model

Within the Leader Approach, LAGs represent the instrument of the 'territorial governance of rural development' (Tola 2010 p. 75) and give a concrete form to local development policies, by formulating the LDP, the planning process which follows three main phases: territorial analysis, animation and concertation (Calabrò et al. 2005). The first activity includes a territorial diagnosis through a knowledge process aimed at pointing out the useful information to define a suitable strategy for the exigencies of territory. Within the territorial animation phase, both activities of listening to the territory in order to detect the requirements and the needs of operators, and activities in search of suitable stakeholders to be involved in the partnership are carried out. The final stage of the LDP design is represented by concertation assemblies among the partnership's components to define, in a shared and participated way, the interventions to promote and carry out on the territory. Specifically, the partnership (governance) can correspond to the character of the LAGs' members (institutional stakeholders), while in other cases it is widened through the involvement of further stakeholders, which could be distinguished between internal and external partners, according to their formally or informally recognized involvement.

The assessment model proposed in this paper uses an integrated method of AHP and VIKOR to provide a framework for the ranking of the rural governance performance dealing with the Leader Approach in the 2007–2013 Rural

Development Programme in Calabria. Specifically, of the fourteen LAGs selected from the Calabria Region eleven were taken into consideration. In detail, the AHP method has been applied to obtain the weight of the single criteria and sub-criteria; on the contrary, the VIKOR method leads to the ranking of LAGs' *performances*. The evaluation process has been conducted with the collaboration of experts, who are technical directors of the examined LAGs. The evaluation tree is composed of four levels: goals, criteria, sub-criteria and alternatives; in the examined case, the last one is coincident with the LAGs. After 'building' the decisional tree, the criteria (7) and sub-criteria (39) have been submitted to pair-wise comparisons through a questionnaire. This has allowed the experts to express their judgement according to the scale given in Sect. 6.3.1 and to compile pair-wise comparison matrices. Regarding the main goal, the different weight vectors of sub-criteria and criteria are determined by using the AHP. The average of weights obtained through the pair-wise comparisons is given in Table 6.3.

From the analysis of the results obtained through the AHP, it emerges that relative to the GG criteria, greater importance is recognized in 'Effectiveness' and 'Sustainability' with a weight equal to 0.17. Effectiveness shows that the LAGs are oriented mainly towards the achievement of the final objective, which is the elaboration of LDP, mainly focusing on the knowledge assets of the partnership. Indeed, having a solid background of shared knowledge requires the advantage of minimizing the informative asymmetries and shortening the decision times. This is why pre-conservation meetings coming before the assembly are important for the LAGs, because they consent to pre-define better the issues on which the partnership will express its own opinion. With respect to sustainability, great relevance is recognized in the necessity to develop strategies overcoming the planning cycle. An example is offered by the plans thought to be like a propulsive element for the start-up of a future district. Moreover, the LAGs recognize the importance of the 'make system' with other forms of planning expressed on the territory.

Regarding 'Participation', it has been given a weight of 0.16. In particular, the participation in initiatives of collective animation and the degree of representativeness of the territory are considered the most important elements to involve the territory itself. In spite of this, the realization of the collective animation through the organization of public meetings on the territory is meant by the LAGs more as a traditional method of involvement limited to the informative function, more than the consultation one. Furthermore, considering the concept of involvement synonymous of inclusivity, instead of encouraging interaction and dialogue with the local communities, for the LAGs it becomes important to have at their disposal partnerships representative of the various categories of stakeholders operating on the territory.

'Transparency' is given 0.15 and the perception of the role the animators have in arranging and making available the acquired knowledge by all stakeholders (partners and not) for the partnership is meaningful. Indeed, either the lack of information or misinformation or the non-homogeneity of the possessed information can cause protests or oppositions within the partnership itself and influence the decision-making process.

Table 6.3 Criteria and sub-criteria weight average values

Criteria/sub-criteria	Weight
C1—social capital	0.10
1.1 Heterogeneity partnership	0.20
1.2 Consensus partnership	0.16
1.3 Construction mode partnership	0.15
1.4 Collaborative capability	0.35
1.5 Agreements with stakeholders	0.15
C2—efficiency	0.13
2.1 Listening to the territory	0.25
2.2 Duration of partnership's assembly	0.10
2.3 Number of animators	0.12
2.4 Number of partnership's assemblies	0.14
2.5 Period of consultation in partnership's assembly	0.21
2.6 Proactive skills in writing of the partnership	0.18
C3—effectiveness	0.17
3.1 Conflict	0.11
3.2 Presence of active actors	0.21
3.3 Contribution knowledge stakeholders	0.22
3.4 Meetings of pre-consultation	0.28
3.5 Informal meetings	0.19
C4—participation	0.16
4.1 Meetings with stakeholders	0.08
4.2 Gathering information for the SWOT	0.12
4.3 Degree of representativeness of the territory	0.16
4.4 Use of participatory techniques	0.11
4.5 Participation in initiatives of animation	0.16
4.6 Stakeholders' involvement in strategy	0.11
4.7 Expression of the views of the partnership	0.14
4.8 Publicizing collective animation activities	0.12
C5—transparency	0.15
5.1 Verbal publication on the WEB	0.11
5.2 Decision support models	0.17
5.3 Accessibility of information to stakeholders	0.22
5.4 Openness of partnership assemblies	0.23
5.5 Communication animators-partnership	0.27
C6—accountability	0.13
6.1 Influence of stakeholders in decision-making	0.23
6.2 Presence of leader	0.14
6.3 Decision support: technical group	0.34
6.4 Skills: private actors	0.29
C7—Sustainability	0.17
7.1 Continuity of planning	0.24
7.2 Involvement of private actors	0.18
7.3 Involvement of private capital	0.11
7.4 Stakeholder analysis	0.13
7.5 Integration with other plans	0.24
7.6 Presentation of the plan to stakeholders	0.10

Source authors' own elaboration

Both ‘Efficiency’ and ‘Accountability’ record a weight of 0.13. For the former, the usefulness of the sub-dimension territorial animation has been considered more incisive, which, above all through work-shops and thematic tables, has allowed to enrich and strengthen the informative wealth for exploitation during the decision process. The partnership assembly is very important for the concentration of the interventions; indeed, it is meant as a moment of integration of the heterogeneous forms of knowledge involved in the process of territory development. It emerges that the different individual perceptions lead to a different definition of the problems and to a different identification of interventions to insert in the plan. As for the latter, the importance given to the Technical Group has to be highlighted, which, besides leading and supporting the partnership in the formulation and arrangement of the project ideas, makes a contribution to the decisions and to bearing them out, thanks to the wide knowledge of the territory acquired over time.

In contrast, ‘Social Capital’ is considered lower, with a weight of 0.10, which is linked more to intangible aspects of the planning process. However, within the above-mentioned criterion, the sub-criterion cooperation ability is considered quite relevant by the partnership, seeing that the spirit of cooperation is determining both to create a positive environment useful to the actors to decide and act together, and to achieve objectives that without any collaborative relationship would be difficult to reach.

Successively, the VIKOR method was used to rank the 11 examined rural governances in accordance with the average weights of the sub-criteria and criteria assessed by using the AHP method. Initially, the original values of the indicators and the evaluative questions associated with each sub-criterion are calculated separately for each LAG. In particular, some indicator values derived from the interviews were conducted with the experts. As the scales of sub-criteria are not equivalent to each other, all values in the decision matrix were normalized by using Eq. (6.9) and weighted by multiplying them by the weight *wj* of the *j-th* sub-criterion obtained with the AHP. For the sub-criteria C31 and C63 they have been considered their reciprocals, because they represent elements to minimize. On the whole, the normalization procedure does not modify the information content of the data. All the weighted values forming each sub-criterion are aggregated and displayed in Table 6.4, where it is also possible to check the best and worst values of the criteria identified according to formulae (6.11) and (6.12).

Subsequently, the values of S_k, R_k and P_k are calculated by using Eqs. (6.13)–(6.15) (Table 6.5). The priority of weights of the criteria with respect to the main goal is calculated as 0.10; 0.13; 0.17; 0.16; 0.15; 0.13; 0.16. Table 6.5 also shows the values of P_k for $v = 0.5$ and the compromise ranking of the LAGs.

In detail, LAG-7 is the one showing the best performance with respect to the GG ideal model, designed around the criteria pointed out. It is interesting to notice the meaningful gap from the LAG in the 2nd position, equal to a score of 0.21. Among the other LAGs, it is possible to observe lower differences and in particular, LAGs 10 and 5 are the ones showing the lowest performances, respectively, with 0.66 and 0.68. Moreover, it must be underlined that the analysis carried out does not want to reach a strict *ranking* on the state of performance of

Table 6.4 Total values of criteria including best (f^*) and worst (f^-) ones

LAGs	C1	C2	C3	C4	C5	C6	C7
LAG-1	0.32	0.23	0.34	0.39	0.15	0.35	0.26
LAG-2	0.21	0.28	0.30	0.30	0.46	0.35	0.31
LAG-3	0.26	0.27	0.24	0.23	0.15	0.29	0.28
LAG-4	0.24	0.21	0.31	0.31	0.25	0.35	0.28
LAG-5	0.27	0.21	0.19	0.17	0.21	0.20	0.28
LAG-6	0.24	0.38	0.30	0.20	0.15	0.26	0.22
LAG-7	0.35	0.29	0.31	0.29	0.32	0.23	0.38
LAG-8	0.30	0.21	0.23	0.24	0.25	0.20	0.27
LAG-9	0.33	0.31	0.19	0.16	0.25	0.17	0.24
LAG-10	0.27	0.19	0.22	0.28	0.15	0.35	0.17
LAG-11	0.26	0.27	0.30	0.30	0.15	0.27	0.34
f^*	0.35	0.38	0.34	0.39	0.46	0.35	0.38
f^-	0.21	0.19	0.19	0.16	0.15	0.17	0.17

Source authors' own elaboration

Table 6.5 Local groups' action rankings according to P_k value

LAGs	S_k	R_k	P_k	Ranking
LAG-1	18.15	6.72	0.21	II
LAG-2	19.86	9.81	0.59	VIII
LAG-3	30.96	6.72	0.48	IV
LAG-4	25.58	7.79	0.49	V
LAG-5	39.38	6.93	0.68	XI
LAG-6	29.47	7.38	0.52	VI
LAG-7	15.59	5.27	0.00	I
LAG-8	32.96	6.74	0.53	VII
LAG-9	32.39	7.46	0.59	VIII
LAG-10	33.91	7.77	0.66	X
LAG-11	24.93	6.72	0.36	III

Source authors' own elaboration

the LDP's planning process adopted by the LAGs. Indeed, its main purpose is to provide input for discussion, in order to point out the elements which could favour or obstruct the achievement of ideal performances. It comes out that the reading of the results can be useful both to single out eventual virtuous behaviours which could be adopted as 'best practices' in other LAGs, and to highlight any criticism. In this sense, LAG-7, the first classified, is the one presenting the best result for the social capital and sustainability dimensions. In particular, social capital differs for the wide recourse to agreements of cooperation with the stakeholders showing, in this way, an open-minded LAG towards the setting-up of relationships which give the opportunity to strengthen the links with the territory and to support the implementation activity of the plan. As regards sustainability, through an intense activity including also public meetings, the LAG has introduced the plan to the whole community, in order to legitimise it even under the profile of social consent.

LAG-1, the second classified, differs because it has obtained the best result in terms of effectiveness, participation and accountability dimensions. Dealing with participation, it has been the only LAG which has experienced the territorial laboratory to allow the participatory planning of the plan. As regards effectiveness, in the partnership a very important element has been the wide presence of active actors both in dialogue and collaboration. Also the informative input given by the stakeholder partners showed itself particularly incisive in the individualisation of the strategy, enriching the technical abilities of the LAG's institutional actors. In the accountability dimension, LAGs 2, 3, 4, 10 and 11 distinguish themselves because they gave their stakeholders (partners) the possibility to take an active part in decision processes. In LAG-11, the third classified, a low value of performance for the transparency dimension can be observed, where the lack of use of the website to make available the minutes of the partnership's assembly is noticed. This is a common aspect for almost all the LAGs in Calabria. Moreover, similarly to LAGs 1, 3, 6 and 10, LAG-11 does not show itself to be favourable in the involvement of the partnership's assembly of subjects external to the LAG itself. On the whole, LAG-11 characterizes itself to be the partnership with the most active actors, indicating a strong sense of responsibility and interest by the involved subjects in the LDP planning. Moreover, it is also the most active for the capacity of 'making synergy' with other projects expressed both as a partner and as a lead partner.

LAG-3 (the fourth classified) records the times of the partnership's assembly that are longer rather than the average ones, showing a low capacity of control and a consequent increase in the times to reach the final decision. In LAG-4 (the fifth classified) the performances both of the dimensions and of the sub-dimensions are more or less in line with the average values, except for the participation, in which it differs positively when experiencing the on-line forum to interact with the territory in its whole. In this way, the local community can formulate and express its own opinion about more relevant issues for them. Instead, LAG-6 (the sixth classified) presents an excellent level for the efficiency dimension which relates to the capacity of acquiring the knowledge which is at the basis of the decision process. On the one hand, the sub-mentioned LAG has been fundamental for the partnership's assembly as a main moment for the concertation of the events to insert in the LDP. On the other, as regards participation, it has had the lowest degree of representativeness in the territory, involving a few categories of actors coming from the agricultural world. With respect to LAG-8 (the seventh classified), both the performances of the dimensions and of the sub-dimensions are more or less in line with the average values.

LAGs 2 and 9 are both in eighth position, but LAG-2 distinguishes itself because it has used an interactive model of support to the group decisions to allocate the resources in the single interventions, while it shows a not so good performance for the social capital dimension. What affects this value is the modality through which the LAG has mobilized the local actors of the partnership, by using mainly friendship nets or interpersonal contacts rather than by stimulating and involving the interest of potential stakeholders towards the activities of territory development. LAG-9, instead, shows low values of performance in the dimensions of effectiveness, participation and accountability. For the effectiveness dimension, the fact that the partnership is only composed of the institutional stakeholders of the LAG is significant, while the involvement of further stakeholders is carried out in a marginal measure and in a completely informal way. This also confirms the meaningful detachment between the territory and LAG-9, testified by the low degree of participation to its initiatives of animation organized in the territory (participation dimension). Dealing with accountability, in the project activity a lack of proactive contribution of the private component is noticed, which should be a propulsive engine for the development of one's own territory, because it is an expression of the needs of the economic and productive world.

LAG-10, which occupies the next to the last position, is interesting for its lowest results concerning the efficiency and sustainability dimensions. The former is characterized by a marginal recourse to the partnership's assembly, which represents the moment of integration of the different forms of knowledge involved in the process of territory development. Indeed, the LAG preferred to build the plan favouring more the confrontation between technicians and single individuals of the partnership through the organization of specific meetings, bypassing the contemporary comparison among all the partnership's actors. The latter is characterized by the lack of integration with other plans, linked to a choice of LAGg-10 itself which wanted to concentrate its own human resources only on the LDP.

In last position, LAG-5 characterizes itself for a strong closure towards an involvement of external subjects in the elaboration of a strategy. Moreover, the role held by the technical group in designing the plan is particularly incisive, because of the past experience in the realization of different planning instruments and because of the well-established nets of relationships among public and private subjects operating on the territory.

Finally, in order to synthesize all the information obtained by the benchmarking study, the main force and weak points of the planning process of the examined partnerships are presented in Table 6.6.

The important role of direction held by the local partnerships in the policies of rural development finds a confirmation also in the future 2014–2020 planning. With respect to the 2007–2013 cycle, on the one hand, new policies aim at involving the local stakeholders, increasing as much as possible their participation in decisions; on the other hand, they return to the centrality of the territory through a new vision: 'one area, one strategy'. Considering such aspects, it emerges that, in general, the rural governances in Calabria have acquired a greater experience in harmonizing different planning instruments operating on the same territory; on the contrary, they show meaningful delays in implementing concrete forms of participatory planning. The LAGs justify their inaction towards more open forms of participation, asserting that they are too expensive both in terms of time, costs and organization.

Table 6.6 Reported strengths and challenges/difficulties of local rural planning of the partnerships

Strengths	Challenges/difficulties
– Presence of a cooperative environment – Prevalence of representative partnerships of organized and widespread interests – Lack of presence of conflicts – High activity of animation addressed to listening to the territory (workshops-thematic tables) – Meaningful contribution of the stakeholder partners' knowledge in the individualisation of the strategy – High recourse to pre-concertation meetings – High proactive ability of the private actors – Synergy of the DLP with other plans	– No recourse to agreements of cooperation – Weak proactive capacity of the written project ideas – Meaningful presence of passive actors – Poor recourse to territorial meetings for the involvement and consultation of the local communities – Marginal use of participatory techniques – Poor recourse of the publication of the minutes in the website – Poor or absent decision role of the stakeholder partners in some partnerships

Source authors' own elaboration

6.5 Conclusion

Approaching the 2014–2020 EU planning period, this study intends to make a contribution to the matter of the assessment of rural policies, providing a fact-finding contribution on the performance of the organizational and interactive modalities adopted during the planning phase within the partnership process of the Leader Approach. The use of the AHP-VIKOR integrated model appears to be an efficient instrument to evaluate how the performance of the rural planning of the LAGs differs with respect to an ideal GG model, elaborated on the criteria of Social Capital, Efficiency, Effectiveness, Participation, Accountability and Sustainability and on the 39 sub-criteria linked to them. The AHP-VIKOR model presents some advantages. First of all, it enables to take into consideration the perception expressed by experts operating in the LAGs, through the weighing of the criteria and the sub-criteria. Second, the individualisation of the synthetic index of the performance score (P_k) allowed the aggregation and comparison of the final results among the single LAGs, which makes more immediate the communication of the results to policy makers at the regional level and to the LAGs. In particular, for the LAGs, the proposed method for the performance measurement can support them both in pointing out possible virtuous behaviours adopted by other LAGs and highlighting eventual elements of criticism of their own planning process. On the whole, the value of the method is seen in its capacity of having allowed the contemporary evaluation of all the seven criteria related to the GG. The future step of the research is to proceed with the comparison between the results of the performance of the LDP planning process phase and the results of the performance carried out in the following implementation phase. Indeed, the informative output gained through the analysis is useful to evaluate, at the end of the present planning cycle, as the different arrangements of the partnership process have influenced the accomplishment of the outcomes expected by rural policies.

Acknowledgments This research has been funded by Regione Calabria—Department 11, Sector 3, through the project 'Decision Support Systems for the participation in governance processes of rural development in Calabria (PAReGO 2011–2013)'.

References

Aa, Vv. (Eds.). (2007). *Il processo partenariale nella nuova fase di progettazione integrata in Sardegna* (p. 79). Rome, Italy: Formez.

Abdi, M., Fathi, M. R., Faghih, A., & Tavakoli, S. (2011). Performance evaluation of metal ore mining firms in Tehran stock exchange using of VIKOR and AHP methodology. *European Journal of Scientific Research, 63*(1), 110–121.

Abrams, P., Borrini-Feyerabend, G., Gardner, J., & Heylings, P. (2003). Evaluating governance. A handbook to accompany a participatory process for a protected area, report for parks canada and CEESP/CMWG/TILCEPA. http://www.ibcperu.org

Annunzi, I. (2006). Assunzioni di responsabilità e dinamiche di governance nelle aree leader. In A. Cavazzani, G. Gaudio, & S. Sivini (Eds.), *Politiche, governance e innovazione per le aree rurali* (pp. 259–270). Naples, Italy: Edizioni Scientifiche Italiane.

Augustyn, M. M., & Knowles, T. (2000). Performance of tourism partnerships: A focus on York. *Tourism Management, 21*(4), 341–351.

Baykasoglu, A., & Kaplanoglu, V. (2008). Application of activity-based costing to a land transportation company: A case study. *International Journal of Production Economics, 116*(2), 308–324.

Bazzazi, A. A., Osanloo, M., & Karimi, B. (2011). Deriving preference order of open pit mines equipment through MADM methods: Application of modified VIKOR method. *Expert Systems with Applications, 38*(3), 2550–2556.

Birolo, L., Secco, L., Da Re, R., & Cesaro, L. (2012). Multi-system governance within the EU policy for rural development: A proposal for LAGs self-evaluation in the LEADER program. *126th EAAE Seminar on New challenges for EU agricultural sector and rural areas. Which role for public policy?* Capri (Italy), 27–29 June.

Börzel, T. A., Pamuk, Y., & Stahn, A. (2008). Good Governance in the European Union, Berlin Working Paper on European Integration No. 7, Berlin, Germany: Freie Universitat Berlin.

Brinkerhoff, J.M. (2002). Assessing and improving partnership relationships and outcomes: a proposed framework. *Evaluation and Program Planning, 25*(3), 215–231.

Calabrò, T., De Luca, A. I., Gulisano, G., & Marcianò, C. (2005). The rural governance system in leader plus: The application of an integrated planning methodology in Calabria (South Italy). *New Medit, 4*(3), 38–46.

Calıskan, H., Kursuncu, B., Kurbanoğlu, C., & Guven, S. Y. (2013). Material selection for the tool holder working under hard milling conditions using different multi criteria decision making methods. *Materials and Design, 45*, 473–479.

Callan, K. (2007). Elements of effective governance. *Measurement, accountability and partecipation.* Public Administration and Public Policy, *126* (p. 363). New York, USA: CRC/Taylor and Francis Group.

Căluşer, M., & Sălăgean M. (Eds.). (2007). *Good governance in multiethnic communities: Conditions, instruments, best practices, ways to achieve and measure good governance at the local level* (p. 70). Belgium: A Joint Publication of the Ethnocultural Diversity Resource Center and the King Baudouin Foundation. http://www.edrc.ro

Chatterjee, P., Athawale, V. M., & Chakraborty, S. (2009). Selection of materials using compromise ranking and outranking methods. *Materials and Design, 30*(10), 4043–4053.

Chia-Chi, S. (2011). A hybrid performance evaluation system for notebook computer ODM companies. *African Journal of Business Management, 5*(3), 987–1000.

Chowdhury, N., & Skarstedt, C.E. (2005). The principle of good governance. *Draft Working Paper, CISDL Recent Developments in International Law Related to Sustainable Development Series* (p. 22). Oxford: UK.

Cundill, G., & Fabricius, C. (2010). Monitoring the governance dimension of natural resource comanagement. *Ecology and Society, 15*(1), 15.

Di Iacovo, F., & Scarpellini, P. (2006). La governance e le aree rurali: Una introduzione critica. In A. Cavazzani, G. Gaudio, & S. Sivini (Eds.), *Politiche, governance e innovazione per le aree rurali* (pp. 161–187). Naples, Italy: Edizioni Scientifiche Italiane.

EC. (2001). *European Governance—A White Paper*. European Commission, COM 428.

EC. (2005). Council Regulation No. 1698/2005 of 20 September 2005 on support for rural development by the European Agricultural Fund for Rural Development (EAFRD). *Official Journal of the European Union L, 277*, 1–40.

Fanizzi, L., & Misceo, S. (2010). L'applicazione dell'Analytic Hierarchy Process (AHP) nella valutazione ambientale iniziale. *L'ambiente Periodico Tecnico-Scientifico di Cultura Ambientale, XVII, 7*, 1–7.

Franceschetti, G. (Ed.). (2009). Capitale sociale e sviluppo rurale. *La potenzialità dell'approccio LEADER e la sua trasferibilità* (p. 72). Padova, Italy: Cleup SC.

Franceschetti, G., Da Re, R., Secco, L. (2012). Misurare la qualità della governance nei territori rurali:un set di indicatori e alcuni risultati applicativi. *1st AIEAA Conference "Towards a Sustainable Bio-economy:Economic Iusses and Policy Challenges*" (pp. 4–5). Trento, Italy: Giugno.

Franceschetti, G., Da Re, R., & Secco, L. (2012b). Un set di indicatori per misurare la qualità della governance nei territori rurali. *Agriregionieuropa, 8*(30), 70–75.

GFI. (2009). *The Governance of Forest Toolkit (Version 1): A draft framework of indicators for assessing governance of the forest sector.* The governance of forest initiative, Imazon/ICV/ WRI. Retrieved September, from www.wri.org/gfi

Governa, F. (2004). Modelli e azioni di governance. Innovazioni e inerzie al cambiamento. *Rivista Geografica Italiana, 1*, 1–27.

Kahraman, C., & Kaya, I. (2010). Realizing policies by projects using fuzzy multiple criteria decision making. In D. Ruan (Ed.), *Computational intelligence in complex decision systems.* Paris, France: Atlantis Press.

Kelly, C., Essex, S., & Glegg, G. (2012). Reflective practice for marine planning: A case study of marine nature-based tourism partnerships. *Marine Policy, 36*(3), 769–781.

Kørnøv, L., & Thissen, W. A. H. (2000). Rationality in decision- and policy-making: Implications for strategic environmental assessment. *Impact Assessment and Project Appraisal, 18*(3), 191–200.

Liou, J. J. H., & Chuang, Y. T. (2010). Developing a hybrid multi-criteria model for selection of outsourcing providers. *Expert Systems with Applications, 37*(5), 3755–3761.

Lopolito, A., Giannoccaro, G., & Prosperi, M. (2011). Efficiency of LEADER Programmes in the creation of tangible and intangible outputs: A Data Envelopment Analysis application to Local Action Groups performances. *122nd EAAE Seminar "Evidence-Based Agricultural and Rural Policy Making: Methodological and Empirical Challenges of Policyevaluation"*, Ancona (Italy), February 17–18.

Marcianò, C., & Palladino, M. (2013). La pianificazione integrata in un'area calabrese nell'ottica di una rete di affiliazione. Analisi della struttura del partenariato locale e del suo sviluppo. In C. Marcianò (Ed.), *Governance rurali in Calabria* (pp. 337–402). Reggio Calabria, Italy: Centro di Stampa di Ateneo.

McCall, M. K. (2003). Seeking good governance in participatory-GIS: A review of processes and governance dimension in applying GIS to participatory spatial planning. *Habitat International, 27*(4), 549–573.

McCall, M. K., & Dunn, C. E. (2012). Geo-information tools for participatory spatial planning: Fulfilling the criteria for 'good' governance? *Geoforum, 43*(1), 81–94.

Nardone, G., Sisto, R., & Lopolito, A. (2010). Social capital in the LEADER initiative: A methodological approach. *Journal of Rural Studies, 26*(1), 63–72.

Opricivic, S., & Tzeng, G. H. (2003). Comparing DEA and MCDM method. In T. Tanino, T. Tanaka, & M. Inuiguchi (Eds.), *Multi-objective programming and goal programming. Theory and applications*. Berlin, Germany: Springer.

Opricovic, S., & Tzeng, G. H. (2002). Multicriteria planning of post-earthquake sustainable reconstruction. *Computer-Aided Civil Infrastructure Engineering, 17*(2), 211–220.

Opricovic, S., & Tzeng, G. H. (2004). Compromise solution by MCDM methods: A comparative analysis of VIKOR and TOPSIS. *European Journal of Operational Research, 156*(2), 445–455.

Opricovic, S., & Tzeng, G. H. (2007). Extended VIKOR method in comparison with outranking methods. *European Journal of Operational Research, 178*(2), 514–529.

Ou Yang, Y. P., Shieh, H. M., & Tzeng, G. H. (2013). A VIKOR technique based on DEMATEL and ANP for information security risk control assessment. *Information Sciences, 232*, 482–500.

Reeb, D. (2004). Towards the development of criteria and indicators to assess the level of participation in NFP processes. The evaluation of forest policies and programmes. *EFI Proceedings, 52*, 87–102.

Saaty, T. L. (1988). *Mathematical methods of operations research.* New York, USA: Dover Publications.

Santiso, C. (2001). Good governance and aid effectiveness: The world bank and conditionality. *The Georgetown Public Policy Review, 7*(1), 1–22.

Schniederjans, M. J., & Garvin, T. (1997). Using the analytic hierachy process and multi-objective programming for the selection of cost drivers in activity based costing. *European Journal of Operational Research, 100*(1), 72–80.

Secco, L., Da Re, R., Birolo, L., & Cesaro, L. (2011). La qualità della governance in ambito rurale: Prime riflessioni sull'auto-valutazione dei GAL nel LEADER. *Agriregionieuropa, 7*(26), 33–37.

Srivastava, M. (2009). *Good governance-concept, meanong and features: a detailed study.* Social Science Research Network, Retrieved from SSRN: http://ssrn.com/abstract=1528449 or http://dx.doi.org/10.2139/ssrn.1528449

Storti, D., Tarangioli, S., & Zumpano, C. (2010). L'attuazione delle politiche di sviluppo rurale in Italia. In D. Storti & C. Zumpano (Eds.), *Le politiche comunitarie per lo sviluppo rurale. Il quadro degli interventi in Italia* (pp. 67–91). Rome, Italy: Inea.

Tavakkoli-Moghaddam, R., Mousavi, S. M., & Heydar, M. (2011). An integrated AHP-vikor methodology for plant location selection. *Transactions B: Applications, 24*(2), 127–137.

Tola, A. (2010). Strategie, metodi e strumenti per lo sviluppo dei territori rurali. *Il modello del GAL dell'Ogliastra (Sardegna) per la valorizzazione delle risorse agro-alimentari e ambientali* (p. 176). Milano, Italy: Franco Angeli.

Vinodh, S., Kamala, V., & Shama, M. S. (2013). Compromise ranking approach for sustainable concept selection in an Indian modular switches manufacturing organization. *International Journal of Advanced Manufacturing Technology, 64*(9–12), 1709–1714.

Vučijak, B., Kupusović, T. S., & Cerić, M. K. A. (2013). Applicability of multicriteria decision aid to sustainable hydropower. *Applied Energy, 101,* 261–267.

Yücenur, N. G., & Demirel, N. C. (2012). Group decision making process for insurance company selection problem with extended VIKOR method under fuzzy environment. *Expert Systems with Applications, 39*(3), 3702–3707.

Zeleny, M. (1975). The theory of the displaced ideal. In M. Zeleny (Ed.), *Multiple criteria decision making* (pp. 153–206). Berlin, Germany: Springer.

Zeleny, M. (1982). *Multiple decision making* (p. 563). New York, USA: Mc Graw-Hill.
Zurla, P. (2010). Governance e riflessività per le politiche locali: Tra sviluppo economico e coesione sociale. In N. De Luigi, A. Martelli, & P. Zurla (Eds.), *Pratiche di governance tra welfare e sistemi locali di produzione* (pp. 15–30). Milano, Italy: Franco Angeli.

Authors' Biography

Giuseppa Romeo graduated in Agricultural Sciences and Technology from the Mediterranean University of Reggio Calabria in 2006. She completed her Master's in Agricultural Economics and Policy from the Centre for Advanced Training in Economics and Policy for Rural Development at the University of Naples Federico II in 2008. She earned her doctoral degree in Agricultural Economics and Policy from Palermo University in 2012. Her research interests are related to the fields of agricultural economics and policy, water resource management and integrated rural development.

Claudio Marcianò is Associate Professor of the Agraria Department at the Mediterranean University of Reggio Calabria. His recent works include Rural Governance in Calabria (editor, 2013, in Italian), Participatory Rural Development experiences in Calabria (editor, 2012), Processes and Politics of Integrated Rural Development in Calabria (co-editor, 2009, in Italian) and Rural Districts in Calabria (co-editor, 2008, in Italian). He has also authored or co-authored numerous articles, book chapters, and papers on agricultural and forest economics and policy, fishery economics, integrated rural development, agro-industrial waste treatment and valorization.

Part III
Surveying and Experimental Designs in Agricultural Policy Analysis

Chapter 7
Consumers' Perception of Wastewater Usage in Agriculture: Evidence from Greece

Foivos Anastasiadis, Fragiskos Archontakis, Georgios Banias and Charisios Achillas

Abstract The need for wastewater usage is increasing, especially in coastal regions with limited freshwater supply. In Greece, the only applications of water reuse projects concern irrigation purposes in the agricultural sector. One of the key issues concerning the adaptation of such projects and further expansion of such initiatives is consumer perceptions. To that end, the aim of this chapter is to explore consumer awareness about the reuse of wastewater for agricultural purposes in order to accept such policies. The study reveals a positive attitude of the public towards recycled water reuse in agriculture. Education is positively correlated with higher awareness regarding agricultural and landscape irrigation. However, the study reveals several obstacles for a wider acceptance of similar practices, especially for older people.

7.1 Introduction

Unquestionably wastewater is a valuable resource of water, especially in coastal regions with limited freshwater supply; for instance, in the Syrian Arab Republic, 67 % of sewage effluent is reused; in Egypt, 79 %; and in Israel, 67 %, mostly for irrigation and environmental purposes (Kay 2011). Moreover, irrigation with wastewater can be an excellent contribution to reducing water demand and recycling nutrients, while it could also improve soil health and reduce the amount of pollutants discharged in watercourses. Especially in agriculture, reuse of wastewater reduces the water footprint of food production on the environment while it simultaneously involves activities such as higher crop yields and changes in cropping patterns, which also reduce carbon footprint (Hanjra et al. 2012). Therefore, adapting sustainable strategies such as reuse of wastewater in

F. Anastasiadis (✉) · F. Archontakis · G. Banias · C. Achillas
School of Economics, Business Administration & Legal Studies,
International Hellenic University, Thessaloniki-N. Moudania, Greece
e-mail: f.anastasiadis@ihu.edu.gr

C. Zopounidis et al. (eds.), *Agricultural Cooperative Management and Policy*,
Cooperative Management, DOI: 10.1007/978-3-319-06635-6_7,

agriculture could fulfil sustainability's triple bottom line with obvious economic, environmental and social benefits.

However, from a consumer perspective there are still acceptability complications mainly due to low levels of awareness resulting in trust issues. Despite the fact that wastewater is widely used worldwide and besides the ground-breaking developments in treatment techniques, securing safety, several studies have shown mixed findings not only with respect to reuse of wastewater per se, but also about consuming products that have been irrigated with such water (Menegaki et al. 2007; Kantanoleon et al. 2007). Public acceptability is a key issue towards further development of such practices in agriculture as well as in the general production level. The importance of the adaptation of such practices is increasing significantly, given that the production systems on the one hand require more and more resources while on the other hand they are among the principal liable sectors for their degradation.

A key objective of this study is to explore consumers' perspective in order to reveal a clearer picture towards the acceptability of such sustainable practices as reuse of wastewater in agriculture. The conceptual approach builds upon investigation of three key aspects: (i) consumer awareness of alternatives for wastewater reuse (ii) how familiar they are about the benefits of wastewater reuse, and (iii) their concerns regarding wastewater reuse. The findings on one hand suggest a generic positive attitude that it is promising for further adaptation of such sustainable strategies, but on the other hand it also highlights the weaknesses towards that direction such as low levels of awareness in certain groups of consumers.

In the material that follows, the key outcome of an extensive literature review on the topic is provided, covering both technical and public acceptability issues. Afterwards the research methodology is presented, including formulation of hypotheses and basic statistics concerning the sample. Finally, the findings take place with a discussion, followed by the conclusion and suggestions.

7.2 Literature Review

The reuse of wastewater is a complex and broad issue; therefore depending on the objective it could be explored from different angles. In the framework of the current study, two are the main themes for a deeper investigation with respect to the consumers' perspective on wastewater usage in agriculture: (1) the importance of the suggested strategy, including some technical justifications, and (2) public acceptability with a special focus on consumers' behaviour. Thus, the literature has been reviewed accordingly and the core outcome is presented below.

7.2.1 Wastewater Importance

Wastewater comprises liquid wastes generated by households, industry and commercial sources, as a result of daily usage, production, and consumption activities (Kontos and Asano 1996). Municipal treatment facilities are designed to treat raw wastewater to produce a liquid effluent of suitable quality that can be disposed to the natural surface waters with minimum impact on human health or the environment. The disposal of wastewater is a major problem faced by municipalities, particularly in the case of large metropolitan areas, with limited space for land-based treatment and disposal. On the other hand, wastewater is also a resource that can be applied for productive uses since wastewater contains nutrients that have the potential for use in agriculture, and other activities (Marecos do Monte et al. 1996).

In both developed and developing countries, the most prevalent practice is the application of municipal wastewater (both treated and untreated) to land. In developed countries where environmental standards are applied, much of the wastewater is treated prior to use for irrigation of fodder, fibre and seed crops and, to a limited extent, for the irrigation of orchards, vineyards, and other crops (Barrett and Segerson 1997). Other important uses of wastewater include recharge of groundwater, landscaping (golf courses, freeways, playgrounds, schoolyards, and parks), industry, construction, dust control, wildlife habitat improvement and aquaculture. In developing countries, though standards are set, these are not always strictly adhered to. Wastewater, in its untreated form, is widely used for agriculture and aquaculture and has been the practice for centuries in countries such as China, India and Mexico (Marecos do Monte et al. 1996).

Agriculture is the largest consumer of freshwater resources, currently accounting for about 70 % of global water diversions (sometimes even up to 80–95 % in developing countries) (Seckler et al. 1998). With increasing demand from municipal and industrial sectors, competition for water will increase and it is expected that water now used for agriculture will be diverted to the urban and industrial sectors (see also Refsgaard and Magnussen 2006).

A number of various examples from Asia, North Africa and Latin America are witness to the above-mentioned fact (Molle and Berkoff 2006). One observed response to this squeeze on agricultural water supply is to promote greater use of treated urban wastewater for irrigation. Discounting the significance of this practice as a partial solution to the freshwater squeeze in agriculture, it is argued that the total volume of treated wastewater available (even if all of it is treated) is insignificant in many countries in terms of the overall freshwater balance and the volumes that will need to be transferred from agriculture to municipal use. While this may be true in most parts of the developing world, in the water-short arid and semi-arid zones of the Middle Eastern, Southern and Northern African regions, the Mediterranean, parts of China, Australia and the USA, domestic water use can represent between 30 and 70 % of irrigation water use (or between 10–40 % of total water use) in extreme cases (Abu-Zeid et al. 2004; Angelakis et al. 1999; FAO 1997; Lallana et al. 2001; Peasey et al. 2000; WRI 2000; UNEP 2002; WHO 2006; AATSE 2004).

Substitution of freshwater by treated wastewater is already seen as an important water conservation and environmental protection strategy, which is simultaneously contributing to the maintenance of agricultural production. In Australia where the share of domestic water use (20 % of total water use) is the second highest in the world, after the USA, the limited total water supply in the country has necessitated careful use of water and recycling. It should be highlighted that in 2000 up to 11 % of wastewater was being recycled in major cities (Vigneswaran 2004).

Another typical example of good practice in this regard where over the past 20 years water reuse has been integrated into the national water resources management strategy is Tunisia (which is a middle-income country with an arid climate). Over 60 wastewater plants in Tunisia produce high quality reclaimed water for use in agriculture, and irrigation of parks and golf courses (Bahri 2002). Currently, about 43 % of the treated wastewater is being recycled for these purposes. A recent comprehensive compilation of data on water reuse (Jimenez and Asano 2008) provides an understanding of common practices around the world, particularly of treated wastewater for municipal and industrial uses, agriculture and groundwater recharge.

7.2.2 Wastewater Public Acceptability

Several factors have been proven to be important as regard to the public acceptability of wastewater in the literature, yet an overall conclusion on whether reuse of wastewater is widely accepted cannot be drawn. More specifically, Menegaki et al. (2007) have found that a similar set of factors (e.g. from environmental awareness to economic factors like income) are significant for both farmers and consumers in accepting reuse of wastewater and more specifically they are willing to use and pay for such water and products produced using it. Nonetheless, there are important differences between consumers' and farmers' attitude since the latter are driven by freshwater scarcity while the former mainly by high levels of environmental awareness.

In a review paper concerning factors influencing public perceptions of water reuse in Australia, Po et al. (2004) highlighted among others the extent of disgust (the "Yuck" factor), perceptions of risk from recycled water, sources of recycled water (e.g. rainwater, toilet water etc.), the issue of choice, trust and knowledge, attitudes towards the environment, environmental justice issues, the cost of recycled water and socio-demographic factors. Education has also been among the factors affecting willingness to reuse wastewater. Tsagarakis and Georgantzis (2003) for instance have shown that the more educated are more willing to reuse such water.

The acceptability of wastewater reuse could be illustrated via the different alternative ways of reusing it. The dominant category is agricultural irrigation, including irrigation of crops, flowers, cotton, fodder and orchards (Janosova et al. 2006). Another category is landscape irrigation, including irrigation of parks, school yards, freeway medians, golf courses, cemeteries, greenbelts and residential areas (Friedler et al. 2006). There is also industrial reuse, i.e. cooling water for thermal power plants, boiler feed, process water (Janosova et al. 2006) and groundwater recharge dealing with groundwater replenishment, saltwater intrusion control and subsidence control (Chu et al. 2004). Moreover, recreational and environmental uses is a category in which wastewater is reused in creating artificial wetlands, sustaining in-stream flows and aquifer recharge (Bixio et al. 2006), enhancing natural wetlands (Chu et al. 2004) and recreational lakes (Friedler et al. 2006). The final category is for non-potable urban uses, which is quite broad including among others urban lawn watering, road cleaning, car washing and toilet flushing (Chu et al. 2004).

Reused wastewater in agriculture has a major advantage in that it is usually a constant and reliable supply with obvious economic benefits. In addition, a huge quantity of water suitable for reuse after a certain treatment usually is discharged into the environment in a way that it could cause severe degradation in water bodies such as lakes, rivers and the coastal marine environments. The degradation is often related to the presence of organic and inorganic nutrients, which can cause problems such as eutrophication and algal blooms (Toze 2006). Furthermore, the reuse of wastewater for purposes such as agricultural irrigation reduces the amount of water that needs to be extracted from environmental water sources (Lopez et al. 2006).

Finally, with respect to the concerns about wastewater, the most important seems to be the level of treatment since it is strongly related with the health impact caused by inadvertent consumption of the treated water and the price of the recycled water (Urkiaga et al. 2006). Other concerns are related to the salinity and the existence of pathogenic microorganisms to harmful effluents and the environmental effect of microbiological agents, quality and cost of treated wastewater (Dolnicar et al. 2010).

7.3 Research Methodology

In order to shed more light regarding consumers' understanding of the issue (i.e. "awareness"), potential gains (i.e. "benefits") and considerations (i.e. "concerns") of wastewater use, we conducted a survey. The sample consists of 355 individuals from the Greater Area of Thessaloniki that were surveyed by means of a structured online questionnaire. Based on the theoretical foundations set out in previous sections of the chapter, the following hypotheses were formulated about the influence of personal factors on awareness, benefits and concerns regarding wastewater reuse:

7.3.1 Consumer Survey Hypotheses

H1a Consumers' awareness of alternatives for wastewater reuse is positively related with education

H1b Consumers' awareness of alternatives for wastewater reuse is positively related with income

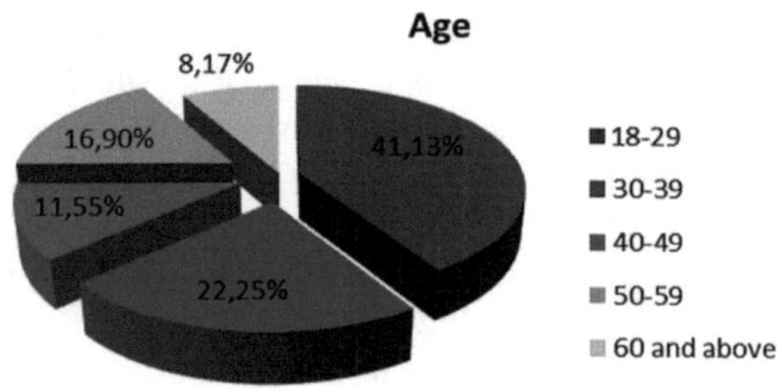

Fig. 7.1 Age

H2a Consumers with higher education are more familiar with more benefits from wastewater reuse

H2b Male consumers are more familiar with more benefits from wastewater reuse

H3 Female consumers are more concerned regarding wastewater reuse

Education is considered a key issue to overcoming public doubts on wastewater reuse. More specifically, the aforementioned doubts mostly relate to public health and water quality. To that end, improved public education to ensure awareness of the technology and its benefits, both environmental and economic, is recommended. Moreover, gender, age and annual income are expected to influence consumer perspective in respect to their attitude towards wastewater reuse. Thus, as dependent variables, the socio-economic & demographic characteristics of the survey's respondents were selected, i.e.: Gender, Age, Income and Education, the main explanatory variables in our models.

A 5-point Likert scale was used for the dependent variables, ranging from completely unaware/strongly disagree/5th choice (1) to very aware/totally agree/ 1st choice (5). As regards income, the sample was grouped into 10 classes, as follows; (i) Less than €5.000 (ii) €5.001–12.000 (iii) €12.001–19.000 (iv) €19.001–26.000 (v) €26.001–33.000 (vi) €33.001–40.000 (vii) €40.001–47.000 (viii) €47.001–54.000 (ix) €54.001–61.000 and (x) More than €61.001. Educational classes were set as (i) Secondary education (ii) Technological education (iii) Holder of a university degree and (iv) Holder of a Master's degree/Ph.D.

The results of the survey are briefly shown in Figs. 7.1, 7.2, 7.3 and 7.4.

The descriptive statistics of socio-economic and demographic variables of the dataset are presented in Table 7.1. According to the summary statistics, 45 % were male (n = 160) whereas approximately 55 % were female (n = 195). The average respondent is middle-aged, while the average annual income is just above EUR 19 K. Finally, in the sample surveyed, the average person has a tertiary education degree.

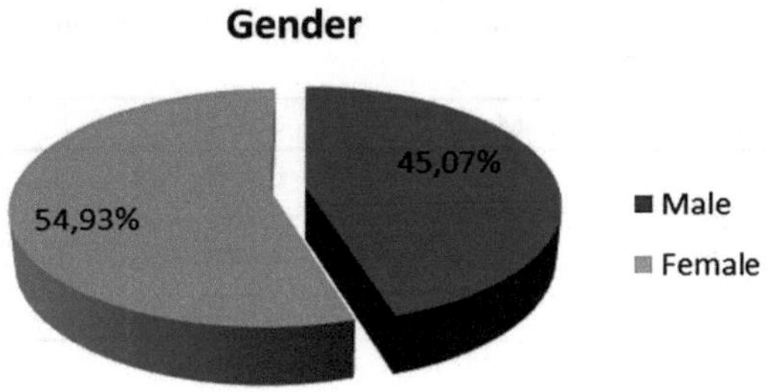

Fig. 7.2 Gender

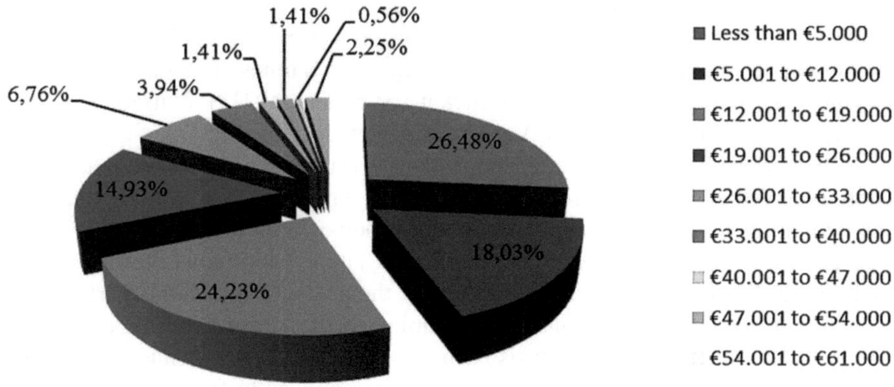

Fig. 7.3 Income

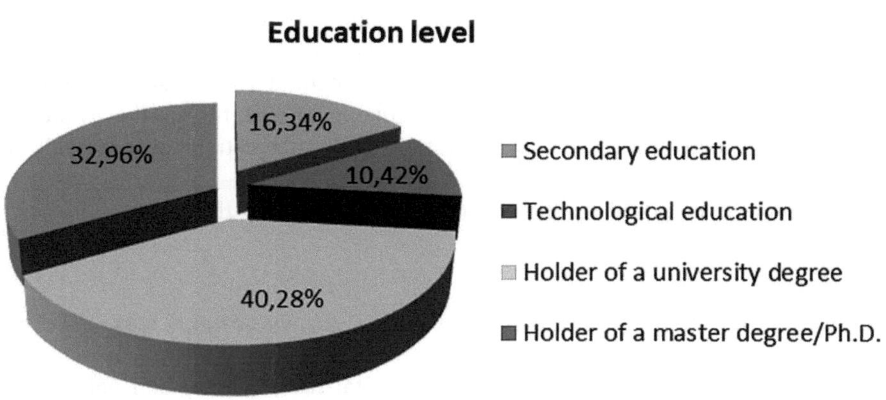

Fig. 7.4 Education

Table 7.1 Descriptive statistics of dependent variables: socio-economic & demographic characteristics of the survey's respondents

Variable	Obs	Mean	Std. Dev.	Min	Max
Gender	355	0.451	0.498	0	1
Age	355	2.287	1.364	1	5
Income	355	3.011	1.977	1	10
Education	355	2.899	1.039	1	4

7.4 Findings and Discussion

Preliminary linear regression analysis (not presented in this chapter—available on request) of the averages of Var1a–Var1f, Var2a–Var2h and Var3a–Var3g shows that Education is the only significant variable for awareness (capturing H1a) and benefits (H2a), while Gender is significant for concerns (H3). Variables Var1a–Var1f, Var2a–Var2h and Var3a–Var3g are presented in Table 7.2.

However, given the nature of the dependent variables (i.e. being ordinal, but not continuous) we will apply here an ordered probit regression, a widely used method for estimating such models. It is an ordered response model which uses the probit link function. The basic idea is that there is a latent continuous metric variable underlying the ordinal responses observed by the analyst. Thresholds partition the real line into a series of regions corresponding to the various ordinal categories. The detailed ordered probit analysis (which was carried out by using the package Stata v.10.0) is not presented in this chapter—we choose to focus on the signs of the explanatory variables that are significant for each of the dependent variables (see Tables 7.3, 7.4 and 7.5).

Indicatively, Table 7.6 presents the results corresponding to one ordered probit regression (for the dependent variable *Var1c*) for clarifying purposes.

Table 7.6 indicates that the only significant explanatory variable for the dependent variable "Wastewater reuse-Industrial reuse" (i.e. *Var1c*) is the level of *Education*. The positive coefficient (highly significant at 1 %, as shown by the *p*-value) indicates that the higher the education level, the higher the awareness regarding the industrial reuse of wastewater. Finally, the estimated thresholds are reported.

The hypotheses posed in this chapter are supported by the data, as discussed below:

H1:

- Higher education implies higher awareness for wastewater reuse (see last column in Table 7.3).
- In certain cases of recreational/environmental uses (Var1e) and non-potable urban uses (Var1f) the consumer's income is positively related with awareness (see last two rows in Table 7.3).

Table 7.2 Definitions of dependent variables

Variable name	Variable definition	
Var1a	1a Wastewater reuse categories:	Agricultural irrigation
Var1b	1b Wastewater reuse categories:	Landscape irrigation
Var1c	1c Wastewater reuse categories:	Industrial reuse
Var1d	1d Wastewater reuse categories:	Groundwater recharge
Var1e	1e Wastewater reuse categories:	Recreational and environmental uses
Var1f	1f Wastewater reuse categories:	Non-potable urban uses
Var2a	2a Wastewater reuse benefits:	Constant, reliable and sustainable source of water
Var2b	2b Wastewater reuse benefits:	Improves nutrient balance of underutilised land
Var2c	2c Wastewater reuse benefits:	Reduces overuse/demand for fresh water sources
Var2d	2d Wastewater reuse benefits:	Reduces effluent discharge to surface waters, lakes
Var2e	2e Wastewater reuse benefits:	Can contribute to economic development & tourism
Var2f	2f Wastewater reuse benefits:	Can reduce coastal marine pollution
Var2g	2g Wastewater reuse benefits:	Nutrients in recycled water used as fertiliser
Var2h	2h Wastewater reuse benefits:	Can reduce dependence on expensive water storage
Var3a	3a Wastewater reuse concerns:	Level of treatment of the recycled water
Var3b	3b Wastewater reuse concerns:	Health problems due to accidental consumption
Var3c	3c Wastewater reuse concerns:	Price of recycled water
Var3d	3d Wastewater reuse concerns:	Presence of pathogenic microorganisms in wastewater
Var3e	3e Wastewater reuse concerns:	Quality of recycled water
Var3f	3f Wastewater reuse concerns:	Specific water activities that involve physical contact
Var3g	3g Wastewater reuse concerns:	Implementation of regulations about wastewater

H2:

- Higher education implies higher benefits for wastewater reuse (see last column in Table 7.4).
- Male consumers see more benefits with respect to wastewater reuse and its contribution to economic development and tourism, and potential nutrients use of recycled water as fertiliser.

Table 7.3 Awareness—Ordered probit regression results (significant signs up to 10 %)

	Gender	Age	Income	Education
Var1a				(+)
Var1b		(+)		(+)
Var1c				(+)
Var1d				(+)
Var1e			(+)	(+)
Var1f			(+)	

Table 7.4 Benefits—Ordered probit regression results (significant signs up to 10 %)

	Gender	Age	Income	Education
Var2a				(+)
Var2b				
Var2c		(−)		(+)
Var2d				(+)
Var2e	(−)			
Var2f		(+)		
Var2g	(−)			
Var2h				(+)

Table 7.5 Concerns—Ordered probit regression results (significant signs up to 10 %)

	Gender	Age	Income	Education
Var3a	(−)			
Var3b	(−)			(−)
Var3c				
Var3d	(−)			
Var3e	(−)			
Var3f	(−)			
Var3g	(−)			

Table 7.6 Awareness—Ordered probit regression results for dependent variable *Var1c*

	Coefficient	Std. error	*P*-value
Gender	0.106	0.113	0.349
Age	−0.039	0.050	0.435
Income	0.044	0.034	0.193
Education	0.199	0.058	0.001
Threshold 1	0.548	0.226	
Threshold 2	0.450	0.222	
Threshold 3	1.157	0.226	
Threshold 4	2.019	0.241	

H3:

- Gender is statistically significant in almost all models (see Table 7.5) in the sense that females are clearly more concerned towards wastewater reuse (with the exception of pricing issues of recycled water where gender was not significant).

In particular, for the variables directly connected with agriculture, such as: (i) wastewater reuse of agricultural irrigation (Var1a) (ii) wastewater reuse of landscape irrigation (Var1b), and (iii) potential nutrients use of recycled water as fertiliser (Var2g), we find that:

- Higher education levels are positively correlated with higher awareness regarding agricultural and landscape irrigation.
- Older consumers may be more aware regarding agricultural and landscape irrigation.
- Female consumers see less benefit with respect to potential nutrients use of recycled water as fertiliser.

7.5 Conclusions and Suggestions

This study aimed to explore consumers' perspective towards reuse of wastewater in agriculture. Key findings suggest a positive attitude, since both variables related to agriculture (Var1a: agricultural irrigation, & Var1b: landscape irrigation), among several variables not connected to agriculture of the construct 'awareness of alternative categories for wastewater reuse' are significant. That indicates a certain familiarity with the concept of reusing recycled water in agriculture, since in any other case it would be easier to select only the other non-agriculture related alternatives available. Moreover, further analysis confirmed previous studies with respect to the positive correlation between education, and both awareness and benefits about recycled water. These results consist of the basic elements of a strong foundation for further adaptation of sustainable practices in agriculture.

However, there is a clear negative correlation between gender and benefits from wastewater reuse as well as gender concerns with respect to recycled water. Females seem to be more anxious about almost every possible issue related to wastewater, such as health problems due to accidental consumption, quality and level of treatment. They are also more sceptical regarding well-defined benefits from using recycled water in agriculture, for example, that it could be used as a fertiliser due to the nutrients it contains. These findings reveal the main bottlenecks of a wider acceptance of recycled water utilisation. Overcoming such barriers is a key stage towards more sustainable strategies not only in agriculture but in the general production sector as well.

Valuable insights concerning awareness issues and specific obstacles related to recycled water should be used for designing targeted campaigns to inform specific groups. Scientific evidence with respect to the safety of recycled products has to be illustrated in a simplified and comprehensive way in public events. Benefits from adopting sustainable practices in general and wastewater reuse in particular must be clarified, properly supported and effectively communicated. Moreover, similar initiatives should be incorporated into the education system, given that the current study shows that older consumers may be more aware regarding recycled water reuse in agricultural and landscape irrigation. A wide range of such awareness-raising campaigns and successful long-term training initiatives is a prerequisite in order for policy makers to consolidate sustainability in their agendas.

Last but not least, limitations of this study combined with the key results provide specific suggestions for future research. Deeper investigation on the motivation behind the scepticism related to females' concerns for reusing wastewater could reveal more insights. Designing separate gender-specific studies in order to compare results should give a better understanding of the different perceptions between genders. Finally, researching related topics in an urban-based (stratified sampled) population would expose the perceptions of those who are actually taking the decisions of adopting or rejecting such practises.

Acknowledgment This research has received funding from the European Union's Seventh Framework Programme (FP7-REGPOT-2012-2013-1) under Grant Agreement No. 316167 (Project Acronym: GREEN-AgriChains). Moreover, the present scientific research was partially conducted in the context of the project entitled "International Hellenic University (Operation—Development)", which is part of the Operational Programme "Education and Lifelong Learning" of the Ministry of Education, Lifelong Learning and Religious Affairs, and is funded by the European Commission (European Social Fund—ESF) and national resources. Special thanks to Mr. Konstantinos Takolas for his contribution in the survey.

References

AATSE (2004). *Water recycling in Australia: A review undertaken by the Australian Academy of Technological Sciences and Engineering (AATSE) water recycling in Australia*, Parkville: Australian Academy of Technological Sciences and Engineering.

Abu-Zeid, K., Abdel-Megeed, A., & Elbadawy O. (2004). Potential for water demand management in the Arab region. *Proceedings of the International Water Demand Management Conference*, Dead Sea, Jordan 2004: Ministry of Water and Irrigation, Jordan and USAID.

Angelakis, A. N., Marecos Do Monte, M. H. F., Bontoux, L., Asano, T. (1999). The status of wastewater reuse practice in the Mediterranean basin: need for guidelines. *Water Research, 33*, 2201–2217.

Bahri, A. (2002). Water reuse in Tunisia—Stakes and prospects. In S. Marlet & P. Ruelle (Eds.), *Vers une maîtrise des impacts environnementaux de l'irrigation*. France: Montpellier.

Barrett, J., & Segerson, K. (1997). Prevention and treatment in environmental policy design. *Journal of Environmental Economics and Management, 33*, 196–213.

Bixio, D., Thoeye, C., De Koning, J., Joksimovic, D., Savic, D., et al. (2006). Wastewater reuse in Europe. *Desalination, 187*, 89–101.

Chu, J., Chen, J., Wang, C., & Fu, P. (2004). Wastewater reuse potential analysis: implications for China's water resources management. *Water Research, 38*, 2746–2756.

Dolnicar, S., Hurlimann, A., & Nghiem, L. D. (2010). The effect of information on public acceptance—The case of water from alternative sources. *Journal of Environmental Management, 91*, 1288–1293.

FAO. (1997). Irrigation in the near east region in figures. Water Reports No. 9, Rome.

Friedler, E., Lahav, O., Jizhaki, H., & Lahav, T. (2006). Study of urban population attitudes towards various wastewater reuse options: Israel as a case study. *Journal of Environmental Management, 81*, 360–370.

Hanjra, M. A., Blackwell, J., Carr, G., Zhang, F., & Jackson, T. M. (2012). Wastewater irrigation and environmental health: Implications for water governance and public policy. *International Journal of Hygiene and Environmental Health, 215*, 255–269.
Janosova, B., Miklankova, J., Hlavinek, P., & Wintgens, T. (2006). Drivers for wastewater reuse: Regional analysis in the Czech Republic. *Desalination, 187*, 103–114.
Jimenez, B., & Asano, T. (2008). *Water reuse: An international survey of current practice, issues and needs* (pp. 648). London: IWA Publishing.
Kantanoleon, N., Zampetakis, L., & Manios, T. (2007). Public perspective towards wastewater reuse in a medium size, seaside, Mediterranean city: A pilot survey. *Resources, Conservation and Recycling, 50*, 282–292.
Kay, M. (2011). Water for food: innovative water management technologies for food security and poverty alleviation, UNCTAD/DTL/STICT/2011/2, United Nations.
Kontos, N., & Asano, T. (1996). Environmental assessment for wastewater reclamation and reuse projects. *Water Science and Technology, 33*, 473–486.
Lallana C., Krinner W., Estrela T., Nixo S., Leonard J., Berland J. M. (2001). Sustainable water use in Europe, Part 2: demand management, European Environment Agency.
Lopez, A., Pollice, A., Lonigro, A., Masi, S., Palese, A. M., et al. (2006). Agricultural wastewater reuse in southern Italy. *Desalination, 187*, 323–334.
Marecos do Monte, M. H. F., Angelakis, A. N., Asano, T. (1996). Necessity and basis for establishment of European guidelines for reclaimed wastewater in the Mediterranean region, *Water Science and Technology, 33*, 303–316.
Menegaki, A. N., Hanley, N., & Tsagarakis, K. P. (2007). The social acceptability and valuation of recycled water in Crete: A study of consumers' and farmers' attitudes. *Ecological Economics, 62*, 7–18.
Molle, F., Berkoff, J. (2006). Cities versus agriculture: Revisiting intersectoral water transfers, potential gains and conflicts. Comprehensive Assessment Research Report 10. Colombo, Sri Lanka: International Water Management Institute.
Peasey A., Blumenthal U., D. Mara, Ruiz-Palacios G. (2000). A review of policy and standards for wastewater reuse in Agriculture: A Latin American Perspective, WELL Study, Task No. 68 Part 11.
Po, M., Kaercher, J., & Nancarrow, B. E. (2004). *Literature review of factors influencing public perceptions of water reuse*. Canberra: CSIRO Land and Water.
Refsgaard, K., & Magnussen, K. (2006). Attitudes, experience and knowledge among farmers to recycling of sludge. *Food Economics—Acta Agriculturae Scandinavica, Section C, 3*, 49–63.
Seckler, D., Amarasinghe, U., Molden, D., de Silva, R., & Barker, R. (1998). *World water demand and supply, 1990 to 2025: Scenarios and issues*. Colombo, Sri Lanka: International Water Management Institute.
Toze, S. (2006). Reuse of effluent water—benefits and risks. *Agricultural Water Management, 80*, 147–159.
Tsagarakis, K. P., & Georgantzis, N. (2003). The role of information on farmers' willingness to use recycled water for irrigation. *Water Science and Technology: Water Supply, 3*, 105–113.
UNEP (2002). Global environment outlook 3—Past, present and future perspectives, United Nations Environment Programme (UNEP).
Urkiaga, A., De las Fuentes, L., Bis, B., Chiru, E., Bodo, B., et al. (2006). Methodologies for feasibility studies related to wastewater reclamation and reuse projects. *Desalination, 187*, 263–269.
Vigneswaran, V. (2004). Acknowledge all approaches: The global outlook on re-use. *Water 21*, 32–7 (IWA Publishing).
WHO (2006). Guidelines for the safe use of wastewater, excreta and grey water: Wastewater use in agriculture, World Health Organization (WHO).
WRI (2000). A guide to world resources 2000–2001: People and ecosystems: The fraying web of life, World Resources Institute (WRI). ISBN: 1-56973-443-7.

Author Biographies

Dr. Foivos Anastasiadis is an Academic Associate at the School of Economics, Business Administration and Legal Studies, International Hellenic University. He holds an M.Sc. from Wagenigen University, The Netherlands (2004) and a Ph.D. from Imperial College, London (2010). Prior to joining IHU, Anastasiadis was employed as a consultant in the market research sector as well as in the supply chain and procurement processes business.

Dr. Fragiskos Archontakis holds an M.A. in Economics and Econometrics from the University of Manchester and a Ph.D. (Econ) from the European University Institute, Florence. Prior to joining International Hellenic University, he was employed among others by the European Commission, the Spanish High Research Council (CSIC), the University of Alicante, RAND Europe and the University of Athens.

Dr. Georgios Banias is an Academic Associate at the School of Economics, Business Administration and Legal Studies, International Hellenic University. He holds an M.Sc. (2005) and a Ph.D. in Mechanical Engineering (2009) from Aristotle University of Thessaloniki (AUTH). He worked as a Scientific Researcher at the Laboratory of Heat Transfer and Environmental Engineering in the Department of Mechanical Engineering. He has been involved in National and European Research Projects and in the educational activities of the Mechanical Engineering Department at AUTH.

Dr. Charisios Achillas is an Academic Associate at the School of Economics, Business Administration and Legal Studies, International Hellenic University. He graduated in 1999 from the Mechanical Engineering Department of Aristotle University of Thessaloniki. He then continued with an M.Sc. in Engineering Project Management from UMIST, Manchester, UK. In 2009, he was granted his Ph.D. in Mechanical Engineering from Aristotle University Thessaloniki in the field of integrated waste management of industrial products and optimisation of reverse logistics networks. Since 2003, he has been involved in educational activities of the Mechanical Engineering Department of Aristotle University of Thessaloniki, as well as in several domestic and European projects.

Chapter 8
Modelling Structural Change in Ex-Ante Policy Impact Analysis

Frank Offermann and Anne Margarian

Abstract Model-based ex-ante policy impact analyses are nowadays widely used in agricultural policy consulting. However, so far very few existing applications try to assess the impact on farm numbers and the re-allocation of resources between farms. Due to data availability, these studies generally use normative or ad hoc decision rules on farm exits. In this chapter, we fill this gap, combining an empirically-based estimation of profit-dependent farm exit probabilities with prospective modelling of farm adjustments and selected factor markets. This study combines farm-individual information from farm-structural surveys for 1999, 2003 and 2007, and economic information from farm accountancy data for Germany. The estimated model explains farm exit probabilities depending on current and expected future profits, the expected development of competitors (e.g., neighbouring farms competing on the land market) and farm and regional structural characteristics influencing farms' strategic decision-making. The econometric exit model is iteratively coupled to a representative farm group model for Germany, facilitating the ex-ante analysis of complex policy reforms. A first application on dairy market reform scenarios highlights the diverging impacts these may have on the developments of the number of dairy farms of different size or region, and their income and output.

8.1 Introduction

The use of models for ex-ante analysis of policy changes is widespread in the domain of agriculture. However, prospective farm level analyses are generally restricted to the modelling of adjustments with respect to the level of production

F. Offermann (✉) · A. Margarian
Thünen-Institute, Braunschweig, Germany
e-mail: frank.offermann@ti.bund.de

C. Zopounidis et al. (eds.), *Agricultural Cooperative Management and Policy*, Cooperative Management, DOI: 10.1007/978-3-319-06635-6_8,

activities, production intensity and the allocation of resources. Very few attempts have been made to model potential impact of future policy changes on farm numbers and the related reallocation of resources between farms. This is partly due to the numerous challenges to modelling structural change (e.g., the complex and often strategic nature of respective decisions; the heterogeneity of farm(er)s; the interlinkages between farms and the complex interaction with policies), but also a consequence of limited data availability.

Most of the existing approaches that model future developments of farm numbers are based on Markov-Chain analyses (Zimmermann et al. 2009). However, the potential for prospective policy analyses is limited by the generally rather aggregated level of the estimated model, ignoring regional specificity of structural change. Furthermore, all existing studies use only indirect proxies for changes in farm profitability, which limits the type of policy scenarios, which can be analysed. In addition, the consistency of total use of fixed resources (e.g. land, quota) is not ensured.

Multi-agent models (e.g. Balmann 1997; Happe 2004) provide an interesting alternative, as they are well suited to capture the key factors of farm structural change in a bottom-up approach by accounting for heterogeneity and interaction between agents while at the same time allowing a detailed representation of farm business. However, at the individual level, personal traits are very important determinants for exit decisions, and respective data availability is very limited. Thus, in existing studies (e.g., Freeman 2005; Kellermann et al. 2008) the decision rules for farm exit are generally based on ad-hoc/normative rules (e.g., a farm is assumed to exit if income falls below a certain normatively set level).

Only few attempts have been made to overcome some of these limitations by combining empirically-based estimates of the impact of economic parameters on structural change and a prospective modelling of farm performance (e.g. Hennessy and Rehman 2006).

Against this background, the objective of this chapter is to develop and apply a model system, which projects future structural change in agriculture under different policy or market scenarios. Specifically, the aim is to combine an empirically-based estimation of profit-dependent farm exit probabilities with a simulation model that provides prospective modelling of farm adjustments and land and quota markets, and to examine the effects that the endogenous modelling of structural change has on the results (e.g., production and income) of ex-ante policy impact analysis. We also want to explicitly evaluate how endogenously taking account of structural change alters results compared to a trend-based extrapolation of structural change.

The rest of the chapter is structured as follows: First, a brief overview of the formulation and data used for the estimation of the exit model and the specification of the simulation model is given, followed by a description of the linkage between the two models. Using a baseline scenario, we evaluate the impact of endogenously taking account of structural change compared to a trend-based extrapolation of structural change. The effects of changes to the economic conditions on farm numbers are then illustrated for two dairy market scenarios. The chapter ends with a discussion of results and future research implications.

8.2 Methods and Data

8.2.1 The Farm Exit Model

Structural change in agriculture is affected by a multitude of factors, e.g. technology, prices, human capital, off-farm income, demographics, market structure, or political environment. Empirical studies of the importance of individual factors on the decision to exit farming highlight that the impact of economic performance criteria strongly depends on farm and farmer characteristics (e.g., Sumner and Leiby 1987; Bremmer et al. 2004; Weiss 1999; Juvancic 2006; and the overview in Mann 2003). There are two further important aspects which contribute to the challenge of understanding and projecting structural change in agriculture: Firstly, land is a key production factor, but is limited and immobile, and thus there is a close interdependency between a farmer's own decision to exit farming and those of her neighbours, giving rise to strategic elements in decision-making (Margarian 2010a). Secondly, sunk costs and the existence of status quo rents can lead to a persistence of 'suboptimal' equilibria, a phenomenon known as path dependency (Balmann 1995). Margarian (2010a) thus found that initial regional farm structure is a key factor determining structural change.

In view of the findings of the literature, the aim was to specify an econometric model, which takes into account farm and farmer characteristics and own and neighbouring farms' (future) economic performance while accounting for the regional specificity of structural change (Fig. 8.1).

The econometric estimation of profit-dependent farm exit probabilities is hampered by data availability. The farm accountancy data network (FADN), which provides extensive information on the economic performance of farms, is organised as a rotating panel, and does not allow determining whether a farm exits the survey due to the closing down of the farm or other reasons. The farm structural survey on the other hand provides information on farm exits but does not include information on the economic performance. Data protection rules prevent a linking of the data on the single farm level (and thus, for example, the identification of exiting farms in FADN). Therefore, as a first step, a detailed profit model was estimated based on German FADN data for the period 1998–2007. In this model, expected profits depend on farm structural characteristics, such as resources (e.g., own and rented land, family and hired labour), animal numbers and cropping areas. Using this model, individual farm data from the farm structural surveys (FSS) from 1999, 2003 and 2007 for West Germany (575,000 observations) were then supplemented with estimated profits.

A logistic regression model (Eq. 8.1) was then estimated, with the exit probability depending on:

- Current profits (profit level),
- Expected profit development (allowing for asymmetric impact of positive and negative development) with fixed resources, and

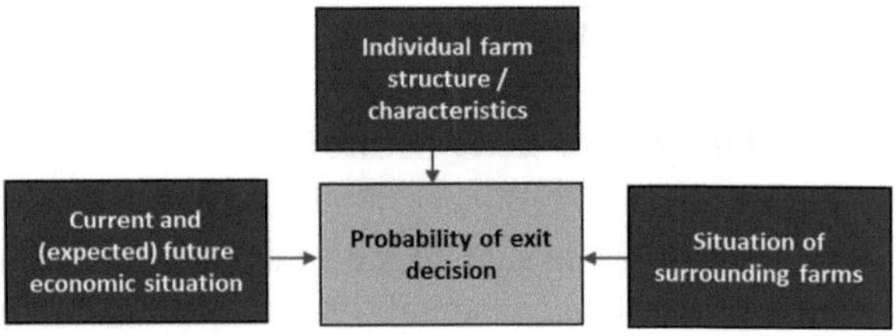

Fig. 8.1 Determinants of farm exit decisions in the estimated exit model

- The development of regional profitability.

$$\ln\left(\frac{p}{1-p}\right) = logit(p) = X_{\text{Type,Size,Region}}$$
$$C + CT_{\text{Type}} + CS_{\text{Size}} + CR_{\text{Region}} + CTS_{\text{Type,Size}} + CRS_{\text{Region,Size}} + CRT_{\text{Region,Type}}$$
$$+ \left(\beta MGEW + \beta MGEWS_{\text{Size}} + \beta MGEWR_{\text{Region}}\right) \bullet MGEW$$
$$+ \left(\beta KGEW + \beta KGEWS_{\text{Size}} + \beta KGEWR_{\text{Region}}\right) \bullet KGEW$$
$$+ \left(\beta PGEW + \beta PGEWS_{\text{Size}} + \beta PGEWR_{\text{Region}} + \beta PGEWT_{\text{Type}}\right) \bullet PGEW$$
$$+ \left(\beta NGEW + \beta NGEWS_{\text{Size}} + \beta NGEWR_{\text{Region}} + \beta NGEWT_{\text{Type}}\right) \bullet NGEW \quad (8.1)$$

MGEW = profit level (difference to average farm profits)
KGEW = expected change in profits of neighbours (= average regional profits)
PGEW = expected change in profits (if positive; else 0)
NGEW = expected change in profits (if negative; else 0)

The parameters are differentiated by region (42), farm type (4) and farm size (3) with the definitions of these characteristics aligned to those used in the farm model FARMIS. The reference farm for the estimation of the coefficients is a small dairy farm in region 1 (Schleswig-Holsteinische Marsch). The model has more than 600 estimated parameters, which not only reflects the complexity of structural change phenomena, but also raises significant challenges for the interpretation and condensation of results. 'Expected profit developments' here refers to the profits expected after a period of four years and were estimated at fixed resources, to reduce issues of endogeneity (profits are influenced by realised growth; growth realised depends on expected profits and thus probability to exit farming). The results of the estimation are documented in Margarian (2010b).

For the ex-ante exit model used in the modelling exercise, the share of farms which exit farming in a specific farm group defined by farm type, size and region can then be calculated as:

$$Exit\,\Pr ob_{\text{Type,Size,Region}} = \frac{e^{X_{\text{Type,Size,Region}}}}{1 + e^{X_{\text{Type,Size,Region}}}} \quad (8.2)$$

Due to the many interaction terms and the heterogeneity of regional results, the influence of changes of single exogenous variables on exit rates cannot easily be deduced from the estimated coefficients. Therefore, the impact of predefined

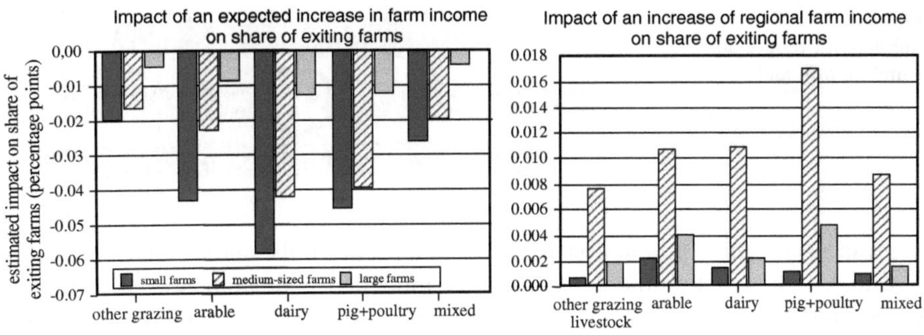

Fig. 8.2 Illustration of the impact of selected changes in economic variables on farm exit rates

isolated changes of profitability was calculated for all combinations of regions, farm sizes and farm types and compared to the results of the model with no changes in exogenous variables. Figure 8.2 provides an overview of the average impact on the share of exiting farms for an expected increase in farm income (of 10,000 EUR/farm), and an expected increase in regional farm income (of 10,000 EUR/farm). These results highlight that an expected increase of profits reduces the share of exiting farms, especially for small arable, dairy, and pig and poultry farms. On the other hand, an increase of regional farm income increases the exit probability due to increased competition, especially for medium-sized farms. However, the diversity of results across regions is often large.

8.2.2 The Farm Model

FARMIS is a comparative-static process-analytical programming model for farm groups (Osterburg et al. 2001; Bertelsmeier 2005; Offermann et al. 2005). Production is differentiated for 27 crop and 15 livestock activities. The matrix restrictions cover the areas of feeding (energy and nutrient requirements, calibrated feed rations), intermediate use of young livestock, fertiliser use (organic and mineral), labour (seasonally differentiated), crop rotations and political instruments (e.g. set-aside and quotas). The model specification is based on information from the German farm accountancy data network covering about 11,000 farms, supplemented by data from farm management manuals. Data from three consecutive accounting years is averaged to reduce the influence of yearly variations common in agriculture (e.g. due to weather conditions) on model specification and income levels. Key characteristics of FARMIS are: (1) the use of aggregation factors that allow for representation of the sectors' production and income indicators; (8.2) input–output coefficients which are consistent with information from farm accounts; and (3) the use of a positive mathematical programming procedure to calibrate the model to the observed base year levels. Prices are generally exogenous and are provided by market models. An exception to this applies to

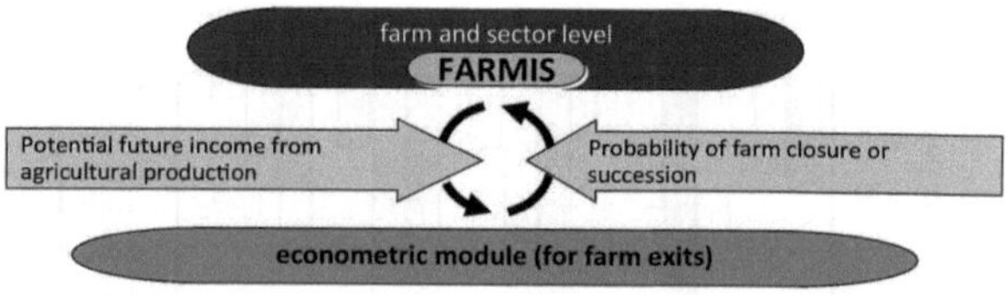

Fig. 8.3 General approach to linking exit model and simulation model

specific agricultural production factors, such as the milk quota, land and young livestock, where (simplified) markets are modelled endogenously, allowing the derivation of respective equilibrium prices under different policy scenarios. FARMIS uses farm groups rather than single farms not only to ensure the confidentiality of individual farm data, but also to increase manageability and the robustness of the model system when dealing with data errors that may exist in individual cases. Homogenous farm groups are generated by the aggregation of single farm data. For this study, farms were stratified by region, type and size resulting in 467 farm groups, which represent the western German agricultural sector.

In the current FARMIS implementation, farm exits are exogenous and the projection of farm numbers to the target year is based on an extrapolation of historical exit rates, which are derived from the FSS differentiated by region, farm type and farm size.

8.2.3 Linking the Models

The basic idea for our approach is to use the simulation model FARMIS to estimate expected future profits, and to use the econometrically estimated exit model to determine profit-dependent exit rates (Fig. 8.3). These exit rates can be implemented in the FARMIS model for the projection of the aggregation factors of the farm groups. As the exit model uses expected future profits at fixed resources as an exogenous variable, as a first step, FARMIS accordingly needs to be run with farms' resources fixed at their current level; then, as a second step, FARMIS is rerun with adjusted farm numbers to reallocate resources and determine new profits.

As the exit rates are estimated for 4 year period and FARMIS is usually applied for a 10–15 year projection horizon, the application for policy impact analyses requires a repeated, iterative application of the two models. FARMIS thus provides the level of current farm profits, the expected change in farm income and the expected change in profits of neighbours, which all enter the exit model as exogenous variables, and allocates resources, which are set 'free' by the exit of farms, the rate of which is determined by the exit model.

8.3 Impact of Dairy Market and Policy Scenarios on Structural Change

8.3.1 Scenarios

For this chapter, scenarios are projected up to the year 2019, with the model base period being 2007 (average of the years 2006–2008). The 4 year projection horizon of the exit model thus implies three iterations of the combined modelling system for the years 2011, 2015 and 2019.

The baseline scenario is based on the Thünen Baseline (Offermann et al. 2012). The 2003 Reform and the Health Check of the CAP are fully implemented, which leads to regional flat rates for first-pillar payments, and the milk quota scheme being abolished in 2015. Furthermore, the sugar market reform decided upon in 2005 is implemented and the set-aside obligations are removed in 2008. For the farm modelling, prices and yields are exogenous and were determined by partial and general equilibrium models. The baseline scenario was also modelled for a model version with a trend-based projection of farm numbers to be able to determine the effect of endogenously accounting for farm exits on the results.

To analyse the impact of changes in the economic environment on farm numbers, two stylised simplified scenarios for dairy market were defined. The first scenario assumes a continuation of the quota scheme at base year levels, with prices being fixed to the values of the baseline scenario (which is clearly unrealistic but here the objective is not (yet) to provide policy impact analysis but rather to examine the principle effect of changes to selected exogenous variables). The second scenario assumes milk prices to be 16 % higher than in the baseline.

8.3.2 Results

8.3.2.1 Impact of Endogenous Versus Trend-Based Projection of Farm Numbers

Table 8.1 provides an overview of the change in farm numbers in the baseline scenario compared to historical exit rates. Overall, structural change is projected to increase slightly, with farm numbers being 5 % lower than under a trend-based projection. In terms of annual exit rates, the difference is comparatively small (annual decrease of farm numbers is 2.6 % compared to 2.2 %). However, differences exist between farms of different types, regions and size. Exit rates are higher than historical values, especially for arable and pig and poultry farms, farms in the southern regions of Germany, and smaller farms. According to the model results, the highest impact of the baseline scenario on farm exit rates is expected for small arable farms.

As more farms exit the sector, the remaining farms can grow more (Table 8.2). On average, farm size in terms of arable area increases by 5 %. The largest

Table 8.1 Development of farm numbers in western Germany in the baseline scenario, endogenous versus trend-based projection of structural change

	Base year	Baseline		Difference
	2007	Trend	Endogenous	Endog. to Trend (%)
All farms (western Germany)	200749	152950	145485	−5
North	71954	55251	53987	−2
South	101455	77122	70862	−8
Centre	27340	20577	20636	0
Arable	41137	24563	21508	−12
Dairy	68667	55942	51899	−7
Other grazing livestock	19632	13021	14201	9
Mixed	44496	33461	33387	0
Pig + Poultry	10839	9984	8511	−15
Arable, <50 ha	22234	9109	5573	−39
Arable, 50–100 ha	9140	6844	7125	4
Arable, >100 ha	9764	8610	8809	2
Dairy, <30 cows	34096	24002	20481	−15
Dairy, 30–60 cows	23652	21611	20764	−4
Dairy, >60 cows	10918	10329	10654	3

Table 8.2 Development of farm sizes in ha UAA in western Germany in the baseline scenario, endogenous versus trend-based projection of structural change

	Baseline		Difference
	Trend	Endogenous	Endog. to Trend (%)
All farms (western Germany)	71	75	5
North	90	92	2
South	58	63	9
Centre	70	70	0
Arable	119	134	13
Dairy	65	69	7
Other grazing livestock	80	77	−4
Mixed	79	80	1
Pig + Poultry	47	53	13
Arable, <50 ha	51	72	40
Arable, 50–100 ha	90	88	−2
Arable, >100 ha	215	211	−2
Dairy, <30 cows	36	41	13
Dairy, 30–60 cows	66	68	3
Dairy, >60 cows	128	126	−2

increase in average size is observed for the small arable farms, which however does not imply that the individual farms in this group grow strongly, but rather is a result of the fact that in this group, especially the very small farms exit, thus increasing the average size of the group. The overall impact of the accelerated structural change on production and farm income is negligible for the baseline scenario, and results are almost identical to the model version with historical exit rates (e.g., difference <1 % for cereal, milk, beef and pork production).

8.3.2.2 Impact of Dairy Market Scenarios on Farm Numbers

The impact of the dairy market reform scenarios is identified by comparing results to those of the baseline scenario with endogenous structural change. Milk production is affected quite differently by the two scenarios: With higher milk prices, total milk production is increasing by 12 %, while the continuation of the milk quota at 2007 levels reduces milk production by 13 % compared to the baseline. Somewhat surprisingly, the impact on the total number of dairy farms is almost identical, with dairy farm numbers being 4–5 % higher than under baseline conditions. The total figure however masks significant differences at the more detailed level. Figure 8.4 illustrates the development of the numbers of small, medium and large dairy farms (here, farm size is related to the number of dairy cows in the base year). The number of small farms decreases strongly under all three scenarios; however, the number of smaller farms is higher in the dairy market scenarios than under the baseline scenario. Especially the continuation of the milk quota scheme seems to slow down the exit rates of smaller dairy farms. The positive impact of higher prices on the number of smaller dairy farms is reduced after the abolishment of the milk quota scheme in 2015. The number of medium-sized dairy farms is positively affected by both dairy scenarios; however, in contrast to the smaller dairy farms, here the effect is larger for the scenario with higher milk prices. The number of larger dairy farms is actually lower than under the baseline scenario, with a continuation of the quota scheme, as the overall limitation of milk production and the higher competition from small and medium-sized farms reduces the chances for growth and continuation. In contrast, higher milk prices with no limitation on sector output provide an opportunity especially for the larger dairy farms, whose number significantly increases compared to the baseline scenario.

8.4 Discussion and Conclusions

For our modelling system, for the baseline scenario, the endogenous modelling of structural change has little impact on aggregated outcome, but affects farm size distribution (depending on type, size and region) compared to a trend-based projection of farm numbers. The impacts may possibly be more pronounced for scenarios with larger changes compared to the base year, e.g. full liberalisation, and in this case endogenously accounting for structural change may affect results for total production. However, for very large changes of profitability of agricultural production, which may not have been observed in the past, the estimated coefficients for the profit-dependent exit rates may not be valid any more (a general problem of all econometric approaches).

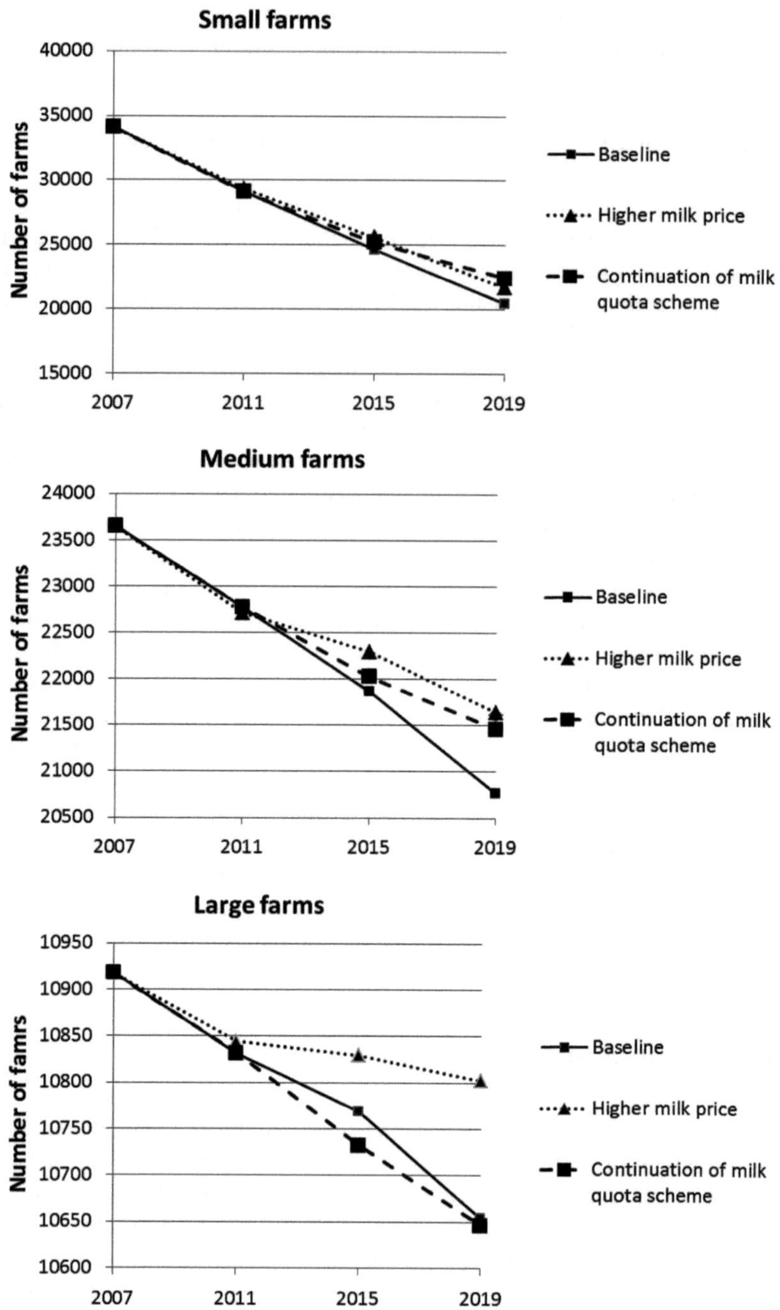

Fig. 8.4 Impact of dairy market scenarios on farm numbers

The evaluation of two stylised milk market scenarios reveals significant differences in the impact of policy or market changes on the number of dairy farms of different sizes. In the application presented here, product prices were exogenously fixed. A coupling to market models would enlarge the range of realistic, policy-relevant scenarios, which could be analysed, e.g. the impact of quota schemes on structural change, or the net effects of investment aid under a quota scheme on farm numbers.

The condensation and interpretation of the diverse and heterogeneous impacts of policies on regional structural change remains a big challenge. Possibly a meta-analysis of model outcomes (see e.g. Happe 2004; Margarian 2010b) may shed more light in this respect.

Acknowledgement The authors acknowledge the financial support provided by the Deutsche Forschungsgemeinschaft (DFG) in the framework of the research group 'Structural Change in Agriculture' (SiAg), which made this work possible.

References

Balmann, A. (1995). *Pfadabhängigkeiten in Agrarstrukturentwicklungen. Begriff, Ursachen und Konsequenzen*. Berlin: Duncker and Humblot.

Balmann, A. (1997). Farm based modelling of regional structural change. *European Review of Agricultural Economics, 25*(1), 85–108.

Bertelsmeier, M. (2005), Analyse der Wirkungen unterschiedlicher Systeme von direkten Transferzahlungen unter besonderer Berücksichtigung von Bodenpacht- und Quotenmärkten. Schriftenreihe des Bundesministeriums für Ernährung, Landwirtschaft und Forsten: Angewandte Wissenschaft 510.

Bremmer, J., Oude Lansink, A. G., Olson, K. D., Baltussen, W. H., Huirne, R. B. (2004). Analysis of farm development in Dutch agriculture and horticulture. *Strategic Decision Making and Firm Development in Dutch Horticulture*, pp. 23–39.

Freeman, T. R. (2005): From the ground up: An agent-based model of regional structural change. Thesis, Department of Agricultural Economics, University of Saskatchewan.

Happe, K. (2004). *Agricultural Policies and Farm Structures—Agent-Based Modelling and Application to EU-Policy Reform*. Halle: Vauk Verlag.

Hennessy, T., & Rehman, T. (2006), Modelling the impact of decoupling on structural change in the farming sector: Integrating econometric and optimisation models. Contributed paper at the International Association of Agricultural Economists Conference, Gold Coast, Australia, August 12–18, 2006.

Juvancic, L. (2006), Determinants of farm survival and growth in Slovenia (1991–2000), Paper presented at the 96th EAAE Seminar in Tänikon, Schweiz.

Kellermann, K., Happe, K., Sahrbacher, C.; Balmann, A., Brady, M., Schnicke, H., Osuch, A. et al. (2008), AgriPoliS 2.1—Model documentation. Technical Report. Halle (Saale), IAMO, http://www.agripolis.de/documentation/agripolis_v2-1.pdf.

Mann, S. (2003). Theorie und Empirie des agrarstrukturellen Wandels? *Agrarwirtschaft, 52*(3), 140–148.

Margarian, A. (2010a), Die regionale Spezifität des Agrarstrukturwandels: Eine theoretische und empirische Analyse. Institutioneller Wandel der Landwirtschaft und Ressourcennutzung (ICAR), Bd. 41. Aachen: Shaker Verlag.

Margarian, A. (2010b), Gewinnentwicklung und Betriebsaufgabe in der Landwirtschaft: Angebotseffekte, Nachfrageeffekte und regionale Heterogenität. Arbeitsberichte aus der vTI Agrarökonomie 11/2010. Braunschweig: Thünen-Institut.

Offermann, F., Kleinhanß, W., Hüttel, S. and Küpker, B. (2005). Assessing the 2003 CAP reform impacts on german agriculture using the farm group model FARMIS. In F. Arfini (ed.), *Modelling agricultural policies: State of the art and new challenges. Proceedings of the 89th European Seminar of the EAAE, Parma, Italy*, February 3–5, 2005. Monte Parma: Universita Parma Editore, (pp. 546–564).

Offermann, F., Banse, M., Ehrmann, M., Gocht, A., Gömann, H., Haenel, H.-D., Kleinhanss, W., Kreins, P., von Ledebur, O., Osterburg, B., Pelikan, J., Rösemann, C., Salamon, P., Sanders, J. et al. (2012), vTI-Baseline 2011–2021: Agri-economic projections for Germany, Landbauforschung, [special issue] 358.

Osterburg, B., Offermann, F., & Kleinhanß, W. (2001). A sector consistent farm group model for german agriculture. In T. Heckelei H. P. Witzke & W. Henrichsmeyer (Eds.), *Agricultural sector modelling and policy information systems* (pp. 152–160). Germany: Verlag Vauk Kiel.

Sumner, D. A., & Leiby, J. D. (1987). An econometric analysis of the effects of human capital on size and growth among dairy farms. *American Journal of Agricultural Economics, 69*(2), 465–470.

Weiss, C. R. (1999). Farm growth and survival: Econometric evidence for individual farms in upper Austria. *American Journal of Agricultural Economics, 81*(1), 103–116.

Zimmermann, A., Heckelei, T., Perez Dominguez, I. (2009). Modelling farm structural change for integrated ex-ante assessment: Review of methods and determinants. *Environmental science & policy, 12*, 601–618.

Author Biographies

Dr. Frank Offermann is a senior researcher at the Thünen-Institute of Farm Economics. He is an agricultural economist with extensive experience in model-based policy assessment, and co-ordinator of the Thünen model network. Contact: frank.offermann@ti.bund.de.

Dr. Anne Margarian is a senior researcher at the Thünen-Institute of Rural Studies. She is an agricultural economist with a specific interest in the economy and job opportunities in rural areas, functional differentiation in regional development and structural change in agriculture. Contact: anne.margarian@ti.bund.de.

Chapter 9
Public Preferences for Climate Change Adaptation Policies in Greece: A Choice Experiment Application on River Uses

Dimitrios Andreopoulos, Dimitrios Damigos, Francesco Comiti and Christian Fischer

Abstract Climate change is a multidimensional issue with serious environmental and socio-economic implications. Mountain areas, in particular, show high vulnerability to climate change. Among others, alterations in temperature and precipitation can severely affect freshwater ecosystems, in terms of both quality and quantity. As a result, services provided by river ecosystems will deteriorate, affecting economic activities and social welfare. This study comprises one of the first attempts to monetize non-market benefits of adaptation to climate change impacts on mountainous rivers. In this direction, a choice experiment was conducted using a face-to-face survey to examine the preferences of Konitsa's residents, a mountain settlement located in the Prefecture of Ioannina (Greece). Simple and extended Conditional Logit models were calibrated in order to analyze trade-offs of choices and to estimate the welfare effects of climate change adaptation measures. The resulting values and reliability considerations indicate that people support adaptation actions, being willing to pay for all river services.

9.1 Introduction

Mountains are critical reservoirs of natural assets and particularly of water resources. Half of the global population depends on the water that is gathered, purified and stored in mountain areas (Gret-Regamey et al. 2012). Nevertheless, mountain areas are sensitive to climate change. Thus, even limited alterations in temperature and precipitation could have massive implications for water resources.

D. Andreopoulos (✉) · F. Comiti · C. Fischer
Faculty of Science and Technology, Free University of Bozen/Bolzano, Bolzano, Italy
e-mail: Dimitrios.Andreopoulos@natec.unibz.it

D. Damigos
School of Mining and Metallurgical Engineering, National Technical University of Athens, Athens, Greece

C. Zopounidis et al. (eds.), *Agricultural Cooperative Management and Policy*, Cooperative Management, DOI: 10.1007/978-3-319-06635-6_9,

In Southern Europe and especially in the Mediterranean region, the impacts of climate change on water reserves are anticipated to accrue from the decrease in annual precipitation (Bates et al. 2008; Philandras et al. 2001). These impacts are largely linked to a decrease in the water use potential, which has already emerged due to the present high demand for water resources.

Bearing in mind the above remarks, this study aims at investigating socioeconomic implications of climate change in the forthcoming decades on different uses of the Aoos River in Ioannina Prefecture, Epirus Region (Western Greece). In Greece, over the last five decades the precipitation patterns have been altered indicating a decrease of 30–150 mm per decade. Respectively, the rivers' runoff has dropped between 5 and 10 % during the last century in the Greek territory (Milly et al. 2005). Additionally, the application of different global emission scenarios (e.g., A1B, A2, B2) recommended by the Intergovernmental Panel on Climate Change (IPCC) for future climate trends implies that the precipitation will drop 3–7 and 14–22 % for the periods 2021–2050 and 2071–2100, respectively. Consequently, a decline is predicted for the total water potential from 7–20 to 30–50 % for the respective time periods, for the entire country (Stournaras et al. 2011). Regarding Western Greece, simulations of the emission scenarios A1B, A2, B2 indicate a decrease of above 20 % (A1B, A2) in water potential (infiltration and surface runoff), while the decrease in water potential for the B2 emission scenario is anticipated to be up to 10 % for the period 2071–2100 (compared with the period 2021–2050) (Stournaras et al. 2011). Furthermore, in the Aoos catchment, which is part of the Vikos-Aoos National Park, the application of daily output data from the RACMO2 regional climate model (RCM) developed within the framework of the ENSEMBLES project (PRUDENCE 2007), with at least a 10 % precipitation decrease for the period 2021–2050 (Giannakopoulos et al. 2011).

Provided that the Aoos River supports a rich ecosystem and a number of human activities, as discussed in the next section, the potential impacts of climate change on river water flow could significantly affect human well-being, especially that of Konitsa's residents, a mountain settlement heavily dependent upon the examined river system. To this end, the aim of this survey is to investigate beliefs and preferences of the general public regarding climate change issues and their reflections on people's willingness to pay for adaptation, in order to provide valuable insight to those involved in adaptation decision-making processes.

More explicitly, the study uses a Choice Experiment (CE) to investigate Konitsa residents' willingness to pay for adaptation interventions to climate change in the local water resources, so as to avoid welfare losses due to possible complications on existing river water uses. CEs rely on people's own stated preferences and have been widely applied for the valuation of environmental goods and services. The preferences are emerged by applying survey techniques in the context of a constructed contingent but realistic market. In the context of this CE survey, the assumptions adopted for constructing the alternative choices were based on the climate change predictions described above, since, to our knowledge, there are no other studies examining climate-related impacts on the water potential of the study area.

9.2 Study Area and Methodology

The mountainous town of Konitsa (40°2′44″N 20°44′56″E) is situated in the Northern Pindos mountain range and lies at an altitude of 700 m. The location of the town is at the southern exit of the Vikos-Aoos canyon, which has been designated as a national park. The Aoos River flows through the homonymous canyon and crosses the town at its southwestern part. The river is 260 km long, 70 km of which are found in the Greek territory. The average flow is 52 m^3s^{-1} and the river is not characterized by major modifications. Close to the river spring there is a hydroelectric plant that produces an average of 10^3 MWh electrical energy per year. The town has about 3,000 inhabitants, living in 750 households (Papageorgiou et al. 2005). Over the years, the local economy was based on the primary sector, since in the southern section of the town there is an irrigated plain area. Currently, interest has also turned to the tourism industry due to the national park, in which there are many hiking and rafting possibilities supported by the services of the Aoos River. In particular, the most characteristic direct and indirect river uses in Konitsa are: (a) irrigation of 1,000 ha of the plain area; (b) rafting in sufficient flow conditions for 7 months per year; (c) electricity production upstream of the town of 10^3 MWh/year. Additionally, the present situation of the Aoos River ecology is "good" and thus meets the requirements of the European Water Framework Directive (WFD) 2000/60.

Forecasting a decrease of above 20 % in the water potential of the area of interest over the forthcoming decades (Giannakopoulos et al. 2011), a respective decrease in the Aoos River runoff is anticipated. In the absence of any adaptation strategy, Aoos River services will significantly decline. More explicitly, the assumed changes in the Aoos River services are as follows: (a) irrigated land will be reduced to 700 ha; (b) the rafting period will decrease to 4 months per year; (c) electricity production will decrease by 25 %; and (d) the ecological state will worsen to "poor."

In order to estimate the non-market benefits associated with adaptation to climate change impacts on river uses, a CE survey was performed. Being a stated-preference technique, CEs enable the inclusion of non-use values in the economic assessment that usually are bypassed by other preference techniques. Non-use values are critical in regard to environmental goods or services valuation since most of them are not considered in normal markets (Pearce and Turner 1990). In this way, CEs are consistent with the Total Economic Value theory (Pearce and Turner 1990). CEs have considerable merit in measuring use and non-use values because they provide a richer description of the attribute trade-offs that individuals are willing to make (Adamowicz et al. 1998). As a result, many recent studies consider CEs as the most suitable technique for environmental valuation (Adamowicz et al. 1998; Alriksson and Oberg 2008; Hoyos 2010).

The underlying basis of a CE is the idea that "any good can be described in terms of its attributes, or characteristics, and the levels that these take" (Bateman et al. 2002). In a CE, respondents are presented with a series of alternative resource use options at a different cost (or "price") and are asked to choose their most preferred one. In this study, the Aoos River uses are assigned as the attributes of the CE, while the levels are defined by the "amount" of services provided prior and posterior to the consideration of climate change effects.

9.3 Theoretical Background

The basis for most of the microeconomic models of consumer behavior is the maximization of consumer utility under a finite income. CE is an application of this theory and the theory for value of Lancaster. According to this theory "the consumers reap the utility of the good from its characteristics and not from the good itself" (Lancaster 1966). As a result, the utility that the consumer obtains from one good or service is equal with the part-utilities (including the "price") combination deriving from the attributes of the good or service. Therefore, the individual has a utility function of the form:

$$U_{jn} = U(x_{jn}, p_{jn}, Z) \tag{9.1}$$

where the individual n receives utility U_{jn} when he/she chooses the alternative j. His choice is influenced by the attributes of the good j, which are presented as vector x, the price p of alternative j and the socioeconomic characteristics of the respondent Z. According to utility maximization theory, the respondent weights the attributes' levels of every alternative choice and opts for the one that offers the highest utility. Therefore, the individual n will choose the alternative j from a choice set C, only if the utility is higher from every other choice i (Pearce and Turner 1990)

$$(\mathrm{j}|\mathrm{C}) = U_{jn} > U_{in} \tag{9.2}$$

The neoclassical economic theory assumes that the individual chooses a good or service after prioritizing the different characteristics with a stable manner and gets the one that offers higher utility (theory of utility maximization) (Hanemann 1984a, b). In practice, the studies of stated preferences are not always consistent with this, since individuals several times make choices that do not maximize their utility. This occurs due to lack of information, market failure in pricing goods including externalities or opportunity costs/benefits (e.g., pollution, jobs created, etc.), nonobservable features (i.e., personal characteristics) and the secondary characteristics of alternative product or service choices which are not included (Louviere et al. 2002). The integration of these undefined components is based on the random utility approach (McFadden 1974), which combines the deterministic of human behavior with the stochastic of unobservable components. So, the utility function is divided into two parts, i.e., the deterministic (V_{jn}) and the nonobservable (ε_{jn}):

Table 9.1 Attributes and levels for various scenarios included in the CE survey

Attribute	Levels
Attr1: Irrigated Area (in hectares)	700, 900, 1000
Attr2: Rafting period (in months)	4, 6, 7
Attr3: Electricity production (% decrease)	0, 10, 25
Attr4: Ecological state	Poor, fair, good
Price: Monthly voluntary subscription for 10 years	2€, 5€, 10€, 15€, 20€ (zero cost for "no action")

$$U_{jn} = V_{jn} + \varepsilon_{jn} \tag{9.3}$$

Hence, the probability of the individual n to choose the alternative j from a choice set C is given by:

$$\begin{aligned}\text{Prob}(j|C) &= \text{Prob}(U_{jn} > U_{in}) \Rightarrow \text{Prob}(V_{jn} + \varepsilon_{jn} > V_{in} + \varepsilon_{in}) \\ &\Rightarrow \text{Prob}(V_{jn} - V_{in} > \varepsilon_{in} - \varepsilon_{jn}) \\ &\Rightarrow \text{Prob}(\varepsilon_{in} - \varepsilon_{jn} < V_{jn} - V_{in})\end{aligned} \tag{9.4}$$

In order to derive an explicit expression for this probability, it is necessary to know the distribution of the random part $\varepsilon_{in} - \varepsilon_{jn}$. According to this form of the distribution, the estimation model is inferred. The typical assumption is that the error component is Gumbel-distributed as well as independently and identically distributed (*iid*) (Train 2003). This condition accommodates the multinomial logit model, which is commonly used since any choice can be correlated with many statistical parameters.

9.4 Experimental Design and Data Collection

The survey design phase is the most important part of the design process, provided that it contains assumptions and decisions that affect and constrain the survey development. Applications of CEs to environmental goods or services mostly encompass three different alternatives. Each of the two first alternatives consists of different attribute level combinations, while the third is defined standardly as the situation that induces no action, change or improvement of one environmental good or service in return of zero price. Based on the particular characteristics of the study area, the involved attributes in the CE refer to the most important direct and indirect river uses, namely irrigation, rafting, hydroelectricity production, and ecological state. In addition, a basic constituent of each design is the hypothetical monetary cost that respondents should be voluntarily willing to pay per month, when selecting the preferred alternative. The attributes and the respective levels are presented in Table 9.1.

The design that permits different combinations to be generated by the product of the attribute levels number is referred to as full factorial and in this study it could give rise to 405 possible sets (3^4*5^1). This number is far from respondents' evaluation abilities and requires large cognitive and time sources. To delineate the number of different combinations a fraction factorial design is used, which statistically is a representative subsample of the total combinations. The latter is created using the principles of orthogonality (the levels of the attributes are altered independently from the other attribute levels), balance (minimizing the probability of one level repetition within the same choice card) and D-efficiency (ordinary least squares efficiency).

The previous conditions for the experimental design were specified into the Sawtooth software CBC version. Focusing only on main effects of the attributes that most conjoint studies are based, the routine "Complete Enumeration" was employed. This routine considers all possible tasks and chooses the ones that lead to nearly orthogonal design, in terms of main effects. Minimal overlap was taken also into account and alternatives are kept as different as possible. Furthermore, given the small sample size, "Complete Enumeration" was deemed to be appropriate to provide more information by combining to the greatest degree different options in every choice situation. Eventually, the aforedescribed strategy in our study yielded 96 different alternatives, which were merged into pairs plus the status quo scenario. The generated 48 choice sets were blocked into 8 versions of 6 choice sets and each respondent was allocated one of each version randomly. A hold-out choice set was also included to introduce the respondents to all of the different attribute levels (the fixed set was drawn up by all the attribute levels) and make clear to them what the choice exercise pertained to. Dominant choice tasks were reconsidered or slightly altered in order to be consistent and utility balanced.

The design report indicates that this strategy is optimally balanced, since all attribute levels are presented in an equal number of times. In terms of orthogonality, the design procedure of Sawtooth uses the ordinary least squares efficiency to impute relative standard errors of main effects in each level. The efficiency derives from the comparison of the present standard errors with those if the design was optimally orthogonal. An empirical rule demonstrates that the closest the OLS efficiency is to 1.0 the better design it is (Orme 2010). In this study, orthogonality was slightly disregarded being around 0.95 for most of the attribute levels. Optimal orthogonal design is achieved by generating a large number of choice versions and sets. Nevertheless, the number of choice versions (8 versions) and the choice alternatives (2 per choice set) were held low in order to reduce both fatigue effects and possible utility inconsistent choice sets that could be produced by a larger experimental design. The logit report of simulated data for a specified number of respondents produced the standard errors in utility estimation for each attribute level. In the present design we achieved for all attribute levels a standard error around 0.05, indicating that our design is acceptable and quite efficient (Orme 2010). An example of a choice set is presented in Table 9.2.

The CE technique is a part of a broader questionnaire, which attempts to reveal various aspects of the examined issue. Preference elicitation is doable by asking

Table 9.2 Choice set (including status quo, moderation alternative, absolute adaptation alternative)

	Status quo (i.e., "no action")	Alternative 1	Alternative 2
Irrigated Area	700 ha	900 ha	1,000 ha
Rafting period	4 months	6 months	7 months
Electricity production	Decrease by 25 %	Decrease by 10 %	No decrease
Ecological state	Poor	Fair	Good
Price	0€	10€	20€

different question types prior to the choice exercises, whilst the choice tasks enable the procedure of trading off on attributes. The attributes of the environmental good are opted to better represent the total utility of the environmental good. Respondents' socioeconomic profile is also of interest in order to acquire data on the individual-level basis. In addition, perceptions about the examined issue and socioeconomic characteristics of the participants may constitute significant components of extended or interacted forms of utility models, which are produced by using variables that stem from attitudinal and/or respondents' socioeconomic profile.

The questionnaire deployed for this study was structured into five parts. First, respondents confronted broad questions about the local environmental status with special regard to the river ecosystem of Aoos. Second, it contained general questions in order to know how and how much people use the Aoos River. Third, participants were asked to provide their opinions about climate change issues in the global perspective and how this may affect water provision in the local watershed. Fourth, people encountered the choice tasks and were allowed to trade off on the main Aoos River uses. Fifth, survey questions were included concerning sociodemographic characteristics and follow-up control questions.

The survey was carried out during the period of January and February 2013 on the residents of the town of Konitsa. Candidates were selected randomly and were personally interviewed. The outcome of the survey was 303 completed questionnaires. Approximately 15 % of the respondents (i.e., 45) opted standardly for the status quo scenario mainly for protest reasons. These individuals were excluded from the welfare analysis in order to reduce the "false zero bias" (Poirier and Fleuret 2010). Protest bidders stated that they would be willing to contribute in management scenarios to alleviate climate change impacts on river uses, in case of more reliable national and local strategic environmental planning.

9.5 Descriptive Statistics

Respondents were 41 years old on average. The average family size was about 3.5 persons (adults and children together). Up to 67 % of the respondents stated that their permanent residence was in Konitsa town, while 21 % were in the town for work purposes. Regarding education, two large groups emerged: one of high-school graduates (25.1 %) and one of university degree holders (27.7 %). The majority was employed (84.2 %) and declared a total annual household income that did not exceed €14,500 on average.

Looking at people's perceptions of their living environment, it appeared that for 75 % the local environmental conditions had been good or very good. The Aoos River was indicated by the respondents as an important ecosystem for most of them (86.8 %), while the opinions about the Aoos River status were divided, since half of the respondents described it as "fair" or "low" and half as "good" or "very good". In the same framework, 59.1 % of the participants noted that the Aoos River has changed for the worse in the last 10–15 years. Questions regarding use or non-use values provided by the river found the majority of people in agreement with the notion that the river does provide all these services. Most of the people who manifested that they do use the Aoos River mentioned recreational purposes. A small group (11.9 %) that uses the river water for economic purposes does so mainly due to irrigation needs.

Almost 90 % of the respondents were concerned about the future condition of the Aoos River, while 21.8 % identified the reduction in water flow as the most possible threat. About 77 % of the respondents were aware about climate change issues. Furthermore, around 70 % were convinced that climate change will affect the river, and about half said that they were concerned about the impacts on water flow (48 %). About one-third of the respondents foresee that the impacts of climate change on the Aoos River will arise in about 20–30 years. Almost all respondents recognized the need for adaptation measures against climate change at a local level. Regarding river water use priorities, respondents opted for the good ecological status of the river (48.8 %), the irrigation water (42.6 %), the hydro-electricity production (5 %) and the rafting activities (3.4 %).

9.6 Econometric Results

The conditional logit (CL) model is a specification of the general multinomial logit model and is defined such that it is a function of choice-specific characteristics only (Poirier and Fleuret 2010). The CL model is basically used in all CE studies and offers an overview of the average preferences, constituting the benchmark for further analysis (Juutinen et al. 2011). With the CL model, the choice set must comply with the "Independence from Irrelevant Alternatives" (IIA) property. The IIA property implies that the relative probabilities of two alternatives being chosen from a choice set are unaffected by the introduction, or removal, of other alternatives in that choice set (Bliem et al. 2012). Furthermore, the CL model assumes homogenous preferences across respondents and all interviewees' answers are pooled. In the context of this study, preference homogeneity allows to focus on the attributes of the Aoos River affected by climate change, as the exclusive factors for the choice decision. The observable component of the utility function reflects the

Table 9.3 Results of the CL and extended CL models

Variable	CL Model	Extended CL model
Irrigation area	0.1016*** (0.026)	0.1022*** (0.026)
Rafting period	0.0628*** (0.026)	0.0637*** (0.026)
Electricity production	0.0159*** (0.032)	0.0159*** (0.003)
Ecological state	0.5670*** (0.042)	0.5674*** (0.042)
Price	−0.0468*** (0.006)	−0.0466*** (0.006)
ASC	0.7200*** (0.126)	2.2719*** (0.504)
Additional variables interacted		
Aoos*ASC	–	−0.2672*** (0.0828)
Age*ASC	–	−0.0308*** (0.007)
Educ*ASC	–	0.0827* (0.0504)
Summary statistics		
Log-Likelihood	−1467.2077	−1447.5333
Pseudo R^2	0.1935	0.2043
AIC*	2946.415	2913.067
BIC*	2985.48	2971.664
Observations	4,968	4,968
Sample Size	276	276

Note standard errors in parentheses *: $p < 0.1$,**: $p < 0.05$ and ***: $p < 0.01$
AIC Akaike Information Criterion, *BIC* Bayesian Information Criterion

sum of the attributes' part-worth utilities of the respondents (Zander and Garnett 2011). The following model depicts the utility function that an individual i gets from alternative n at choice situation t:

$$U_{\text{nit}} = \beta_j^C \text{ASC}_j + \beta^{\text{Ir}}\text{IrrigationArea}_{\text{nit}} + \beta^{\text{Raf}}\text{RaftingPeriod}_{\text{nit}} \\ + \beta^{\text{El}}\text{ElectricityProduction}_{\text{nit}} + \beta^{\text{EC}}\text{EcologicalStatus}_{\text{nit}} + \beta^{P}\text{Price}_{\text{nit}} + \varepsilon_{\text{nit}}$$

where $\beta_j^C \text{ASC}_j$ denotes the "alternative specific constant" (ASC) and is equal to 1 for alternatives other than status quo, and β^{Ir}, β^{Raf}, β^{El}, β^{EC}, β^{P} represent the vector of coefficients describing attributes associated with the different uses of the Aoos. The term ε_{nit} displays the error component incorporated in random utility models.

An extended conditional model was also estimated with the inclusion of interacted opinions and socio-demographic variables in order to overcome a possible violation of the IIA property. The extended form model permits unbiased estimation of the conditional coefficients (Birol et al. 2006a; De Valck et al. 2012), since it takes into account some sources of preference heterogeneity, accounting for the relative impact of respondents' beliefs. The interacted variables were created by respondent-specific social economic and attitudinal characteristics with choice-specific attributes and/or with the ASC coefficient.

The results of the models are reported in Table 9.3. Turning to the general CL model, the log-likelihood value achieved (−1467) and Pseudo R^2 ($\sim$0.2) are comparable with those reported by other studies (Goibov et al. 2012; Juutinen et al.

2011; Tait et al. 2012) and are interpreted as a good fit for our model (Hensher and Johnson 1981). The coefficients are highly significant at the 1 % level and their signs are as expected. More explicitly, the positive sign of the ASC coefficient indicates that respondents prefer moving away from the status quo scenario accounting the river ecosystem services as contributors to their utility deriving from the Aoos River. Furthermore, it is proved that all the attributes used (i.e., water uses) in the CE are significant factors in the choice of the river management scenario under climate change impacts. Higher levels of any single attribute except for the price increases the probability that a management scenario is selected. In particular, river management scenarios with the higher levels of irrigation area, rafting period, hydroelectricity and ecological state are preferred by the respondents, *ceteris paribus*. In line with expectations, the price attribute has a negative sign; thus it poses a negative utility effect in the case that scenarios with higher payment levels are chosen. The price attribute, apart from the willingness to pay estimations about river uses, also enables the trading-off process on the other attributes, since it allows the respondents to prioritize the river uses at a certain level of price each time. Overall, these results point out that positive and significant economic values exist for higher levels of ecological, economic and recreational attributes of the Aoos River, as discussed in the next section.

The second column of Table 9.3 depicts the extended form of the general CL model with the inclusion of interacted variables. After testing various interactions of the four river uses with the respondents' socio-demographic profile and perceptions toward the Aoos River ecosystem, a model that includes education, age and perceptions about the Aoos River state was found to fit the data reasonably well. The interaction term Aoos*ASC shows a significant negative sign, meaning that those who believe that the Aoos River is generally in a good or very good condition are more positive about selecting river management scenarios toward climate change adaptation and therefore higher levels of river use attributes. As per interaction term Age*ASC, the negative sign indicates that young people are more likely to move away from the status quo option, selecting alternative schemes, i.e., policies that promote adaptation measures. A similar attitude is observed concerning respondents with higher education, i.e., they are more likely to select adaptation policies. The coefficients of the CE attributes representing the main river uses remain approximately as in the "simple" CL model. The log-likelihood and Pseudo R^2 values were slightly improved, indicating a better model fit with the extended CL model.

An additional analysis to understand better people's behavior about climate change adaptation scenarios was undertaken by imputing CL models for subsets of data on the basis of socio-demographic or attitudinal variables. The comparisons between the groups created by these subsets are given as follows:

- Women were on average more willing to move away from status quo scenarios but more reluctant to choose scenarios with high price levels. Men valued irrigation higher than women. In contrast, ecological status and rafting were valued higher by women.

- Younger respondents were more likely to move away from the status quo. Nevertheless, older respondents are willing to contribute more money when they choose adaptation scenarios. Furthermore, older respondents are willing to contribute twice as much as younger respondents to irrigation.
- Respondents who do not live permanently in Konitsa are willing to pay less than permanent residents. In addition, it is found that their choices are dominated by the ecological state of the river.
- Unemployed respondents are more willing to move away from the status quo, but they are willing to contribute less money than employed respondents. It is interesting though that they valued all the river uses, except for the ecological state, more than employed people, implying that these river uses could provide job opportunities in the future.
- People with higher incomes (>13,000 €/year) have a higher willingness to pay than people with lower incomes. The former are willing to pay much more for the ecological state of the river, while the latter are willing to pay more for economic uses of river water.
- Respondents who are unaware of climate change issues are willing to contribute more money for adaptation, probably due to an overwhelming sense of insecurity relating to climate change-imposed risks.
- As mentioned, respondents who believe that the Aoos River is generally in a good or very good condition are less likely to select river management scenarios toward adaptation for climate change. Nevertheless, these people indicate a willingness to pay more for adaptation than those who stated that the condition of the Aoos River is fair or bad.
- Both users and non-users of the river are equally likely to support adaptation scenarios. However, river water users are willing to pay more toward this direction.

9.7 Welfare analysis

Once the parameter estimates have been obtained, the WTP values for the marginal change in an attribute (known as "implicit price") are estimated by dividing the estimated coefficient on the attribute of interest by the negative coefficient on the monetary variable. In other words, the value of a marginal change in any of the attributes in terms of welfare measurements accrues from the ratio of the coefficient of attribute j and the price coefficient (Hanley et al. 2002), as follows:

$$\mathrm{WTP} = -\frac{\beta_j}{\beta^P} \tag{9.5}$$

where WTP stands for willingness to pay. As given in Table 9.4, the benefit estimates for all river attributes indicate that Konitsa's residents are willing to

Table 9.4 Marginal WTP for the CE attributes

Attribute	Implicit price (€) CL model	Implicit price (€) Extended CL model
Irrigation area	2.17	2.19
Rafting period	1.34	1.37
Electricity production	0.34	0.34
Ecological state	12.12	12.17

contribute according to the CL model every month per household €2.17 for every 100 ha preserved irrigated area, €1.34 for having an extra month of sufficient flow for rafting activities, €0.34 for 10 % more hydroelectricity production and €12.12 for improving the state of the river ecosystem to the next better level.

The above-mentioned implicit prices do not provide estimates of compensating surplus (CS) for alternative adaptation scenarios. Welfare measures derive from the marginal rate of substitution between the residual of the initial utility state and the alternative utility state divided by the marginal utility of income, which is represented by the coefficient of the payment attribute. Thus, in order to estimate WTP for adaptation to climate change, three distinct hypothesized scenarios were defined, as follows:

- Scenario 0 represents the "do-nothing" case, in which no adaptation actions are considered. As a result, river water uses deteriorate due to climate change with subsequent loss of utility. More explicitly, the irrigated land will be reduced from 1,000 to 700 ha, the rafting period will be confined to 4 months per year, the electricity production will decrease by 25 %, and the ecological state will experience a decline from "good" to "poor."
- Scenario 1 stands for a moderate adaptation policy. In this case, all river water uses are preserved to some extent from climate change-induced impacts. More specifically, the irrigated land will decrease by 10 % (i.e., from 1,000 to 900 ha), the rafting period will be shorter from 7 to 6 months per year, and the electricity production will decrease by 10 %. Finally, the Aoos River ecology will be characterized as "moderate."
- Scenario 2 foresees a strong adaptation policy that maintains the present river status in the future, i.e., irrigation land will remain the same as today (i.e., 1,000 ha), river water levels will support rafting activity for 7 months per year, electricity production will not decrease, and the present situation of the Aoos River ecology will be characterized as "good," meeting the requirements of the European water directive 2000/60.

To find the CS associated with each of the above-described scenarios, the difference between the welfare measures under the status quo and the alternatives scenarios are estimated. Welfare changes are then obtained by using the compensating surplus formula described by Hanemann (1984), as in Eq. 9.6.

Table 9.5 Compensating surplus for each scenario (€/month)

Scenario	CL model	Extended CL model
Scenario 1	40	52
Scenario 2	59	71

$$CV = -\frac{1}{\beta^P}(V^1 - V^0) \quad (9.6)$$

where β^P is the parameter estimate of price, and V^0 and V^1 represent a representative respondent's utility before and after the change under consideration. The estimates of WTP for the alternative scenarios are given in Table 9.5.

As expected, the CS increases, moving from the status quo situation to the adaptation scenarios considered. For the best-fit Extended CL model, the results indicate that households are willing to pay almost €52 per month (i.e., approximately €620 per year) for moderate adaptation. The voluntary contribution increases to €71 per month (i.e., about €850 per year) for an all-inclusive solution for adaptation, which will preserve all human and ecosystem services of the Aoos River to current levels.

9.8 Conclusions

This study comprises one of the first attempts to monetize non-market benefits associated with adaptation to climate change impacts on mountainous water resources. Toward this direction, a CE was conducted using a face-to-face survey on a sample of Konitsa's residents in order to analyze trade-offs of choices and to estimate the welfare effects of adaptation measures.

The results of the CE highlight that there are positive and significant economic benefits associated with different river ecosystem services (both use and non-use values). In particular, respondents were willing to pay to move away from the "do-nothing" situation toward a new direction that could preserve the current levels of the use of Aoos River by means of appropriate climate change adaptation measures. As regards the particular attributes of river water, the "ecological state" was ranked relatively highly in terms of implicit price order, implying that the local community's well-being depends heavily on river-supporting services rather than provisioning (i.e., irrigation water, hydroelectricity generation) or cultural (i.e., rafting) services. This is consistent with the ecosystem services theory where supporting services are necessary for the production of all other ecosystem services (Millenium Ecosystem Assessment 2005) and the findings of other studies, e.g., (Zander and Garnett 2011), where respondents were willing to preserve river water primarily for its intrinsic value. On the other hand, the relatively low marginal WTP for hydroelectric power generation is likely related to a latent NIMBY ("not in my back yard") behavior against the large hydroelectric plant that has been constructed at the spring of the Aoos River, as the dam decreases the volume of water flowing downstream.

The economic analysis performed in this study attempts to assist mountain communities to integrate cost-effective adaptation into their long-term sustainable water resources management and fulfills in general the needs of efficient management policies emphasized in the WFD 2000/60/EC, the first legislative action in which the interrelation between water aquifers and socioeconomic values has been acknowledged (Birol et al. 2006b). Thus, it fills an important gap in information in Greece, regarding environmental/resource costs of river water (Kanakoudis and Tsitsifli 2010). Furthermore, since this is the first effort in Greece and one of the few in the world to establish non-market values of adaptation benefits on a local scale, it is anticipated that communities that undergo similar climate change trends could make use of the present monetary estimates through value transfer approaches. Finally, the observed influence of the individual-related characteristics on choice preferences is a matter that can lead to a better deliberation process among decision makers and the stakeholders, so as to develop an optimal climate change adaptation plan for the local water resources.

References

Adamowicz, W., Boxall, P., Williams, M., & Louviere, J. (1998). Stated preference approaches for measuring passive use values: Choice experiments and contingent valuation. *American Journal Agricultural Economics, 80*(1), 64–75.

Alriksson, S., & Oberg, T. (2008). Conjoint analysis for environmental evaluation a review of methods and applications. *Environmental Science and Pollution Research, 15*(3), 244–257.

Bateman, I., Carson, R., Day, B., Hanemann, M., Hanley, N., & Hett, T. (2002). *Guidelines for the use of stated preference techniques for the valuation of preferences for non-market goods*. UK, Cheltenham: Edward Elgar.

Bates, B. C., Kundzewicz, W. Z., Wu, S., & Palutikof, J. P. (2008). Climate change and water. *Technical Paper VI of the IPCC*. Geneva, Switzerland: IPCC Secreteriat.

Birol, E., Karousakis, K., & Koundouri, P. (2006a). Using a choice experiment to account for preference heterogeneity in wetland attributes: The case of cheimaditida wetland in Greece. *Ecological Economics, 60*(1), 145–156.

Birol, E., Karousakis, K., & Koundouri, P. (2006b). Using economic valuation techniques to inform water resources management: A survey and critical appraisal of available techniques and an application. *Science Total Environment, 365*(1–3), 105–122.

Bliem, M., Getzner, M., & Rodiga-Lassnig, P. (2012). Temporal stability of individual preferences for river restoration in Austria using a choice experiment. *Journal Environmental Management, 30*, 65–73.

De Valck, J., Vlaeminck, P., Liekens, I., Aertsens, J., Chen, W., & Vranken, L. (2012). Valuing landscape preferences for nature restoration: Drongengoed case study (Belgium). *129 EAAE Seminar*, Castelldefels, Barcelona.

Giannakopoulos, C., Kostopoulou, E., Varotsos, K. V., Tziotziou, K., & Plitharas, A. (2011). An integrated assessment of climate change impacts for Greece in the near future. *Regional Environmental Change, 11*(4), 829–843.

Goibov, M., Schmitz, P. M., Bauer, S., & Ahmed, M. N. (2012). Application of a choice experiment to estimate farmers preferences for different land use options in northern Tajikistan. *Journal of Sustainable Development, 5*, 5.

Gret-Regamey, A., Brunner, S. H., & Kienast, F. (2012). Mountain ecosystem services: Who cares? *Mountain Research and Development, 32*, 23–34.

Hanemann, W. (1984a). Discrete continuous models of consumer demand. *Econometrica, 52*(3), 541–561.

Hanemann, W. (1984b). Welfare evaluations in contingent valuation experiment with discrete responses. *American Journal of Agricultural Economics, 66*(3), 332–341.

Hanley, N., Wright, R., & Koop, G. (2002). Modelling recreation demand using choice experiments: Climbing in Scotland. *Environmental Resource Economics, 22*(3), 449–466.

Hensher, David, & Johnson, Les. (1981). *Applied discrete choice modeling*. New York, USA: John Wiley.

Hoyos, D. (2010). The state of the art of environmental valuation with discrete choice experiments. *Ecological Economics, 69*(8), 1595–1603.

Juutinen, A., Mitani, Y., Mantymaa, E., Shoji, Y., Siikamaki, P., & Svento, R. (2011). Combining ecological and recreational aspects in national park management: A choice experiment application. *Ecological Economics, 70*(6), 1231–1239.

Kanakoudis, V., & Tsitsifli, S. (2010). On-going evaluation of the WFD 2000/60/EC implementation process in the European Union seven years after its launching: Are we behind schedule? *Water Policy, 12*, 70–91.

Lancaster, K. J. (1966). A new approach to consumer theory. *The Journal of Political Economy, 74*(2), 132–157.

Louviere, J., Street, D., Carson, R., Ainslie, A., Deshazo, J., & Cameron, T. (2002). Dissecting the random component of utility. *Marketing Letters, 13*(3), 177–193.

McFadden, D. 1974. Conditional logit analysis of qualitative choice behavior. In: P Zerembka (ed). *Frontiers in econometrics* (Vol. 45: pp. 105–141). New York, USA: Academic press.

Millennium Ecosystem Assessment. (2005). *Ecosystems and human well-being: A framework for assessment*. Washington, USA, DC: Island Press.

Milly, P., Dunne, K., & Vecchia, A. (2005). Global pattern of trends in streamflow and water availability in a changing climate. *Nature, 438*, 347–350.

Orme, B. K. 2010. *Getting started with conjoint analysis. Strategies for product design and pricing research*. Mandison, USA, WI: Research Publisher.

Papageorgiou, K., Kassioumis, K., Blioumis, V., & Christodoulou, A. (2005). Linking quality of life and forest values in rural areas: An exploratory study of stakeholder perspectives in the rural community of Konitsa, Greece. *Forestry, 78*(5), 485–499.

Pearce, David, & Turner, Kerry. (1990). *Economics of natural resources and the environment*. London, UK: Harvester Wheatsheaf, Hemel Hempstead.

Philandras, C. M., Nastos, P. T., Kapsomenakis, J., Douvis, K. C., Tselioudis, G., & Zerefos, C. S. (2001). Long term precipitation trends and variability within the Mediterranean region. *Natural Hazards Earth System Sciences, 11*, 3235–3250.

Poirier, J., & Fleuret, A. (2010). Using the choice experiment method for valuing improvements in water quality: A simultaneous application to four recreation sites of a river basin. *59th conference of association francaise de science economique*, nanterreuniversite.

PRUDENCE (Prediction of Regional scenarios and Uncertainties for Defining European Climate change risk and Effects) (2007). http://prudence.dmi.dk/.

Stournaras, G., Skourtos, M., Kontogianni, A., Yoxas, G., Machleras, A., & Nastos, P. (2011). Climatic changes and impacts on Greece's water systems (pp. 127–152). The environmental, economic and social impacts of climate change in Greece. National Report. Bank of Greece Printing Works, Athens, Greece.

Tait, P., Baskaran, R., Cullen, R., & Bicknell, K. (2012). Non-market valuation of water quality: Addressing spatially heterogeneous preferences using GIS and a random parameter logit model. *Ecological Economics, 75*, 15–21.

Train, Kenneth. (2003). *Discrete choice methods with simulation*. Cambridge, USA: University Press.

Zander, K. K., & Garnett, S. T. (2011). The economic value of environmental services on indigenous-held lands in Australia. *PLoS ONE, 6*(8), 23–54.

Author Biographies

Dimitrios Andreopoulos Dimitrios Andreopoulos obtained his Master's degree from the National Technical University of Athens (Greece) in 2009. He is currently a Ph.D. student under the supervision of Prof. Fischer Christian. His research focuses on economic impacts of climate change on ecosystem services derived from river systems.

Dr. Dimitris Damigos is an Assistant Professor in the School of Mining Engineering—Metallurgy of the NTUA. He holds a Diploma in Mining Engineering and a Ph.D. in Environmental Economics. His research interests include environmental economics, mineral economics, financial and cost-benefit analyses, waste management, and risk assessment and management. He teaches a number of under- and postgraduate courses. He has participated in more than 30 research projects. He has published about 80 papers in the abovementioned fields and he is a reviewer for 14 academic journals

Dr. Francesco Comiti is an Assistant Professor at the Faculty of Science and Technology of the Free University of Bozen-Bolzano since 2009, where he teaches Rural Hydrology and Hydraulics. He obtained the Ph.D. in "Environmental Watershed Management" at the University of Padova in 2004. His current research interests include river hydromorphology, hydrological and sediment transport monitoring, flood and debris flow hazard assessment in mountain basins. He has published 55 papers on these topics, 34 of which on international journals. He is currently an Associate Editor for the journal "Water Resource Research" and has been a reviewer for 20 academic journals.

Dr. Christian Fischer is an Associate Professor of agro-food economics, management, and marketing at the Free University of Bozen-Bolzano. Being specialized in the analysis of agricultural markets he has also developed an interest in the evaluation of non-market goods such as certain ecosystem services.

Chapter 10
The Stakeholder Analysis: A Contribution Toward Improving Impact of Rural Policy

G. Benedetto, D. Carboni and G. L. Corinto

Abstract Since more than a decade agricultural economists pay more attention to CAP's effects evaluation as a consequence of a larger social request for understanding what impacts are generated by the adoption of Rural Development Policy. The EU 2020 strategy fosters a political shift from the market liberalization processes toward policies promoting stability and equity, in addition to environment protection and social inclusion. Thus, CAP will have to be better evaluated by both quantitative and qualitative tools in order to understand the role of local communities in RDP implementation. In studying the role of wine tourism, in particular the successful case of Young Wines Exhibition of Sardinia and the failure of the "Verdicchio di Matelica" Wine Road (Ancona), the use of stakeholder analysis has shown positive contribution in evaluating the role of social actors in success/failure of RDP implementation. This work aims at discussing the necessity to use a more holistic evaluation method, mainly focusing on possibilities and difficulties of involving local social actors and policymakers.

10.1 Introduction

The future Common Agricultural Policy (CAP) scenario will be characterized by financial constraints. Recently, in October 2011, the European Commission gave CAP a declared objective to meet the challenges of food security, sustainable use

G. Benedetto (✉)
Department of Science for Nature and Environmental Resources,
University of Sassari, Sassari, Italy
e-mail: gbenedet@uniss.it

D. Carboni
Department of Social and Human Science, University of Sassari, Sassari, Italy

G. L. Corinto
Department of Education, Cultural Heritage and Tourism,
University of Macerata, Macerata, Italy

C. Zopounidis et al. (eds.), *Agricultural Cooperative Management and Policy*, Cooperative Management, DOI: 10.1007/978-3-319-06635-6_10,

of natural resources and growth within the necessity to maintain CAP spending at 2013 levels in nominal terms. Thus, the importance of policy evaluation is increased to achieve goals of public interest, and for more than a decade, agricultural economists are being urged to pay more attention to CAP's effects evaluation. Pillar II of CAP will obtain close to one-third of the total budget, further posing an essential question to policymakers and evaluators in fair decisions.

This chapter will discuss the possibility to help policy decisions by using the stakeholder analysis (SHA) as a qualitative tool, in addition to quantitative evaluation of rural policy. Section 10.2 is dedicated to a brief discussion about the debate on the policy analysis based on alternative "prescriptive" and "descriptive predictive" approaches, and the possibility to use either quantitative and/or qualitative methods for a better evaluation of (rural) policies. Section 10.3 gives the background on Rural Development Policy, stakeholder analysis and qualitative studies. Section 10.4 sketches the methodology, reporting two case studies conducted using the stakeholder analysis (SHA) in the wine tourism sector, while Sect. 10.5 concludes the discussion.

10.2 Prescriptive and Descriptive Predictive Approaches in Policy Analysis

The analysis of CAP and Rural Policy Development, like the analysis of other policy areas, has suffered from competing theories, paradigms, and evaluation criteria. The main confrontation is between qualitative and quantitative methods, even when researches claim a major integration of different approaches.

More in general, the public policy literature would appear as a source of scarce clarity for policymakers, and the relationships between science and policymaking seem to be unproductive, as politicians rarely read scientific papers and researchers live in their ebony tower.

The vast literature about policy analysis can be approximately divided into two categories (each of them further divisible) (Rossell 1993):

(i) prescriptive analyses;
(ii) descriptive predictive analyses.

Under (i), scholars (Rossell 1993) intend analyses that seek to improve the policymaking process rationally, selecting policy alternatives so that policymakers can best achieve their goals. This is a branch of policy sciences and can follow two lines of doing: qualitative theoretical issues and qualitative works relying on such techniques as cost-benefit analysis, public choice models, decision analysis, and mathematical programming.

Under (ii), works emphasize the attempt to make sense of and evaluate policies already implemented (Rossell 1993), encompassing works related to classifying policies (distributive, redistributive, regulatory); quantitative analysis of factors linked to public expenditures in specific sectors or areas; qualitative case studies in specific policy areas; case studies of the implementation of public policies; and theory of the policy process.

Rossell (1993, p. 156) enucleates a common focus in the very different approaches "that distinguishes policy analysis from political analysis of the policymaking process. Policy analysis is interested in policy content above all else [...] Political analysis, by contrast, is interested in the process and political behavior almost as an end in and of itself."

Policy analysts view the government quite as an instrumental tool to improve—for instance—the welfare of society, tying to integrate theory and practice for human benefit. And, in this sense, they are proactive, but give also the rationale for descriptive predictive analyses and for most of the policy analyses, including the qualitative case studies of one or more policy areas, such as health, energy, and the environment, not excluding agricultural policy.

The debate on positivism is still alive in social sciences (Miller and Yang 2007, p. 110) and scholars such as Bobrow and Dryzek (1987) stated that positivism provided little help in determining public policy, as the most likely important problem in decision-making is acquiring the necessary knowledge. This means that in the approach of results of future policy actions, i.e., if X_1 through X_n cause Y, than the policy decision must include all the factors of X. Bobrow and Dryzek (1987), reviewing the mathematical side of policy analysis, concluded it is of limited value because it does not fit well with political reality. Policy analysts and researchers must assist policymakers not in a technical way but in a political one, designing robust basic structure able to resist the constant shifting coming from political and social actors and lobbies. The mathematical side of policy could be labeled naive because it is apolitical and the political science literature is, by contrast, accused of lacking rigor because it is quite descriptive and atheoretical (Rossell 1993, p. 158) or taking us into uncharted territory (Sadoulet and de Janvry 1995, p. 1). We can argue that many aspects of the two lines might join to facilitate any policy analysis.

Many concepts used in formal models, besides their mathematical rigor, are useful criteria for organizing the information found in case studies and for evaluating policy alternatives. Combining information from case studies and concepts from political science, you can constitute a more useful framework of criteria for ex-ante or ex-post evaluating alternative policies.

The theoretical discussion is not the aim of this chapter. At any rate, it is important to underline the necessity to improve knowledge of most factors determining the results of public policy, especially of intervention in agriculture and rural society, as it is half the part of the EU financial budget.

10.3 Backgrounds

10.3.1 Backgrounds: Rural Development Policy

The revolution in RDP has been essential for reforming the CAP. Many real experiences demonstrate that results are strictly joined to the actual unity of programs and cross-relations with other socio-structural policies. This evolution results from the rationalization of programming and financing based on evaluation processes that have implemented more knowledge for decision-making and policy implementation. The European Commission (2006), OECD (2009), and scholars (Leeuw and Vaessen 2009) have highlighted the valuation process as an integral part of the policymaking (Midmore et al. 2008; Kinsella et al. 2010; Mortimer et al. 2010; Hill and Wojan 2010), even more so when sectoral environments are highly complex (Leon 2005) such as agriculture and rural society (Sali 2010).

As the territory is a strong component of the RDP, many models try to adopt a local scale of analysis, subnational, regional, and even subregional ones. Or, rural policies can be evaluated from the viewpoint of diverse economic categories (i.e., consumers, producers, tax payers, etc.) or using macro-indicators (growth, employment, wages) (Midmore et al. 2008), also introducing economic, social, cultural, and environmental parameters (Kinsella et al. 2010). This latter is the scheme adopted by the EU for the RDP, for the utmost importance of the environment and society in farming (Robinson and Liu 2006; Kilkenny 1999; Balamou et al. 2008).

It is impossible to give a detailed and exhaustive review of the vast literature on evaluation of policies and intervention programs. Several different approaches and proposed classifications aim at defining the boundaries, contexts, and practices of the evaluation processes. In the case of complex policies, such as RDP, the evaluation process has been divided immediately in different and very hard-working paths. For the sake of brevity, we try to describe the eventual complexity, but only considering a few works that adopted quantitative methods. Table 10.1 reports authors and years of publication, area, evaluated policy, research goals, methods, economic categories, and main results.

At any rate, even though progress in the use and implementation of quantitative and computational models has been very interesting, also in Italy, the actual possibility of representing in a satisfactory manner the results of the PAC are to be considered limited (Anania 2005). Regarding this, the main issue appears to be the very low flexibility of the PAC measures to diversify local communities and territories (Midmore et al. 2008), omitting consideration of social and cultural histories, disparities, as well as levels of local endowments. Actually, the RDP measures are diverse and more complex, aiming at fostering the nonfarm activities; but these being only a small fraction of rural activities, farming is still the motor of rural economies and household revenue.

Table 10.1 Quantitative Methods in CAP and RDP Evaluation, some examples

Authors, year	Area	Evaluated policy	Research goals	Methods	Economic categories	Main results
Moro and Skokai (1999)	Northern Italy	CAP arable crop regime	To measure in terms of elasticities the response of production and land allocation decisions to the compensatory payments scheme	Profit function extended to include the effect of aid	RICA farms	The estimated model allows measuring the area considered, supply and land allocation responses to the recent 'Agenda 2000' reform package, as well as to analyze its degree of decoupling
Judez et al. (2004)	Spain	Wetlands plan policy	Effects on income and crop distribution of the implementation of the new Wetlands Plan policy approved in 2003 for four farm types	Positive Mathematical Programming (PMP)	Irrigated vineyards	The findings show that the CAS adherence formula adopted by farm types is unrelated to the type of aid envisaged for COP crops. Moreover, the new CAS measures to which the holdings adhere depend on the CAS formula in force for each one in the base year and on the intensity of vineyard irrigation

(continued)

Table 10.1 (continued)

Authors, year	Area	Evaluated policy	Research goals	Methods	Economic categories	Main results
Helming (2005)	Netherlands	EU Common Agricultural Policy (CAP) and Dutch manure and nutrients policies	To describe the current state-of-the-art of the Dutch Regionalized Agricultural Model (DRAM). Two model applications	Dutch Regionalized Agricultural Model (DRAM)	Regional and sector levels (dairy farming)	It is concluded that the model is a flexible tool for integrated scenario and policy analyses at the agricultural sector level
Arfini et al. (2006)	Italy	CAP reform	To describe the main features of a model that has as an objective the estimation of the effects of agricultural policy measures at national, regional, and subregions	FIPRIM (Fadn Iacs PMP Regional Integrated Model)	Agriculture (RICA farms)	The model results show a general decrease in grain surfaces, which is also a significant increase in forage areas and areas not cultivated in accordance with good agricultural practice. The FIPRIM model can be applied in many contexts to assess agricultural policy and changes in market variables

(continued)

Table 10.1 (continued)

Authors, year	Area	Evaluated policy	Research goals	Methods	Economic categories	Main results
Schmid and Sinabell (2006)	Austria	CAP reform 2003 and the introduction of the new program for rural development	To estimate the impact of the 2003 CAP reform on selected agricultural and environmental indicators to measure rural and agricultural development	Positive Agricultural Sector Model Austria (PASMA)	Conventional and organic production systems (crop and livestock)	Results show that the single farm payment, which was introduced in 2005 will have an important effect on land allocation. Because this payment is only granted if land is maintained in good agricultural and ecological condition, farmland is kept in production which otherwise would most likely have been abandoned
Balamou et al. 2008	Greece and Scotland	Impact of CAP reform	To assess the magnitude and distribution of effects associated with a change in agricultural policy in two predominantly rural regions of the EU	CGE model	Urban and rural areas	The results show that the impacts of CAP reform spread from rural to urban areas within regions. Importantly, they suggest that while coupled price support sustains rural GDP in both study areas, the removal of such support will not lead to drastic negative effects through price and substitution adjustments in the regions

(continued)

Table 10.1 (continued)

Authors, year	Area	Evaluated policy	Research goals	Methods	Economic categories	Main results
Schader et al. (2008)	Switzerland	Agro-environmental policies	1) To explore the general suitability of static programming models as a tool to assess both economic and ecological impacts of single policy measures and complex schemes 2) To discuss an approach to quantify the economic and ecological effect of organic farming and alternative extensification policies and to determine the cost-effectiveness of these measures	Nonlinear programming models (FARMIS)	Organic and nonorganic farms	The model comparison showed that currently, static PMP models are the most widespread sector modeling approach integrating environmental concerns. The approach outlined in this chapter has some advantages and limitations, which are discussed below. By employing CH-FARMIS for the evaluations, sector-level results can be generated, differentiating between farm types and farming systems and regions. On the other hand, the model can only indirectly take into account structural change and conversion to organic agriculture

(continued)

Table 10.1 (continued)

Authors, year	Area	Evaluated policy	Research goals	Methods	Economic categories	Main results
Howitt et. al. (2010)	California	Environmental policies	To estimate the effects of particular water-related policies and external shocks to agriculture in California	Positive Mathematical Programming method for modeling agricultural production and water use called the Statewide Agricultural Production Model (SWAP)	Three categories of land use, including environmental agricultural, and urban, in California	These applications highlight the importance and robustness of SWAP for studying the economics of irrigated agriculture as part of a complex system

10.3.2 Background: Stakeholder Analysis and Qualitative Studies

The EU's institutionalized formal policy evaluation models actually cannot trace the chain causality from sustained actions to their impacts (Baslé 2006). Rigid models are able to assess and measure the extent to which a selected policy intervention realizes the targeted objectives, but are not feasible to grasp some important questions such as how and why that policy operates in the way it has done (Midmore et al. 2008).

Furthermore, the necessity to increase the contribution of social capital, in terms of capacity to establish and manage relationship marketing to improve revenues and the capacity to face globalization, claims policies more grounded in territorial reality (Murdoch 2000). It is clear that deeper insights are necessary. And these are possible only by investigating a small number of cases in a very intensive manner, adopting the suggestions of Yin (1994), i.e., using empirical investigation immersed in a real-life context, especially when the context and the investigated phenomenon are reciprocally influenced. Moreover, the case study method suggested by Yin (1994) gives the opportunity to make some theoretical generalizations when the survey is deep, allowing to make an analytical interpretation of the empirical results. Further insights will be possible by adopting a comparison between different individual cases that are able to reveal the "causal" patterns differently rooted in different places. Places can face the same phenomenon with very different factors of influence actually emerging in separate contexts (Zadek et al. 1997).

The stakeholder analysis (SHA) is a method of investigation that, through the systematic collection of qualitative information and subsequent analysis of this information to understand what the interests are that must be taken into account in the design of a policy, program, or any other action, identify the key players, and interact with them effectively. The SHA allows an understanding of the values, interests, aptitudes, and aspirations of stakeholders favoring a more transparent and coherent dialog between the parties. In this sense, it can be crucial in conflict resolution, as it provides a platform on which the committed people can identify and express their interests converging and/or diverging, in order to build a shared vision between the parties, encouraging them to find a method of negotiation and to recognize the value of joint cooperation (Parnell 2007). The SHA is a highly flexible application; it was initially used in business management and political science, now being used in any field where the stakeholders do act, from the economy to health care and social reform, and in the management of natural resources as evidenced by numerous contributions in the literature (Lindblom 1959; Reich 1994; Grimble et al. 1995; Mitchell et al. 1997; Varvasovszky and Brugha 2000; Schmeer 2000).

The fundamental aim of the application of SHA is to understand the interests of the various parties and find a compromise between the potential conflict of interests identified during the investigation, thus reducing the risk of failure due to the opposition of one or more parts, and thus facilitating the work of policy implementation.

In this direction, the strengths of the tool seem to be self-evident. Actually, its weaknesses or disadvantages are embedded in the practical conduction of the survey. First of all it is expensive, because, to be sound and exhaustive, it requires a workgroup for the formulation of the objectives, the development of tools, the conducting of interviews, and the transfer of results. Moreover, the list of (main) stakeholders and the selection of the key ones are to be consistently taken into account and discussed within the research group. The position of the individual stakeholder has to be well analyzed and assessed in detail as it is fundamental to enucleate the power and leadership emerging between the interviewed people. One particularly sensible point to be evaluated is the position of any respondent in an organization, in order to understand any egoistic and strategic behavior. Also in this sense, the analysis, intrinsically, gives best results if repeated on a continuous basis. Thus, the analyst needs to revise and deepen earlier levels of the analysis, and continuously renew research efforts (Varvasovsky and Brugha 2000).

The SHA can help policymakers follow a systematic process for collecting and analyzing data about key stakeholders in a more feasible framework even though it is impossible to absolutely prevent subjective information, as the analysis is based on what stakeholders declare to analysts.

The policy process is a circular wheel where policy legitimization has the same rank importance of the policy formulation, and being a technical function it is still essential for the politically dominated decisions (Schmeer 1999). The use of the SHA allows policymakers to identify actual key players or effective "stakeholders," and understand whether they might support or contrast or even nullify any public intervention. Policymakers should consider developing strategies to "promote supportive actions and decrease opposing actions before attempting to implement major reform at the national, regional, local, or facility level" (Schmeer 2000, p. 2).

10.4 Methodology

In order to test the stakeholder analysis in two different Italian regions and its usability in support of the process of policymaking, we collected data from January 2011 to December 2012, regarding two case studies related to wine tourism and territorial promotion, interviewing nearly 50 people in total for both cases.

The first case we considered is the "Sardinian Young Wines Exhibition" in the village of Milis, in the Italian region of Sardinia. The second one is "The Verdicchio of Matelica Wine Road" in the Marches, with the main wine production area located in the village of Matelica and its geographic neighborhood. Ex-ante, we knew that the wine exhibition has a history of success and the wine road of failure.

We surveyed press news and conducted face-to-face interviews, recorded in both places. Before making the face-to-face interviews and taping the stories, all the interviewees had been informed that their answers would be useful both for research goals and policymaking.

In making the two field surveys, we duly adopted the guidelines stated by Schmeer (2000) describing the network of interests that surround the conception, organization, and upgrading of the "Sardinian Young Wines Exhibition" and "The Verdicchio of Matelica Wine Road." As the SHA aims at enlightening advantages and disadvantages for the stakeholders and contribution to local development, we followed the following steps: (i) identifying key stakeholders; (ii) collecting and recording the information; (iii) filling in the stakeholder table; (iv) analyzing the stakeholder table; (v) using the information.

In defining the exact stakeholder information or characteristics to be considered we detected name, position and organization, internal/external position to the organization body, and surveyed: (i) knowledge of the wine event and wine route; (ii) position for or against them; (iii) interest; (iv) alliances; (v) resources; (vi) power; and (vii) leadership, to be intended as the willingness to engage, call people or conduct in person an action for or against the event.

The method allows assessment of the "importance" of each stakeholder, stating the power and leadership of each actor, and thus identifying the most important stakeholders and classifying "groups of power" for or against.

10.4.1 Results of the Sardinian Young Wines Exhibition

The identification phase of the key stakeholders was preceded by the construction of the event network, through the examination of paper documents (newspapers, websites, and posters prepared for the event from its birth in 1988 until 2012), and through an informal discussion carried out with the inventors of the festival, who still run the exhibition and who belong to the Pro-Loco of Milis.

The following nine categories of stakeholders were identified:

- Institutions: the Region of Sardinia and Province of Oristano as financial and human supporting resources; the Municipality, managing the organization and logistics; the Pro-Loco of Milis as the factual organizer.
- Associations: the Italian Sommeliers Association (A.I.S.), for the sensory analyses; the Protection Consortium of "Vermentino di Gallura."
- Media: on a national and regional level, the latter acting also as the role of organizing support.
- Exhibitors: wine producers (traditional and young wines); food producers and sellers; nonprofit bodies; and others, as guests of the Festival.
- Hospitality: local accommodations, night-bed suppliers for visitors and exhibitors.

- Sponsor: regional (Vino Novello[1] producers), as Young Wines free supplier.
- Volunteers: local workforce for services and assistance.
- Residents: young, adult, senior, as representatives of the local community.
- Visitors: coming quite exclusively from the province of Oristano, as users and clients.

Around 25 people belonging to different categories were interviewed with the aim of gathering different opinions regarding the Young Wines Exhibition and its success/failure and possible improvement by taping the whole conversation. Following Schmeer's guidelines we enucleate the success reasons of the event (Carrus 2012; Benedetto and Corinto 2014), as follows:

- Unlike other food and wine events, the exhibition promotes a product, the Young Wine, which is not made on site but in the rest of the region. In fact, Milis is best known for other agro-food production (e.g., citrus);
- It has contributed to spreading the technique of production of Young Wine (the former owner of the event was a well-known nationally acclaimed sommelier) that in 1988 was mostly unknown to the whole Sardinian winemakers;
- The origin of the event is absolutely endogenous to the local community; from the beginning the Leadership of the event is still in the hands of local stakeholders, "Pro-Loco" and the Municipality;
- The organizational network of the event, built and characterized by strong bonds of friendship and family nature, fostered social cohesion within the local community, as well as the creation of a strong territorial identity;
- The regional and provincial Institutions do not have the Leadership of the event, and in the absence of public financial support, the exhibition has been carried out by "private" financial support;
- Local stakeholders have invented and pioneered a formula through which the event is self-financing;
- The event was taken as a model for many other regional initiatives born in imitation of the same format of the exhibition;
- Despite the crisis in the product "young wine" (drop in consumption and production of Young Wine since 2003 at national and regional levels), the exhibition has continued to record a progressive increase in the participants coming from extra-regional and extra-national spots (within 30 km around the Milis hotels, bed and breakfasts are full during the 4 days of the event);
- For these reasons, the Regional Institutions have chosen Milis and the exhibition as a backdrop for the launch of the new event "Vip Wine Sardegna" for the promotion of the overall regional excellent food and wine products.

From these findings, we argue that the essential importance of the cultural/local dimension in the process of implementation of rural development policies, and that, often, the success of a territorial marketing plan endogenously arisen and supported, is not directly related to the presence/absence of public funding.

---

[1] Official Italian name for the category of young wines.

10.4.2 Results of the Verdicchio of Matelica Wine Road

The case study of "The Verdicchio of Matelica wine road" tells the story of a failure (Conti 2012). The promotion initiative started locally, in the absence of a regional law about wine roads. Since 2000, the main promoter, the Municipality of Matelica, repeatedly asked for a regional intervention in view of the constitution of the official promotion committee, and continued on the path of a local initiative, which, as mentioned above, failed. In defining the key stakeholders committed to the initiative, we identified the following three broad categories:

- Institutions: the Region of the Marches, the Provinces of Ancona and Macerata, the Municipalities of Camerino, Castelraimondo, Cerreto d'Esi, Esanatoglia, Fabriano, Gagliole and Matelica (territorial area of production of the "Verdicchio" wine).
- Control, Promoting, and Protecting Wine Committees: the producers of the "Verdicchio" wine, the Italian Sensorial Centre of Matelica, the Marchigian Regional Institute for the Promotion of Wines (IMT), the Promoting Committee of the "Verdicchio" road.
- Local businesses: wine producers, wineries, wine shops, tourist farms, food and wine sellers, restaurants, tourism businesses, and B&B.

As in the preceding exposed case, among the above categories we have interviewed around 25 people, using for the analysis the same guidelines by Schmeer.

We found the essential topics as follows:

- The main cause of the failure is to be identified in the scarce public financial funds as stated in the Regional Development Plan;
- Private businesses, pertaining mainly to the agri-food chain and the tourism sector, aiming to gain eventual financial support, started the wine road, even establishing a formal promotion committee in 2003, according to the national law N.185/1999 which allows Regions to adopt specific regulations and financing;
- Over the period 2000–2010, the promotion Committee repetitively stressed the Regional Administration for adopting regulations and financing;
- In 2010, the Committee resigned due to the impossibility of becoming an effective Managing Board given the absence of a regional law on wine roads and the unwillingness of private financiers to sufficiently support the initiative;
- All the local stakeholders declared their willingness to promote the wine road but they regard the lack of regional support as blocking future effective initiatives, and furthermore, the local divisions produce a lack of power and weak or totally lacking leadership;
- The Regional Institute of Wine Promotion is a determinant key stakeholder acting as a strong opponent of wine roads as a means of wine promotion;
- Local economic actors and policymakers cultivate a very different idea of promotional policy in wine tourism.

For all the above-mentioned findings, we argue that this case study illustrates a story of policy failure in a framework of strong local divisions among social actors, lacking private–public coordination and a waste of finance.

10.5 Discussion and Conclusions

The aim of the chapter was to discuss if the SHA could be helpful in policymaking with particular regard to the RDP. In this line, the two cases we have presented are to be intended as an example of an exploratory research aiming at comparing the same phenomenon (wine promotion through wine tourism) in different local situation patterns. We chose these cases because their histories are evident examples of success or failure of the RDP in its regional implementation, moreover shedding some light on the relations between private entrepreneurs and the regional policy agency. We are aware that the importance and full credibility of a comparative case study should be based on more numerous places to be investigated. Nevertheless, we can realize the validity of using the SHA in order to assess local patterns of entrepreneurial behavior and the importance of private initiative in ensuring the actual effectiveness of the public policy intervention. When a local community has some grounded common values it is able to "bargain" with the public agency and shape the final and actual financial intervention, and this is the case of the wine festival in Saridinia. When the local community lacks shared vision and economic intentions, the up-down policy is thus failing, the local initiative will soon fade, and this is clearly the case of the wine road in the Marches.

The effectiveness of a policy depends on actual understanding and support from key stakeholders about policymakers' intentions and the long-run horizon of public financial expenditure.

The background literature shows a major integration between quantitative and qualitative research and a less sticky confrontation among researchers of different specializations. This is not an academic discourse, but an essential point if policymakers want to be faced with fewer policy failures. The role of local communities in determining the final results of policies is essential as well as the strong need not to waste public money. The SHA is a tiring, and sometimes boring tool, because it necessitates the inventory of primary data sources, whose collection is always expensive. Furthermore, its intrinsic subjectivity can be resolved by composing a plural research group and repeating the survey constantly, as the actual presence in the local community of an economic "turbulence" coming from political and social actors and lobbies must be considered. The analysis of the actual willingness and participation in supporting or contrasting policies by key stakeholders appears very useful for the effective impact of Rural Development Policy at any level. The SHA can be used also to contact local communities, demonstrating the public care of private opinions in a more communicative behavior. Economists can play an essential role in "lubricating" this channel and the SHA is a very flexible tool, well adaptable to the solution of different problems.

References

Anania, G. (2005). Alcune considerazioni sui problemi connessi all'uso dei modelli quantitativi nei processi relativi alle politiche agricole. In *La riforma delle politiche agricole dell'UE ed il negoziato WTO: Il contributo di alcune ricerche quantitative alla valutazione dei loro effetti sull'agricoltura italiana*, Milano, Italia: Franco Angeli, pp. 7–26.

Arfini, F. (2006). AGRISP: Un modello per simulare l'impatto delle politiche agricole. *Agriregionieuropa, 7*(2), 1–5. http://www.agriregionieuropa.univpm.it.

Balamou, E., Poulikas, K., Roberts, D., & Psaltopoulos, D. (2008). Modeling the rural-urban effects of changes in agricultural policies: A bi-regional CGE analisys of two case study regions. Paper presented 107th EAAE seminar: *Modeling of agricultural and rural development policies*, 29 gennaio-1 febbraio Siviglia, Spagna.

Baslé, M. (2006). Strengths and weaknesses of European Union policy evaluation methods: Ex-post evaluation of Objective 2, 1994–99. *Regional Studies, 40*, 225–235.

Benedetto, G., & Corinto, G. L. (2014). Food and wine events in Sardinia: The role and performance of institutions and local governments in the Sardinian Young Wines Exhibition of Milis. In A. Cavicchi & C. Santini (Eds.), *Food and wine events in Europe: A stakeholder approach* (pp. 157–171). Routledge, London, UK: Taylor and Francis.

Bobrow, D. B., & Dryzek, J. S. (1987). *Policy analysis by design*. Pittsburgh, PA: University of Pittsburgh Press.

Carrus, S. (2012). *Gli Eventi Enogastronomici ed il loro valore socio-economico: l'applicazione della Stakeholder Analysis alla "Rassegna dei Vini Novelli di Sardegna"*. Sassari, Italia: Unpublished degree dissertation Tesi di Laurea in Economia e Gestione dell'Impresa Agro-Industriale, corso di Laurea Magistrale in Scienze e Tecnologie Alimentari, Università degli Studi di Sassari.

Commissione delle Comunità Europee. (2006). *Manuale del quadro comune per il monitoraggio e la valutazione, Direzione generale dell'Agricoltura e dello sviluppo rurale.*

Conti, S. (2012). *Stakeholder Analysis per la Strada del Vino Verdicchio di Matelica.* Unpublished degree dissertation, University of Macerata.

Grimble, R., Chan, M., Aglionby, J. & Quan, J. (1995). Trees and Trade-Offs: a stakeholder approach to natural resource management. *International Institute for Environment and Development*, Gatekeeper Series 52.

Helming, J. F. M. (2005). *A model of Dutch agriculture based on Positive Mathematical Programming with regional and environmental applications.* Ph.D. thesis, Wageningen University.

Hill, B., & Wojan, T. R. (2010). Evaluation of rural development policy. *EuroChoices, 9*(1), 22–23.

Howitt, R. E., MacEwan, D., Medellín-Azuara, J., & Lund, J. R. (2010). *Economic modeling of agriculture and water in California using the statewide agricultural production model.* Davis, CA: University of California.

Júdez, L., Amorós, J. A., Piniés, M., De Andrés, R., & Urzainqui, E. (2004). Effects of agroenvironmental measures on holdings in a region in the "Mancha" in the context of the Mid-Term-Review of the CAP. Colloque OEnométrie XI, Dijon 21–22 may.

Kilkenny, M. (1999). Explicitly Spatial Rural-Urban Computable General Equilibrium. *American Journal of Agricultural Economics, 81*(3), 647-652.

Kinsella, J., Goetz, S. J., Partridge, M. D., Deller, S. C., & Fleming, D. (2010). Evaluating RD policies for social and human capital development. *EuroChoices, 9*, 42–47.

Leeuw, F., & Vaessen, J. (2009). Impact evaluations and development. NONIE Guidance for Impact Evaluation, NONIE, Washington. http://www.worldbank.org/ieg/nonie/guidance.html.

Leon, Y. (2005). Rural development in Europe: A research frontier for agricultural economists. *European Review of Agricultural Economics, 32*(3), 301–317.

Lindblom, C. E. (1959). The science of muddling through. *Public Administration Review, 19*, 78–88.
Midmore, P., Langstaff, L., Lowman, S., & Vaughan, A. (2008). *Qualitative evaluation of European rural development policy: Evidence from comparative case studies*. In 12th Congress of the European Association of Agricultural Economists—EAAE 2008, Ghent, Belgium.
Miller, G. J., & Yang, K. (Eds.). (2007). *Handbook of research methods in public administration*. Baton Rouge: CRC Press.
Mitchell, R. K., Agle, B. R., & Wood, D. J. (1997). Toward a theory of stakeholder identification and salience: Defining the principle of who and what really counts. *Academy of Management Review, 22*(4), 853–866.
Moro, D., & Sckokai, P. (1999). Modelling the CAP arable crop regime in Italy: Degree of decoupling and impact of Agenda 2000. *Cahiers d'Economie et Sociologie Rurales, 53*, 49–73.
Mortimer, S., Mauchline, A., Park, J., Finn, J., Edwards, D., & Morris, J. (2010). Evaluation of agri-evironment and forestry schemes with multiple objectives. *EuroChoices, 9*(1), 48–54.
OECD. (2009). *OECD rural policy reviews*. Italy: OECD.
Murdoch, J. (2000). Networks—a new paradigm of rural development? *Journal of Rural Studies, 16*, 407–419.
Parnell, J. (2007). *A stakeholder analysis of Macquarie Island. Identifying opportunities and constraints, and facing the future*. Honours thesis, University of Tasmania.
Reich, M. (1994). *Political mapping of health policy; a guide for managing the political dimension of health policy*. Boston, MA: Harvard School of Public Health.
Robinson, D. P., & Liu, Z. (2006). The effects of interregional trade flow estimating procedures on multiregional social accounting matrix multipliers. *The Journal of Regional Analysis & Policy, 36*(1), 94–114.
Rossell, C. H. (1993). Using multiple criteria to evaluate public policies. The case of school desegregation. *American Politics Quaterly*, *21*(2), 155–184 (Thousand Oaks, California, USA: Sage Publications).
Sadoulet, E., & De Janvry, A. (1995). *Quantitative development policy analysis*. Baltimore and London: The John Hopkins University Press.
Sali, G. (2010). "Le politiche agricole strutturali", in Atti del XL Incontro di Studio del Ce.S.E.T., Napoli 25–26 Novembre, pp. 85–101.
Schader, C., Sanders, J., Offermann, F., Lampkin, N., & Stolze M. (2008). *Cost-effectiveness evaluation of swiss agri-environmental measures on sector level*. Paper presented at the 107th EAAE Seminar, January 29–February 1, 2008. Sevilla, Spain: Modelling of Agricultural and Rural Development Policies.
Schmeer, K. (1999). *Guidelines for conducting a stakeholder analysis*. Bethesda, MD: Partnerships for Health Reform, Abt Associates Inc.
Schmeer, K. (2000). *Stakeholder analysis guidelines*. Washington, DC: Partners for Health Reform.
Schmid, E., & Sinabell, F. (2006). The Austrian agricultural sector in 2013 –Management and environmental perspectives. *Jahrbuch der Österreichischen Gesellschaft für Agrarökonomie, 15*, 35–43. http://www.boku.ac.at/oega.
Zadek, S., Pruzan, P., & Evans, R. (Eds.). (1997). Why count social performance? *Building corporate accountability: The emerging practices in social and ethical accounting, auditing and reporting* (pp. 12–34). London, UK: Earthscan Publications.
Varvasovsky, Z., & Brugha, R. (2000). How to do (or not to do)… A stakeholder analysis. *Health Policy and Planning, 15*(3), 338–345 (Oxford University Press).
Yin, R. K. (1994). *Case study research: Design and methods*. Beverly Hills, CA: Sage Publishing.

Authors' Biography

Graziella Benedetto is Associate Professor in Agricultural Economics at the University of Sassari (Italy). She graduated in Agricultural Science and her main fields of interest and research are: local development and agriculture, promotion of rural areas and models of socio-economic assessment, economics of tourism and wine tourism, territorial marketing, sustainability of agro-food production, life cycle assessment. Her works have been published in several books and journals, including Journal of Cleaner Production, Science of the Total Environment, Wine Economics and Policy, Rend. Online Soc. Geo., Franco Angeli. She is a reviewer for Rivista di Economia Agraria, Rivista di Economia Agro-Alimentare, Agricultural and Food Economics, Journal of Wine Research, Journal of Cleaner Production, Journal of Economics World.

Donatella Carboni is Associate Professor in Human Geography at the University of Sassari. She has published several works on qualified national and international journals. She has contributed to the organization of national and international scientific conferences. Her main fields of interest and research are: protected areas, the study of cultural heritage, tourism, immigration, the changing coastline in relation to archeological sites, territorial transformations made by man, water.

Gian Luigi Corinto is Associate Professor in Agricultural Economics at the University of Macerata (Italy). He graduated in Agricultural Science and his main fields of interest and research are: sustainable and rural tourism, local and regional development, innovation in ITC use in rural areas. His works have been published in several books and journals, including Tourism Today, New Medit, Global Bioethics, Enometrica, Cambridge Scholars Publishing.

Graziella Benedetto, Donatella Carboni, and Gian Luigi Corinto have contributed to the conception and design. All the authors were involved in the analysis and interpretation of the data. Graziella Benedetto wrote paragraphs 10.3.2, 10.4.1; Donatella Carboni wrote paragraph 10.3.1 and table 10.1; Gian Luigi Corinto wrote paragraphs 10.1, 10.2, 10.4, 10.4.2, and 10.5. All the authors approved the final version.

The authors wish to thank Silvia Carrus and Sara Conti for having collaborated in collecting data respectively for the Sardinian Young Wines Exhibition and the Wine Road of Matelica case studies.

Part IV
The Influence of Climate Change and Constraints

Chapter 11
Impacts of Climate Change on Agriculture Water Management: Application of an Integrated Hydrological-Economic Modelling Tool in a Semi-Arid Region

A. Scardigno, D. D'Agostino, D. El Chami and N. Lamaddalena

Abstract An integrated hydrological-economic modelling tool—applied to the Apulia region (southern Italy)—is proposed to define water balance components and water use in the agricultural sector. The hydrological model allows assessing the crop irrigation requirements and the water availability, expressed in terms of river flow, groundwater recharge and abstraction, while the integration with the economic model allows simulating the real farmers' decision process in response to any changes both in the constraints and in the boundary conditions. The tool provides a comprehensive information framework including water balance components, crop irrigation requirements, farmers' choices in terms of land use and irrigation techniques, economic results (costs and incomes), and environmental impacts. Climate, land cover and soil datasets have been implemented as thematic maps into a GIS-based model, and integrated with the main economic parameters at the farm and crop level. Future scenarios of climate change have been simulated and their impacts on water balance taken into account. The aim of the results is optimizing the use of water resources and addressing the policies for an efficient water management under severe drought conditions that are likely to occur in the region according to climate change projections.

A. Scardigno (✉) · D. D'Agostino · N. Lamaddalena
Mediterranean Agronomic Institute of Bari, CIHEAM-IAMB,
Via Ceglie 9, 70010 Valenzano, Bari, Italy
e-mail: scardigno@iamb.it

D. El Chami
School of Applied Sciences, Cranfield Water Science Institute (CWSI),
Cranfield University, Bedfordshire MK43 0AL, UK

C. Zopounidis et al. (eds.), *Agricultural Cooperative Management and Policy*,
Cooperative Management, DOI: 10.1007/978-3-319-06635-6_11,

11.1 Introduction

Warming of the climate system has been unequivocal since the mid-twentieth century, with significant increase in global mean air temperature, sea surface temperature and sea level rise, and a considerable part of this warming is occurring despite the global efforts undertaken to reduce emissions of GHGs (Pachauri and Reisinger 2007; Schlesinger and Zhao 1989; Hansen et al. 1988).

Agricultural consequences of this change are manifold. Climate change may affect agricultural productivity and the consequent economic responses may be expressed by a change in the regional distribution and intensity of farming. Therefore, the long-term productivity and competitiveness of agriculture may be at risk, farm communities could be disrupted and conflicts over environmental impacts of agriculture on land and water resources could become increasingly contentious (Calow et al. 2010; Darwin et al. 1995).

Thus, a new generation of modelling is emerging to model impacts of climate change projections on the hydrological cycle and the socio-economic conditions at regional scales. This type of model emphasizes the adaptive capacity of ecosystems, as well as the ability of populations and ecosystems to access available resources such as water (Carpenter et al. 2005).

Multidisciplinary problems require multidisciplinary solutions (Howden et al. 2007) and coupled hydrological-economic modelling has been used in the literature for the potential of integrating the interdisciplinarity of climate change impacts and incorporating the socio-economic behaviour of people into the physical water balance framework and solving the problems of intersectoral supply and demand for water resources (Young et al. 1994).

This modelling approach which has been first reported by Noel and Howitt (1982) to determine optimal spatial and temporal water allocation now covers a range of water resource problems, locations and innovations (Harou et al. 2009; Brouwer and Hofkes 2008). Recent applications in the last decade analysed profit optimization for different water use scenarios with respect to groundwater and surface water at the watershed and basin level (Barthel et al. 2012; Dono et al. 2012; Varela-Ortega et al. 2011; Ahrends et al. 2008; Lanini et al. 2004; Frede et al. 2002; Gillig et al. 2001).

Especially for the Mediterranean basin classified between semi-arid and arid areas and particularly exposed to the impacts of climate change on freshwater (surface and groundwater), some authors (Palatella et al. 2010) showed different climate change signals in different areas and confirmed the need for more assessments and analysis to understand projections, variability and impacts on water resources, which is the primary vital resource weighting directly on people, ecosystems and economies (Calow et al. 2011; Hunter et al. 2010; Carter and Parker 2009; Parry et al. 2007).

Calow et al. (2011) suggested more socio-economic mitigation and ecosystem adaptation studies to be done in order to be able to define the fitness of predictions and how close to the realized change they turn out to be.

Therefore, the objective of this chapter is to assess the quantitative effects of climate change on water balance components and water use in the agricultural sector of the Italian Apulia region in order to support the adoption of adaptation measures. To reach this purpose, a coupled hydro-economic model was developed integrating a hydrological GIS-based model implemented in visual basic and MapWindow and a nonlinear optimization model encoded in GAMS (General Algebraic Modelling System). The model allowed to evaluate the existing trade-offs for the long-term sustainability of agricultural ecosystem services of this particular case study with a changing regional water balance; while the empirical model built for the purpose of this chapter could be generalized to different case studies with different hydro-agronomic conditions and under different climatic zones to reach a global understanding of this local climate variability and its impacts.

11.2 The Case Study

The proposed case study focuses on the Apulia region (Fig. 11.1) located in the southern part of Italy and characterized with wet and mild winters and long dry and hot summers (Ferchichi 2009), very typical for semi-arid climate conditions (Fig. 11.2). The main physical and socio-economic features of this area are common to many Mediterranean parts with similar climate type, where agriculture, a primary resource for the local economy, and water-related issues are highly correlated: the increasing gap between water demand and supply, and the water quality deterioration. In Apulia, the relatively flat landscape is largely occupied by agriculture; up to 72 % of the total area is agricultural lands equivalent to a total of 1,462,785 ha (ISTAT 2007). The largest part is occupied by cereals (33 %) and olive trees (25 %), while the rest is mainly dominated by vineyards, fruit trees and vegetable crops.

Cereals and vegetables are mainly grown in the fertile northern zone of the region, where water is made available to the farmers through a large irrigation infrastructure managed by the Consortium of Capitanata, while olive trees and vineyards dominate the central and southern parts of the region, where surface water is almost completely absent and irrigated crops rely exclusively on groundwater abstraction. As a result, groundwater levels are dramatically abating and the monitoring data for some wells managed by the 'Agency for the Irrigation Development and the Land Transformation in Apulia and Lucania' proved that piezometric values are showing a decline with time (De Girolamo et al. 2002). In addition, in some coastal areas sea water intrusion is observed as a direct effect of groundwater abatement (Polemio and Limoni 2001).

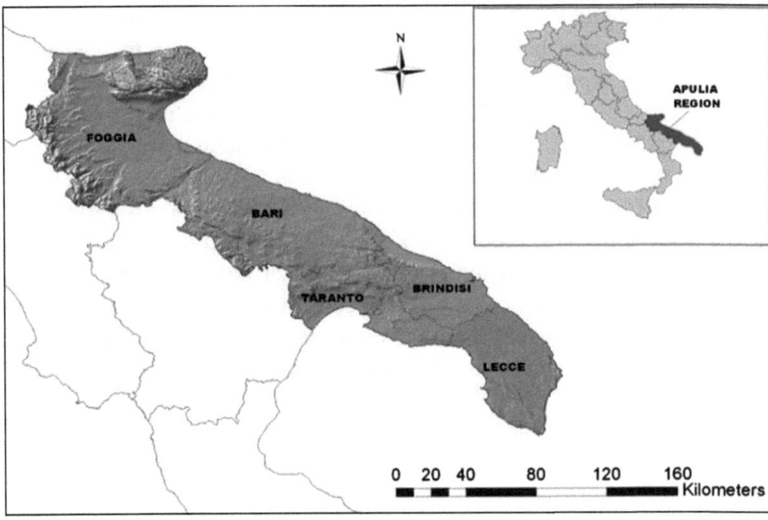

Fig. 11.1 Location of Apulia region and its provinces

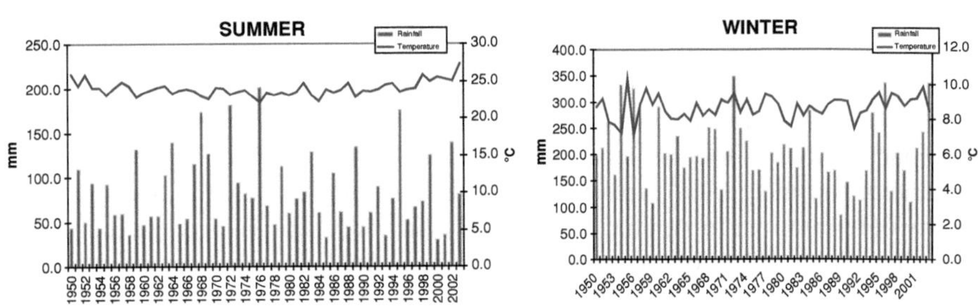

Fig. 11.2 Annual trend of rainfall and temperature in Apulia region for summer and winter seasons

11.3 Methodology

The following figure (Fig. 11.3) is a schematic representation of the proposed integrated model:

11.3.1 Economic Model

In order to estimate the water needs of the agricultural farms in the Apulia region under future scenarios, the identification of the future land use is required. Farmers' decisions in terms of cropping patterns and irrigation techniques are clearly affected by numerous factors such as climate conditions, water availability

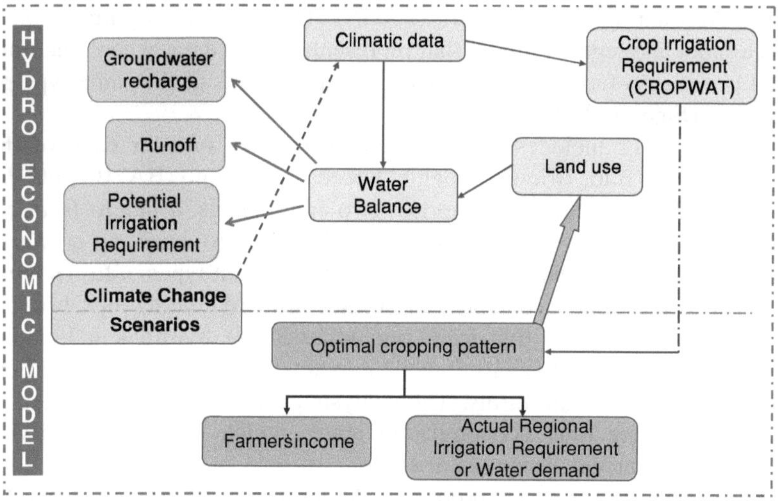

Fig. 11.3 Framework of the proposed integrated model

and market situation. Farmers' decision-making is a very complex process that should take into account the production methods, the seasonal periods and the quantities produced. Thus, farmers' decisions are subject to the prevailing farm's physical and financial constraints, and often to a considerable uncertainty related to yields, prices and input costs. All these factors have been considered in the simulated scenarios. A model written in GAMS (General Algebraic Modelling System) language was developed, integrating agricultural and institutional parameters such as crop production systems, cropping season, irrigation requirements, and water and labour costs. It aims at the identification of the optimal land use of the Apulia agricultural surface as a basis for estimation of the potential irrigation requirement.

The adopted methodology is based on mathematical programming of a farm model widely applied in socio-economic analysis of agricultural farms and irrigated agriculture (El Chami et al. 2011; Varela-Ortega et al. 2011; Louhichi et al. 2010; Arfini and Donati 2008; Marchand et al. 2008; Blanco Fonseca 2007; Janssen and Van Ittersum 2007; Saraiva and Pinheiro 2007, Scardigno and Viaggi 2007; Blanco Fonseca and Iglesias Martinez 2005; Borresh et al. 2005; Buisson 2005; Gómez-Limón and Berbel 2000).

A nonlinear, stochastic, single-year static mathematical programming model, named Climaware_2012, has been used. The model maximizes expected farmers' utility taking into consideration several conditions such as climate conditions; irrigation requirements and management techniques; monthly and total water availability/supply; prices of the products and agricultural input cost; water tariffs.

Agricultural farms in the Apulia region have been distinguished by farm typologies according to their structural specifications—average size, capital and labour availability—as well as socio-economic features as defined and identified in the Farm Accounting Data Network (FADN). Therefore, the optimization model used is a block aggregated model that represents all the area, where each block is referred to a macro-farm standing for the group of farms of the same type present in the area (Dono et al. 2008).

Climaware_2012 includes five subregional models, one for each of the following provinces: Bari, Brindisi, Foggia, Lecce and Taranto (BA, BR, FG, LE and TA). Each subregional model comprises two components: *a macro-level* component that corresponds to all the farms located in the province and *a micro-level* component that corresponds to the blocks of the different types of farms present in the area. Eight specialist farm types have been considered: field crops, horticulture, wine grapes, table grapes, olive, fruits (including citrus), livestock and mixed. The adopted approach allows to analyse the macro area and to highlight the differences and the specificities of the farms. The analysis of the agricultural system is performed by pursuing total economic efficiency, which leads to identifying the optimal solution for the system as a whole.

Climate conditions and soil properties define the set of crops that can be cultivated in the area, while technical and agronomic considerations allow defining the possible combinations among crops (*C*), irrigation techniques (*T*) and irrigation method (*I*). Twenty-six crops have been considered and grouped in six main classes: field crops including durum and soft wheat, barley, oat, broad bean, sugar beet, maize, sunflower and grass meadows; vegetables including broccoli, cabbage, celery, artichoke, lettuce, potato, tomato, watermelon and zucchini; orchards comprising cherry, peach, almond and citrus; vineyards including both table and wine grapes; pasture and olive trees. Four irrigation techniques (dry, complementary, partial and full irrigation) were considered, where full irrigation is relative to the full satisfaction of the crop water requirements; partial irrigation is relative to the satisfaction of 75–85 % of the total crop water requirements; and complementary irrigation is relative to the satisfaction of 45–55 % of the total crop water requirements. Two irrigation methods (drip and sprinkler) with different field application efficiencies were also considered. Crops are differently distributed over the year according to the planting and harvesting dates.

For each crop, the gross irrigation requirements were calculated considering the net irrigation requirements (NIR) estimated by CROPWAT divided by the field efficiency of the different methods of irrigation used (sprinkler 85 % and trickle 90 %). Gross irrigation requirements per crop and technique and per province are expressed on a monthly basis in $m^3\ ha^{-1}$ and are included in the model.

The CROPWAT model is a tool designed by the Food and Agriculture Organization of the United Nations (FAO) to carry out standard calculations for the design and management of irrigation schemes, and for improving irrigation practices. It is based on an approach of the daily water balance calculations used in the Irrigation and Drainage Paper Series 33 and 56 (Doorenbos et al. 1986; Allen et al. 1998).

Yields per crop and per irrigation technique have also been estimates and included in the model as well as data on prices of the different crops and the variable costs.

The model maximizes farmers' utility defined as the expected revenue minus its variability due to yields and price variations (Eq. 11.1).

$$\mathrm{Max}U = \Sigma_f Z_f - \phi \cdot \sigma \tag{11.1}$$

where U is the expected utility; Z_f is the average net income per farm type (€); ϕ is the risk aversion coefficient and σ is the standard deviation of the income distribution (€).

The average net farm income per farm type Z_f is defined as the difference between the gross margins and fixed and variable costs, except for the cost for irrigation water (Blanco Fonseca 2007). It is equal to the summation of the incomes resulting from different farm activities. The value of production refers to the product sold for final consumption or processed. Existing coupled crop subsidies have also been considered. Variable costs are given by the specific cropping expenses. They correspond to the summation of specific crop expenses with costs for temporary labour and mechanization. Specific crop expenses include costs for seeds, fertilizers and pesticides, hire charges and so forth (fuel, insurance and electricity). Labour costs exclusively include costs for wage-earning labour and not implicit costs relative to the family work.

Prices were collected from records of the wholesale local market, and were integrated to generate the 'endogenous prices' of different crops. The equilibrium price of a good in a supply and demand model is actually endogenous because it is set by a producer in response to consumer demand. Therefore, an 'endogenous prices' formula taking into consideration the elasticity of crop price for any change in the demand has been considered.

Costs of irrigation water are not included in the variable costs but were considered separately. According to the water tariff scheme most used in the region, cost of irrigation included in the model consists of two different components: (i) 'the cost of water' given by the volume of water used multiplied by the price of water per cubic metre of water used for the considered crop; (ii) a fixed water tariff to be paid for each hectare of irrigable land. Two different water sources were considered: in addition to the water supplied by the collective distribution network, the possibility of self-supply through wells was also taken into consideration. For the public water source, a binomial water tariff consisting of a fixed fee per hectare of irrigable land and a volumetric fee depending on the consumption was considered, while the cost for private water follows a similar structure with the cost of extraction rising with quantity, due to higher pumping costs for the progressive lowering of the water table level.

The obtained income (Z_f) is the remuneration to factors of production of the family (Eq. 11.2), i.e. land property, labour and capital. It is given by the following equation:

$$Z_f = \sum_{c,t} (Pr_c \cdot Y_{c,t} - V\cos\ t_{c,t}) \cdot X_{c,t,i} + \sum_{c,t} (\text{Dpay}_c \cdot X_{c,t}) - \sum_{c,t} (K\cos\ t_c \cdot X_{c,t}) - (\text{Wtarif} \cdot \text{iLand}) - (\text{PrWat} \cdot \text{QWat}) - (\text{PrLab} \cdot \text{QLab}) \quad (11.2)$$

where, for each farm typology component of the model, $X_{c,t,i}$ are the decision-making variables representing the area cultivated by crop type (c), irrigation technique (t), irrigation method (i); Pr_c: the average crop price (€ ql^{-1}); $Y_{c,t,}$: the crop yield (ql ha^{-1}); dpay_c: the coupled payments under the CAP (€ ha^{-1}); $Vcost_{c,t,}$: the variable costs (€ ha^{-1}); $Kcost_c$: plantation cost (€ ha^{-1}); Iland. irrigable land (ha); Wtarif: the fixed water tariff per unit area (€ ha^{-1}); PrWat: volumetric water tariff (€ m^{-3}); QWat: the annual amount of used water (m^3); PrLab: the labour salary (€ hr^{-1}); QLab : the annual amount of used labour (hr).

Two different sources of risk were considered: a 'market risk' affecting commodity prices and a 'nature risk' affecting yields. Standard deviation of farm income (€) is given by the following (Eq. 11.3):

$$\sigma = \sqrt{\sum_{kp,ky} \frac{(ZK_{kp,ky} - Z)^2}{N_{kp,ky}}} \quad (11.3)$$

where $ZK_{kp,ky}$ is the random income (€); $N_{kp,\ ky}$: number of states of market and nature kp: states of market [1–10] and ky: state of nature [1–10].

The random net income $Zk_{kp,\ ky}$ is calculated using the same equation applied for calculating the expected income Z; the unique difference was the average price (price_c) replaced by the random price ($\text{price\_}k_{c,kp}$) over state of market (*kp*) where $\text{price\_}k_{c,kp}$ is the vector of independent random numbers normally distributed (i.e. they are calculated using a normal distribution function based on the average and the standard deviation of price) and average yield (yield_c) is replaced by the random yield ($\text{yield\_}k_{c,ky}$) over state of nature (*ky*) where $\text{yield\_}k_{c,ky}$ is the vector of independent random numbers normally distributed (i.e. they are calculated using a normal distribution function based on the average and the standard deviation of yield). The main constraints adopted by the model include total and irrigated land constraint; agronomic constraints; labour and water constraints. A specific constraint is considered for olive trees since they are protected by law for the historical and cultural value they reflect.

The constant relative risk aversion coefficient was used also for calibration using the mean standard deviation approach. The model was run for different values of the coefficient in a range between 0 and 1.65 and the simulated results compared with the observed data. The percentage absolute deviation (PAD) parameter was used to validate the parameter between observed and predicted values (Eq. 11.4)

$$\text{PAD} = \sum_{i=i}^{n} \left| X_i^O - X_i^P \right| \div \sum_{i=i}^{n} X_i^O \tag{11.4}$$

where X_i^O is the observed value of the variable and X_i^P the predicted value.

The value of risk aversion that gives the lower PAD value is used for scenario testing (Janssen et al. 2010). The PAD values obtained are lower than 9 % for all the provinces, indicating a good level of the model calibration. The regional Farm Accountancy Data Network database has been used as a primary source for the structural and economic farm data and combined with the data from the Ministry of Labour and Social Security on labour requirements for the different crops. All the data used were adequately integrated and checked both with experts and stakeholders and based on the scientific literature during the preparation and calibration of the model. The model has been built and calibrated towards 2007 data and validated towards 2006.

For each simulation, optimal farmers' choices related to cropping patterns and agri-techniques have been identified, and the effects of such choices on farm revenues, costs and incomes have been estimated. Results have been obtained for each provincial model and then aggregated on the regional scale.

11.3.2 The Hydrological Model

The model used is a geographic information system (GIS)-based model implemented in visual basic and MapWindow (Lamaddalena et al. 2008). It has a simple structure, since as stated by Dooge (1977), keeping the number of parameters as low as possible allows both a more accurate determination of the parameters and a more reliable correlation of the values obtained. The model combines information coming from different cartographic layers, and as outputs it produces thematic maps illustrating the parameters of the water balance and the volumetric irrigation needs.

The model simulates the soil water balance as composed of two connected subsystems: the first represents the water dynamics in the root zone while the second represents the phenomenon of the natural groundwater recharge.

The soil water content variation estimated on a monthly basis is summarized in Eq. 11.5 and represented in the scheme in Fig. 11.4.

$$\delta w/\delta t = P - \text{ETc} - \text{RO} - \text{RO}_{\text{sub}} - \text{GWr} + \text{NIR} \tag{11.5}$$

where P is the monthly precipitation (mm); ETc is the crop monthly evapotranspiration (mm); RO is the monthly surface run-off (mm); RO_{sub} is the monthly subsurface run-off (mm); GWr is the monthly groundwater recharge (mm), and NIR is the monthly net irrigation applied (mm).

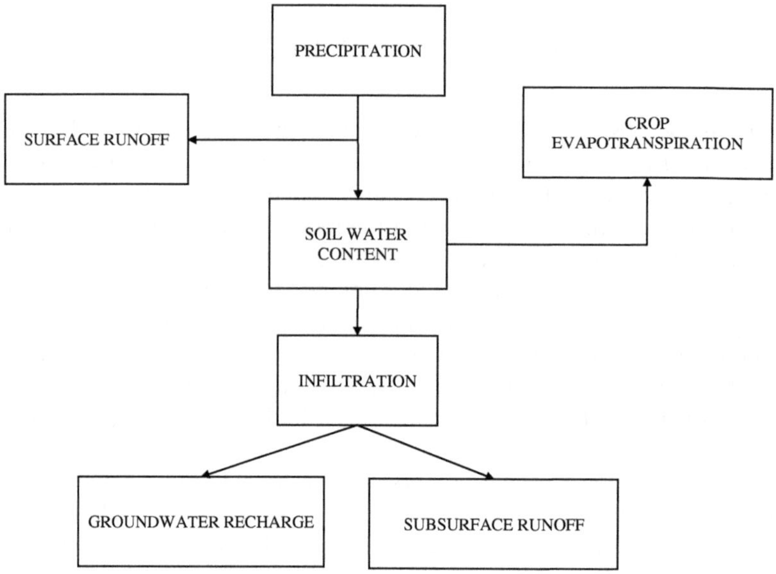

Fig. 11.4 Simplified scheme of the soil water balance

Crop evapotranspiration (ETc) is estimated using Allen et al. (1998) methodology (Eq. 11.6), where reference evapotranspiration[1] (ETo) is adjusted by a correction factor known as crop coefficient (*Kc*) which depends mainly on the crop type, variety and growth stages:

$$\text{ETc} = Kc * \text{ETo}. \tag{11.6}$$

For the simulation, the climatic data and the reference evapotranspiration were imported into the GIS and transformed into a continuous surface of 2 km^2 resolution. These maps were finally intersected with the map containing the land use information, referring to the land cover classes where each class encloses a series of crops with different growing season, cropped area and water requirements. An average value of weighted crop coefficient and root depth according to the area

[1] Eto is the evapotranspiration rate from a reference surface, a hypothetical grass reference crop with specific characteristics, not short of water. In the present study, it has been calculated using the modified version of the Hargreaves–Samani equation (Razieia and Pereira 2013):

$$\text{ETo} = 0.0135 \cdot k_{Rs} \cdot \frac{Ra}{\lambda} \sqrt{(T_{\max} - T_{\min})} \cdot (T + 17.8) \tag{11.7}$$

where *Ra* is the extraterrestrial radiation, and λ is the latent heat of vaporization (MJ kg^{-1}) for the mean air temperature T (°C), that is commonly assumed equal to 2.45 MJ kg-1. 0.0135 is a factor for conversion from American to the international system of units and kRs is the radiation adjustment coefficient, commonly equal to 0.17 (Samani 2004).

occupied by each crop within the corresponding class was then assigned to each class.

The surface run-off is calculated based on the Soil Conservation Service curve number method (US Soil Conservation Service—SCS 1972) (Eq. 11.8):

$$Q = \frac{(P - 0.2 \cdot S)^2}{(P + 0.8 \cdot S)} \tag{11.8}$$

where Q is the run-off rate (mm); P is the precipitation (mm); and S is the potential maximum retention after run-off begins, calculated with the following equation:

$$S = \frac{1000}{\mathrm{CN}} - 10 \tag{11.9}$$

where CN is the curve number that has to be defined based on ground cover type and hydrological conditions. Tables that give the CN to use in the run-off equation for the various cover types and hydrological soil groups are available in the literature. The cover types describe not only what is on the land, but also in some cases its condition from a hydrological standpoint (good, fair or poor). The soils for the site are classified into one of four hydrological soil groups, depending on the soils' ability to infiltrate water. The soil groups are called A, B, C and D, which indicate the greatest infiltration capacity to the least, respectively. Curve number is computationally efficient and does not require detailed information on soil surface conditions or rainfall (Connolly 1998).

The second subsystem of the soil water balance considers the infiltration below the root zone that is partitioned into groundwater recharge (GWr) and subsurface run-off (ROsub), using the coefficients of potential infiltration) which depend on the lithology of the soil (Celico 1986).

In order to define the crop irrigation requirement, an important role is covered by some hydraulic characteristics of the soil that are wilting point and field capacity. In the present study, they have been defined using the Hydraulic Properties Calculator (Saxton et al. 1986) and the soil data (slope, stoniness, morphology, texture, parent material, soil colour, layer depth, permeability, etc.) coming from the ACLA project (Steduto et al. 1999; Steduto and Todorovic 2001).

Finally, net irrigation requirements (NIR) are estimated through the following equation:

$$\mathrm{NIR} = \mathrm{ETc} - P_{\mathrm{eff}} \tag{11.10}$$

where ETc is the crop evapotranspiration, calculated as described in Eq. 3.6; P_{eff} is the effective precipitation (mm), i.e. the amount of precipitation effectively used by crop excluding the run-off and deep percolation losses, calculated as follows (USDA Soil Conservation Service 1967):

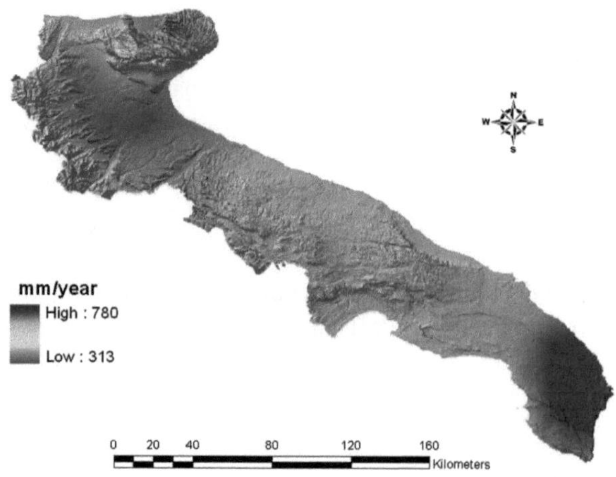

Fig. 11.5 Annual precipitation for the TMY

$$P_{\text{eff}} = \left(\frac{P}{125}\right) \cdot (125 - 0.2 \cdot P). \tag{11.11}$$

As for precipitation and temperature to be included in the model as input, a typical meteorological year (TMY) has been used: starting from the available monthly data of precipitation and temperature of the period 1950–2007 (Figs. 11.5 and 11.6), for the 107 gauging stations distributed in the Apulia region, a representative database of weather data for a 1-year duration has been generated. The hydrological model was calibrated by varying the coefficients of potential infiltration (Celico 1986) that directly influence the rate of the infiltration below the root zone. A satisfactory calibration was considered to be when the coefficient minimized the difference between the simulated and the observed level of groundwater recharge, while at the same time it yielded the least square regression equation of simulated versus observed monthly recharge. Monthly groundwater recharge volumes for the representative aquifer of Tavoliere, situated in the province of Foggia, and for three different years of observations 2008–2009–2010 were selected for calibration. The simulation results were compared to the observed recharge volumes and each of the coefficients' values were varied as necessary to reach a good fit. The model was run after calibration and validation for the whole Apulia region to simulate the water balance components under actual conditions and future scenarios of climate change. Finally, the land use generated by the economic model has been used as input.

11.4 Climate Change Scenario

In order to analyse water balance components under different climate conditions, a climate change scenario was adopted and applied to the case study. In particular, the selected climate change scenario was based on the results of the CIRCE project

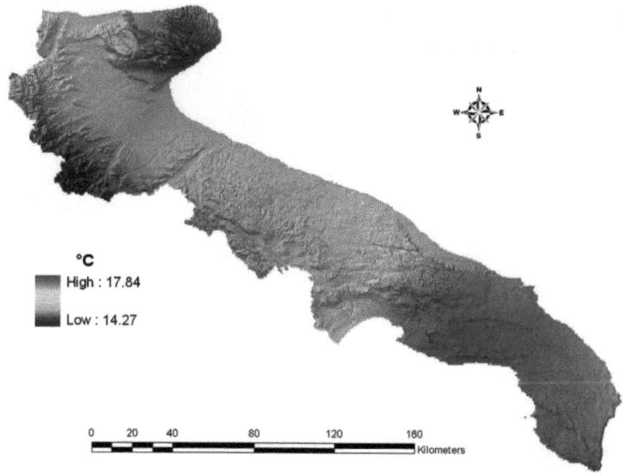

Fig. 11.6 Annual mean temperature for the TMY

(www.circeproject.eu). The global high-resolution[2] model used to perform the climate change projections generated in CIRCE is the Atmosphere-Ocean Global Circulation Model (AOGCM). This model simulates the atmospheric and ocean processes, and the interactions between them through the atmospheric component ECHAM 5.4 (Roeckner et al. 2003) and the oceanic component OPA 8.2 (Madec et al. 1998). The atmospheric model is implemented with a horizontal resolution of about 80 km. The data of anthropogenic aerosols concentration and distribution are specified according to the emission scenario A1B (IPCC-SRES) that refers to a balanced emphasis on the use of all energy sources.

Climate change projection of precipitation and temperature (Figs. 11.7 and 11.8) referring to a long-term average (2036–2065) have been extrapolated and data have been spatialized over the entire region using the geostatistical technique of Kriging. Future precipitation and temperature were used as a climatic input in the hydrological model.

11.5 Results and Discussions

Simulation results for the net irrigation requirements of the main crops are presented in the following table (Table 11.1) for both the baseline and cc_scenario.

The comparison of NIR of the cultivated crops between the baseline and cc_scenario shows a significant increase in the amount of water required for the full irrigation technique for almost all crops in all the region as an effect of the increase in the crop evapotranspiration, due to the increase of temperature, and

[2] A global high-resolution model is a tool used for modelling climate on a global scale.

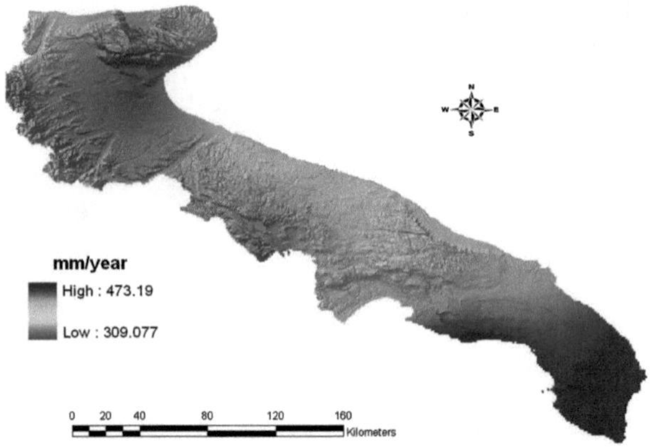

Fig. 11.7 Projection of annual precipitation

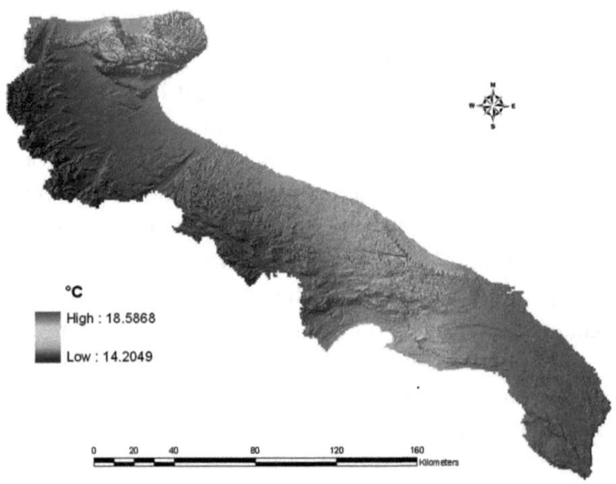

Fig. 11.8 Projection of annual mean temperature

the decrease in the precipitation that affects the water content available for the crop in the soil.

The following figures (Figs. 11.9 and 11.10) show the olive and the citrus change in response to water in the two simulated scenarios: for the optimal production in the cc_scenario higher amounts of water are required; while maintaining the same amount of water given in the baseline scenario reduced levels of production are obtained.

As for the land use, the results obtained for the baseline and climate change scenario in the region show that the total cultivated land is reduced by 8.5 % in the cc_scenario and the percentage of irrigated land decreases from 31 to 22 % of the total agricultural land. Comparing land use in the baseline and cc_scenario, results reveal that the total area planted with field crops declines by 15 %, with vineyard and vegetables by 54 and 22 %, respectively. This change is coupled with a rise of pastures by 31 % and orchard by 12 %. Olive trees remain constant given the specific

Table 11.1 Net Irrigation Requirements of the main crops, m^3/ha

Crop	Scenarios	Province				
		BA	BR	FG	LE	TA
Durum wheat	t3_base	2,655	2,718	2,275	2,719	2,625
	t3_cc	2,820	3,994	2,672	3,875	2,785
Potato	t3_base	5,075	5,076	4,567	5,085	5,198
	t3_cc	5,480	6,793	5,039	6,703	5,295
Tomato	t3_base	5,400	4,996	4,987	5,323	5,297
	t3_cc	5,516	6,855	5,082	6,764	5,339
Olive tree	t3_base	3,184	3,973	3,165	3,980	4,003
	t3_cc	5,124	6,446	4,785	6,226	4,947
Citrus tree	t3_base	3,391	4,450	4,083	4,448	4,818
	t3_cc	5,454	6,882	5,123	6,738	5,294
Table grape	t3_base	3,368	3,842	3,472		3,713
	t3_cc	4,858	5,916	4,432	–	4,630

Fig. 11.9 Change in water response of olive tree

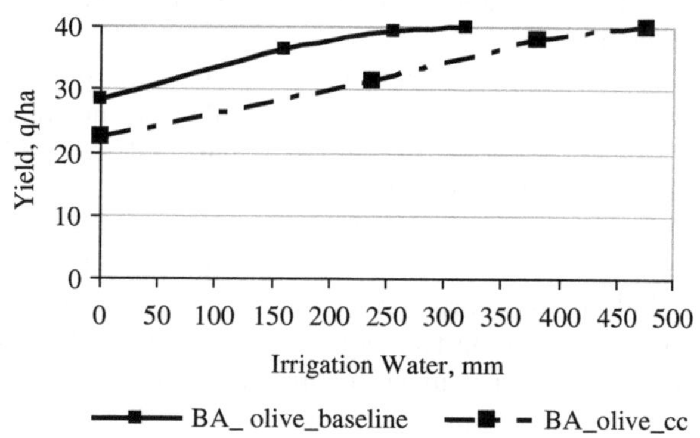

Fig. 11.10 Change in water response of citrus

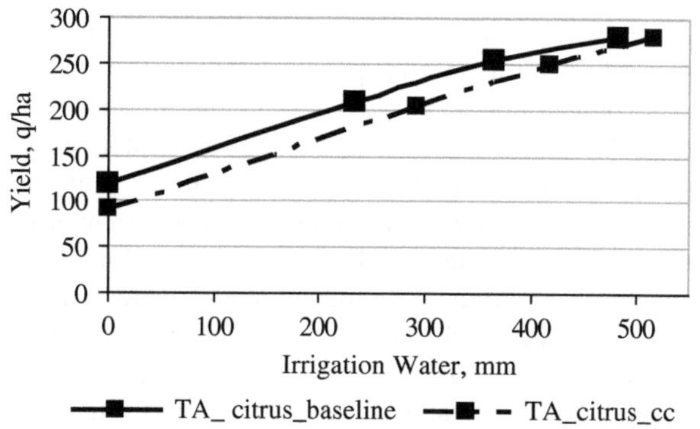

surface constraint included in the model (Fig. 11.11). In terms of irrigation techniques in the cc_scenario, complementary and partial irrigation are applied on 50 % of the total irrigated surface in comparison with 42 % registered in the baseline.

Considering an unchanged number of farms, the average net income per farm over the year calculated at around 29,300 € in the baseline scenario is reduced up to 18,500 € in the new climatic scenario: in both scenarios farms specialized in table grapes, horticulture and fruit show better performances even though the occurring changes in climate weigh on farms specialized in olive, wine grapes and fruit more than other farm types. The reduction in the net farm income is mainly due to a decrease of total revenue not compensated by a decrease of variable costs. Total revenue declines as an effect of a drop in crop sales while payment of Common Agricultural Policy remains quite stable since they are quite completely decoupled from production (Fig. 11.12).

The land use generated by the economic model has been spatialized over the entire region using an automatic program developed in Excel and implemented in GIS for the production of the land use map. Such maps have been used as an input for the hydrological model.

Potential irrigation requirements for the current conditions and for the future scenario of climate change are represented in Figs. 11.13 and 11.14, respectively. The highest water consuming crops in the region are orchards, vineyards and vegetables.

As shown in the maps, the values of water required for irrigation (in mm/year) increase in the future with respect to the current situation but the irrigated areas are drastically reduced. In particular, the total irrigated area in the region declines from 480,712 to 362,055 ha.

The following table (Table 11.2) shows the results of the irrigation requirements per province of the region for the current conditions and the future scenario.

Indeed, the variation in the cropping pattern and in the application of the different irrigation techniques results in the reduction of irrigated areas and consequently water demand for irrigation all over the region. This reduction is more evident in Lecce province that shows a radical change from high value crops such as vegetables and vineyards, to less water demanding crops.

Results of the hydrological model obtained for the entire region in terms of groundwater recharge and surface run-off under current conditions and future scenario of climate change are shown in the following table (Table 11.3).

Groundwater recharge refers to the amounts of water filling the groundwater reservoir and is a useful tool for groundwater resource planning and management (groundwater protection area delineation, pumping network design, etc.). The main recharge areas correspond to the geological units more permeable with quite thin soil cover and shallow rooted vegetation cover (thus reducing the field capacity and augmenting the amount of water available for deep infiltration). In the provinces of Bari and Lecce, where overlying deposits are very thin, the groundwater recharge is maximized in the region.

The previous table (Table 11.3) shows that water availability expressed in terms of groundwater recharge and surface run-off appears to be reduced all over the region in the future scenario of climate change compared to the current situation and this is due to the reduced precipitation.

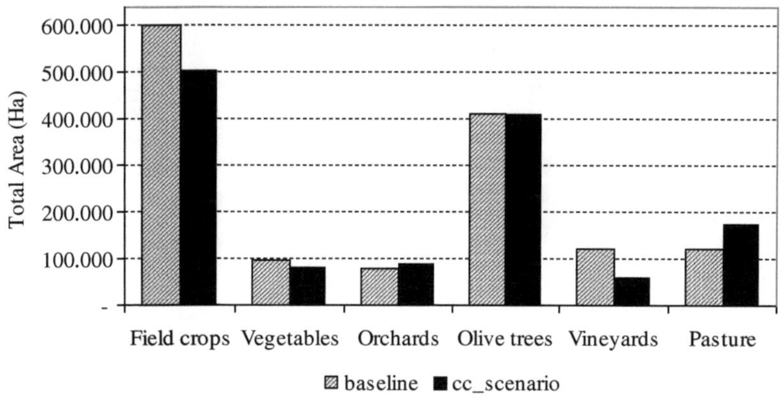

Fig. 11.11 Land use change in the future with respect to the baseline scenario

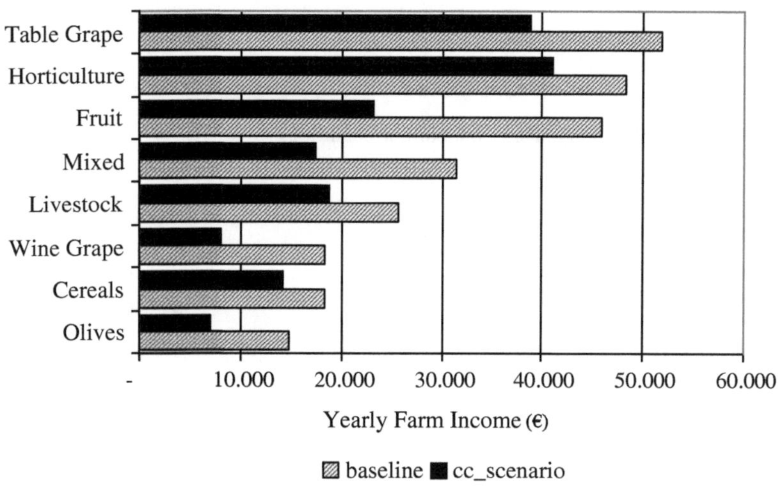

Fig. 11.12 Change in net farm income due to climate change scenario

The climate change scenario combined with the new land use allocation allows noticing that farmers, in order to adapt to the future minor availability of water, tend to reduce the irrigated areas by changing the crops.

11.6 Conclusions

This chapter has fixed the objective to assess the quantitative effects of climate change on water balance components and water use in the agricultural sector of the Italian Apulia region. A coupled hydro-economic model was developed, calibrated

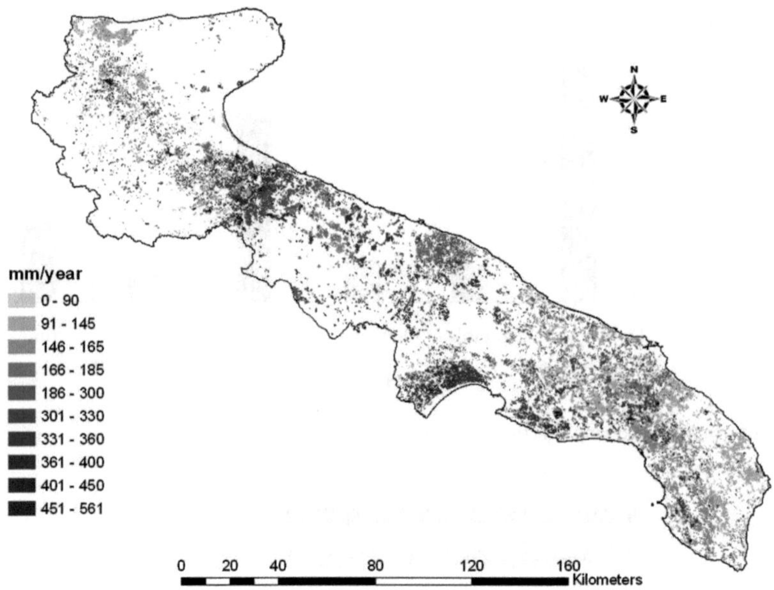

Fig. 11.13 Crop IWR (mm/year), at the current conditions

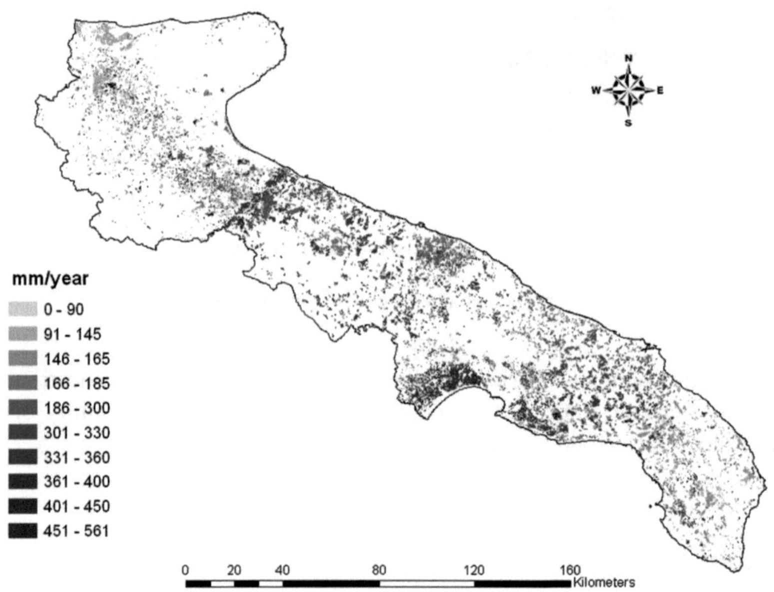

Fig. 11.14 Crop IWR (mm/year), for the future scenario

Table 11.2 Provincial irrigation water requirements (IWRs)

Province	Irrigated area (ha)		IWR—present		IWR—future	
	Present	Future	Mm^3	$m^3\ ha^{-1}$	Mm^3	$m^3\ ha^{-1}$
Bari	120,917	118,204	307	2541	271	2,292
Brindisi	76,057	50,568	144	1,894	122	2,418
Foggia	117,514	88,274	240	2,050	174	1,974
Lecce	100,026	45,817	183	1,839	90	1,976
Taranto	66,198	59,192	191	2,889	176	2,973

Table 11.3 Results of the hydrological model

Provinces	Current conditions		Future scenario	
	Run-off Mm^3	GW recharge Mm^3	Run-off Mm^3	GW recharge Mm^3
Bari	552	711	441	603
Brindisi	309	325	273	266
Foggia	1,200	377	994	320
Lecce	452	590	441	430
Taranto	299	316	242	267

and simulated integrating a hydrological GIS-based model implemented in visual basic and MapWindow and a nonlinear optimization model encoded in GAMS. The simulations evaluated the trade-offs existing in the agricultural ecosystem of Apulia region and their importance for the long-term sustainability of agricultural activity.

Results prove that climate change is likely to reduce agricultural productivity, production stability and incomes in this case study in southern Italy which would increase food insecurity levels. From another side, in a changing climate scenario changes in practices at the farm management level confirm to be a key component in adapting agriculture (FAO 2010; Howden et al. 2007).

Apulian farmers adopt different strategies to manage irrigation water in order to adapt and/or mitigate adverse effects. First of all, they reduce the irrigated surface. Secondly, they shift towards less water-intensive techniques: in the cc_scenario, complementary irrigation is applied on 31 % of the total irrigated surface in comparison with 16 % registered in the baseline. Furthermore, there is an effect of crop substitution and dangerously there is a serious phenomenon of land abandonment since a substantial area of the region traditionally or recently used by agriculture will not be cultivated any more. Results also show that, notwithstanding the complex farm strategies adopted, farm income is seriously affected by future climate conditions.

These results put in question the overall sustainability of the agricultural systems that are supposed to increase productivity to meet food security. Accordingly, the socio-economic sustainability is also vulnerable with an income reduction for farmers of about 37 %.

It is important to stress that the obtained results refer to a future scenario in which water availability for the agricultural sector is unaltered with respect to baseline conditions and more adverse effects are likely to occur if water available for the agricultural sector is reduced in order to satisfy the domestic sector.

Future research should address the evaluation of present and future water balance of the region taking into consideration the variation of nonagricultural demand and a potential reduction in available water for irrigation as well as an increased uncertainty in water availability.

Acknowledgements The authors would like to thank the French Ministry for Ecology, Sustainable Development, Transport and Housing, who funded this research in the framework of the CLIMAWARE project.

References

Ahrends, H., Mast, M., Rodgers, Ch., & Kunstmann, H. (2008). Coupled hydrological-economic modelling for optimised irrigated cultivation in a semi-arid catchment of West Africa. *Environmental Modelling and Software, 23*, 385–395.

Allen, R. G., Pereira, L. S., Raes, D., & Smith, M. (1998). *Crop evapotranspiration. Guidelines for computing crop water requirements*. Food and Agricultural Organization of the United Nations, Irrigation and Drainage Paper 56, FAO, Rome, p. 300.

Arfini, F., Donati, M. (2008). Health Check e efficienza delle aziende agricole: una valutazione comparativa su quattro regioni agricole europee. XLV Convegno Annuale SIDEA, Portici, 7–9 settembre 2006.

Barthel, R., Reichenau, T. G., Krimly, T., Dabbert, S., Schneider, K., & Mauser, W. (2012). Integrated modeling of global change impacts on agriculture and groundwater resources. *Water Resources Management, 26*, 1929–1951.

Blanco Fonseca, M., & Iglesias Martinez, E. (2005). Modeling new EU agricultural policies: Global guidelines, local strategies. *New policies and institutions for European Agriculture, 80th EAAE Seminar*. Ghent, 24–26 September.

Blanco Fonseca, M. (2007). Advanced modeling tools for integrated assessment of water and agricultural policies. *Water and Sustainability Development in the Mediterranean. 3rd Regional Workshop on water and sustainable development in the Mediterranean*, Zaragozza 19–21 March, p. 10.

Borresh, R., Kavallari, A., & Schmitz, P. M. (2005). CAP Reform and the Mediterranean EU-Member States, Paper presented at 80th EAAE Seminar.

Brouwer, R., & Hofkes, M. (2008). Integrated hydro-economic modelling: Approaches, key issues and future research directions. *Ecological Economics, 66*, 16–22.

Buisson, G. (2005). Etude sur les effets de la réforme de la PAC de juin 2003 sur la consommation d'eau par l'agriculture. Document de travail série Etudes, N(05 E06. Direction des études économiques et de l'évaluation environmentale du Ministère de l'écologie et du développement durable.

Calow, R., Bonsor, H., Jones, L., O'Meally, S., MacDonald, A., & Kaur, N. (2011). Climate change, water resources and WASH—A scoping study. Natural Environment Research Council—British Geological Survey (BGS), p. 58.

Calow, R. C., MacDonald, A. M., Nicol, A. L., & Robins, N. S. (2010). Ground water security and drought in Africa: Linking availability, access and demand. *Ground Water, 48*(2), 246–256.
Carpenter, S. R., Pingali, P. L., Bennet, E. M., & Zurek, M. B. (2005). *Ecosystems and human wellbeing: Scenarios* (vol. 2). Findings of the Scenarios Working Group of the Millennium Ecosystem Assessment—United Nations Environment Programme (UNEP), p. 560.
Carter, R. C., & Parker, A. (2009). Climate change, population trends and groundwater in Africa. *Hydrological Sciences Journal 54*(4), Special Issue, 676–689.
Celico, P. (1986). Prospezioni Idrogeologiche. Ed. Liguori, Napoli, vol. I–II.
Connolly, R. D. (1998). Modelling effects of soil structure on the water balance of soil-crop systems: A review. *Soil and Tillage Research, 48*, 1–19.
Darwin, R., Tsigas, M., Lewandrowski, J., & Raneses, A. (1995). *World agriculture and climate change: Economic adaptations*. Natural Resources and Environment Division, Economic Research Service, U.S. Department of Agriculture. Agricultural Economic Report No. 703, p. 86.
De Girolamo, A. M., Limoni, P. P., Portoghese, I., & Vurro, M. (2002). Il bilancio idrogeologico delle idrostrutture pugliesi. Sovrasfruttamento e criteri di gestione. *L'Acqua, 3*, 33–45.
Dono, G., Cortignani, R., Giraldo, L., Doro, L., Ledda, L., Pasqui, M., & Roggero, P. P. (2012). Evaluating productive and economic impacts of climate change variability on the farm sector of an irrigated Mediterranean area. *New challenges for EU agricultural sector and rural areas*. Which role for public policy? 126th EAAE Seminar, 27–29th of June, Capri, Italy.
Dono, G., Marongiu, S., Severini, S., Sistu, G., & Strazzera, E. (2008). Studio sulla gestione sostenibile delle risorse idriche: analisi dei modelli di consumo per usi irrigui e civili. Ricerca Integrata per l'Applicazione di tecnologie e processi innovative per la lotta alla DEsertificazione (RIADE), p. 257.
Dooge, J. C. I. (1977). Problems and methods of rainfall-runoff modelling. In T. A. Ciriani, U. Maione, & J. R. Wallis (Eds.), *Mathematical models for surface water hydrology* (pp. 71–108). New York: Wiley.
Doorenbos, J., Plusje, J. M. G. A., Kassam, A. H., Branscheid, V., & Bentvelsen, C. L. M. (1986). Yield response to water. Food and Agriculture Organization of the United Nations (FAO), *Irrigation and Drainage Papers 33*, 192.
El Chami, D., Scardigno, A., & Malorgio, G. (2011). Impacts of combined technical and economic measures on water saving in agriculture under water availability uncertainty. *Journal of Water Resources Management, 25*(14), 3911–3929.
Ferchichi, A. (2009). Climate variability over Apulia region: A pluviothermometric base assessment of the period 1950–2003. Ph.D. Thesis, University of Basilicata, p. 139.
FAO. (2010). *Climate-smart agriculture policies, practices and financing for food security, adaptation and mitigation*. Food and Agriculture Organization of the United Nations (FAO), p. 41.
Frede, H.-G., Bach, M., Fohrer, N., & Breuer, L. (2002). Interdisciplinary modeling and the significance of soil functions. *Journal of Plant Nutrition and Soil Science, 165*(4), 460–467.
Gillig, D., McCarl, B. A., & Boadu, F. (2001). An economic, hydrologic, and environmental assessment of water management alternative plans for the South Central Texas region. *Journal of Agricultural and Applied Economics, 33*(1), 59–78.
Gómez-Limón, J. A., & Berbel, J. (2000). Multi-criteria analysis of derived water demand functions: A Spanish case study. *Agricultural Systems, 63*, 49–72.
Hansen, J., Fung, I., Lacis, A., Rind, D., Lebedeff, S., Ruedy, R., et al. (1988). Global climate changes as forecasted by the Goddard Institute for Space Studies Three-Dimensional Model. *Journal of Geophysical Research, 93*(D8), 9341–9364.
Harou, J. J., Pulido-Velazquez, M., Rosenberg, D. E., Medellín-Azuara, J., Lund, J. R., & Howitt, R. E. (2009). Hydro-economic models: Concepts, design, applications, and future prospects. *Journal of Hydrology, 375*, 627–643.

Howden, S. M., Soussana, J. F., Tubiello, F. N., Chhitri, N., Dunlop, M., & Meinke, H. (2007). Adapting agriculture to climate change. *Proceedings of the National Academy of Sciences of the Unites States of America (PNAS), 104*(50), 19691–19696.

Hunter, P. E., MacDonald, A. M., & Carter, R. C. (2010). Water supply and health. *PLOS Medicine, 7*(11), 1–9.

ISTAT. (2007). Survey on agricultural holding structure and output. Italian National Institute for Statistics (Istat). Retrieved March 2013, from http://www.istat.it/agricoltura/datiagri/coltivazioni.

Janssen, S., Louhichi, K., Kanellopoulos, A., Zander, P., Flichman, G., Hengsdijk, H., et al. (2010). A generic bio-economic farm model for environmental and economic assessment of agricultural systems. *Environmental Management, 46*(6), 862–877.

Janssen, S. J. C., & Van Ittersum, M. K. (2007). Assessing farm innovations and responses to policies: A review of bio-economic farm models. *Agricultural Systems, 94*(2), 622–636.

Lamaddalena, N., Caliandro, A., Daccache, A., D'Agostino, D. R., Scardigno, A., & De Santis, S. (2008). Riorientamenti produttivi del territorio agricolo pugliese per uno sviluppo rurale sostenibile. Regione Puglia and CIHEAM-IAMB, p. 93. ISBN: 2-85352-395-0.

Lanini, S., Courtois, N., Giraud, F., Petit, V., & Rinaudo, J. D. (2004). Socio hydrosystem modelling for integrated water-resources management—the Hérault catchment case study, southern France. *Environmental Modelling and Software, 19*(11), 1011–1019.

Louhichi, K., Kanellopoulos, A., Janssen, S., Flichman, G., Blanco, M., Hengsdijk, H., et al. (2010). FSSIM, a bio-economic farm model for simulating the response of EU farming systems to agricultural and environmental policies. *Agricultural Systems, 103*, 585–597.

Madec, G., Delecluse, P., Imbard, M., & Levy, C. (1998). OPA 8.1 Ocean General Circulation Model reference manual. Institut Pierre-Simon Laplace Internal Report, 11, p. 91.

Marchand, F. L., Buysse, J., Campens, V., Claeys, D., Fernagut, B., Lauwers, L., Vander Straeten, B., & Van Huylenbroeck, G. (2008). Effects of a flat rate introduction: Shifts in farm activity and impact on farmers' income, 107th EAAE seminar, January 29th–February 1st, Sevilla.

Noel, J. E., & Howitt, R. E. (1982). Conjunctive multi-basin management an optimal control approach. *Water Resources Research, 18*(4), 753–763.

Pachauri, R. K., & Reisinger, A. (Eds.). (2007). *IPCC climate change 2007: Synthesis report*. Contribution of Working Groups I, II and III to the Fourth Assessment Report of the Intergovernmental Panel on Climate Change (IPCC), p. 104.

Palatella, L., Miglietta, M. M., Paradisi, P., & Lionello, P. (2010). Climate change assessment for Mediterranean agricultural areas by statistical downscaling. *Natural Hazards and Earth System Sciences, 10*, 1647–1661.

Parry, M., Canziani, O., Palutikof, J., Van Der Linden, P., & Hanson, C. (Eds.). (2007). *Climate change 2007: Impacts, adaptation and vulnerability*. Contribution of Working Group II to the Fourth Assessment Report of the Intergovernmental Panel on Climate Change (IPCC), p. 976.

Polemio, M., & Limoni, P. P. (2001). L'evoluzione dell'inquinamento salino delle acque sotterranee della Murgia e del Salento. *Memorie della Società Geologica Italiana, 56*, 327–331.

Razieia, T., & Pereira, L. S. (2013). Estimation of ETo with Hargreaves-Samani and FAO-PM temperature methods for a wide range of climates in Iran. *Agricultural Water Management, 121*, 1–18.

Roeckner, E., Bauml, G., Bonaventura, L., Brokopf, R., Esch, M., Giorgetta, M., Hagemann, S., Kirchner, I., Kornblueh, L., Manzini, E., Rhodin, A., Schlese, U., Schulzweida, U., & Tompkins, A. (2003). *The atmospheric general circulation model ECHAM5, Part I*. Max-Planck-Institut fuer Meteorologie Report N. 349, ISSN 0937-1060, p. 127.

Samani, Z. (2004). Discussion of history and evaluation of Hargreaves evapotranspiration equation by George H. Hargreaves and Richard G. Allen. *Journal of Irrigation and Drainage Engineering, 130*(5), 447–448.

Saraiva, J. P., & Pinheiro, A. C. (2007). A multi-criteria approach for irrigation water management. *Agricultural Economics Review, 8*(1), 64–76.

Saxton, K. E., Rawls, W. J., Romberger, J. S., & Papendick, R. I. (1986). Estimating generalized soil water characteristics from texture. *Soil Science Society of America Journal, 50*, 1031–1036.

Scardigno, A., & Viaggi, D. (2007). *Water demand management in the Mediterranean: Progress and Policies*. Water and Sustainability Development in the Mediterranean. 3rd Regional Workshop on water and sustainable development in the Mediterranean, Zaragozza 19–21 March, p. 24.

Schlesinger, M. E., & Zhao, A. C. (1989). Seasonal climatic changes induced by doubled CO_2 as simulated by the OSU atmospheric GCM/Mixed-Layer Ocean Model. *Journal of Climate, 2*(5), 459–495.

Steduto, P., & Todorovic, M. (2001). Agro-ecological characterization of the Apulia region: Methodology and experience. In P. Zdruli, P. Steduto, C. Lacirignola, & L. Montanarella (Eds.), *Soil resources in Southern and Eastern Mediterranean Countries* (vol. 34, pp. 143–158). Options Méditerranéennes, Series B: Studies and Research. Bari: CIHEAM.

Steduto, P., Todorovic, M., Caliandro, A., Catalano, M., Quaglino, E., Rusco, E., Stellati, M., Nerilli, E., De Santis, S., Madaro, A., Pezzato, A., Filomeno, O., D'Amico, O., Fanelli, D., & Cainarca, C. (1999). Agro-ecological characterization of the Apulia region: GIS applications and modelling. In: A. Musy, L. S. Pereira, & M. Fritsch (Eds.), *Emerging technologies for sustainable land use and water management*. 2nd Inter-Regional Conference on Environment-Water, Lausanne (Switzerland), 1–3 September 1999, Ppur. European Society of Agricultural Engineering (p. 13). CD ROM publications.

USDA—Soil Conservation Service. (1967). Irrigation water requirements. Technical Release, 21. Washington, D.C.

US Soil Conservation Service—SCS. (1972). *National engineering handbook* (p. 593). Washington, DC: Agriculture USDA.

Varela-Ortega, C., Blanco-Gutiérrez, I., Swartz, C. H., & Downing, T. E. (2011). Balancing groundwater conservation and rural livelihoods under water and climate uncertainties: An integrated hydro-economic modelling framework. *Global Environmental Change, 21*, 604–619.

CIRCE—Climate change and impact research: The Mediterranean environment www.circeproject.eu.

Young, G. J., Dooge, J. C. I., & Rodda, J. C. (1994). *Global water resource issues* (p. 216). Cambridge: Cambridge University Press.

Author Biographies

Scardigno Alessandra is an agricultural economist, with a Ph.D. in Rural Development and Economic Systems. Since 1997 she has been working for the CIHEAM-IAMB where she accomplishes research and teaching activities in the fields of farm economics, agricultural policy and environment, and economics of water resources in irrigated agriculture. She has been developing a study of the bio-economic modelling approach as a tool to analyse the impacts on agricultural and water policies on the complex interactions between agriculture, natural resources and environment. Involved in several European projects on water resources management in Mediterranean countries, she is actually working on the sustainability evaluation of water policies in the Mediterranean countries and impacts of climate change on water resources and use in Mediterranean agriculture.

D'agostino Daniela is a Researcher at the Department of Land and Water Resources Management of CIHEAM-IAMB, with a Ph.D. in Agro-Forestry and Environment Engineering from the University of Bari. Her scientific interests include surface hydrological modelling at the catchment and regional scale; water resource management and use in agriculture; semi-quantitative modelling for the analysis and management of qualitative data in the participatory and stakeholder-driven processes; elaboration, management and analysis of cartographic data using modern GIS techniques; statistical analysis and management of climatic data. She has been involved in many international and regional research projects and has authored and co-authored about 20 publications in scientific journals, books and conference proceedings.

El Chami Daniel is a researcher at the Cranfield Water Science Institute (CWSI). He graduated as an agricultural engineer (M.Sc. in Irrigation). He received a master's degree in Land and Water Resources Management and a Ph.D. in Agricultural and Food Economics and Politics. He is mainly interested in economics and the sustainable management of agricultural water resources including water resources in agro-food systems. He has also been a consultant for different EU projects and international organizations. He is currently working on different economic aspects of the UK irrigation sector: optimization of rainwater harvesting techniques, economic evaluation of climate change effects on UK agriculture and changes in production costs due to the change in irrigation patterns. Some aspects of his research include sustainable agriculture and water resources management, and agro-food economics.

Lamaddalena Nicola a Civil Engineer at the Polytechnic of Bari (Italy) since 1980, obtained his Ph.D. at the Technical University of Lisbon (Portugal) in 1997. He is the Head of the Land and Water Department at MAIB and, since 2006, an Adjunct Professor at the Polytechnic of Bari (Italy) for the course on 'Advances in Hydraulic Structures'. He is the coordinator of a number of water resources management projects financed by international organizations like FAO, World Bank and IFAD, and by national organizations like the Italian Ministry of Foreign Affairs. His main research interests are in the field of design, rehabilitation and modernization of large-scale irrigation systems, performance analysis of irrigation systems, interactions between distribution and on-farm networks, new technologies for irrigation, water balance at the regional scale and energy saving in pressurized irrigation systems. He has published more than 80 scientific articles in international and national journals. He is also the author of the FAO Irrigation and Drainage Paper n. 59, published in 2000, on Design and Performance Analysis of On-demand Irrigation Systems.

Chapter 12
Expanding Agri-Food Production and Employment in the Presence of Climate Policy Constraints: Quantifying the Trade-Off in Ireland

Ana Corina Miller, Trevor Donnellan, Alan Matthews, Kevin Hanrahan and Cathal O'Donoghue

Abstract This chapter explores the trade-off between competing objectives of employment creation and climate policy commitments in Irish agriculture. A social accounting matrix (SAM) multiplier model is linked with a partial equilibrium agricultural sector model to simulate the impact of a number of GHG emission reduction scenarios, assuming these are achieved through a constraint on beef production. Limiting the size of the beef sector helps to reduce GHG emissions with a very limited impact on the value of agricultural income at the farm level. However, the SAM multiplier analysis shows that there would be significant employment losses in the wider economy. From a policy perspective, a pragmatic approach to GHG emissions reductions in the agriculture sector, which balances opportunities for economic growth in the sector with opportunities to reduce associated GHG emissions, may be required.

12.1 Introduction

The global financial crisis which emerged in 2008 had a considerable adverse impact on the Irish economy. The Irish state was forced to absorb the debts of a number of banks and this led to rapid increase in its debt to GDP ratio. Ireland is a

A. C. Miller (✉)
Agri-Food and Bioscience Institute, Belfast, UK
e-mail: Corina.Miller@afbini.gov.uk

T. Donnellan · K. Hanrahan · C. O'Donoghue
Rural Economy Development Programme, Teagasc, Athenry, Co. Galway, Ireland

A. Matthews
Department of Economics, Trinity College Dublin, Dublin, Ireland

A. C. Miller · A. Matthews
Institute for International Integration Studies, Trinity College Dublin, Dublin, Ireland

C. Zopounidis et al. (eds.), *Agricultural Cooperative Management and Policy*, Cooperative Management, DOI: 10.1007/978-3-319-06635-6_12,

small open economy and several economic sectors are therefore very much export-focused. However, those economic sectors which are mainly dependent on the domestic economy have been adversely affected by a significant drop in consumer spending. Unemployment has increased, as have migration and emigration.

The Irish government has been under pressure to produce employment-focused initiatives. As a small open economy, and as a member of the euro zone, there are relatively few economic policies that can be adopted by government to rapidly promote growth. Increased government spending, which has been a strategy used by some governments internationally, would have little benefit for the domestic economy, since the economic stimulus would quickly leak out of a small open economy. In many respects prospects for the Irish economy remain dependent on a recovery in international demand.

Among the limited suite of options available to the Irish government one of the policy initiatives that has been advanced is a medium-term growth strategy for the agriculture and food processing (agri-food) sector. Given that Irish agriculture is very export-focused and that the medium-term outlook for global agricultural demand is quite positive, the sector is seen as providing growth opportunities, particularly for the rural economy.

The Government's plan of action for the agri-food sector published in 2010, known as Food Harvest 2020 (FH2020) (Department of Agriculture Fisheries and Food 2010), contains a variety of ambitious targets to be achieved by 2020. A mix of volume-based and value-based growth targets are specified in the report. Under these targets, by 2020 milk production in Ireland would grow by 50 % in volume terms when compared with a base period of 2007–2009. The target growth in the value of output from the beef and sheep sectors is 20 % by 2020 and the target for growth in pig sector output value by 2020 is 50 %.

However, Ireland also has stringent greenhouse gas (GHG) emission targets under the EU Commission's Climate and Energy Package Effort Sharing Decision (European Parliament 2009). Under this agreement Ireland is required to reduce its GHG emissions from those sectors not covered by the Emissions Trading System (ETS) by 20 % by 2020, relative to 2005. The proportion of emissions from agriculture in total national GHG emissions is quite high in Ireland at about 30 % in 2012 (Environmental Protection Agency 2013). It reflects the large volume of dairy and beef production and the relative absence of heavy manufacturing industry in Ireland. By way of contrast, the corresponding average EU27 figure for the share of GHG emissions represented by agriculture is just 9 % (Breen et al. 2010). Agriculture's share of non-ETS emissions for which there is a binding target is even higher, at about 40 % of non-ETS emissions in 2012 (Environmental Protection Agency 2013). In the non-ETS sector it is left to national governments to decide how these GHG reduction targets are to be achieved. Although agriculture in Ireland does not have a specific GHG reduction target, if GHG emissions from agriculture remain unchanged or increase, then the required rate of GHG emissions reduction for the other sectors in the non-ETS sector (mainly transport, industrial processes and household energy) will be even higher. According to the Environmental Protection Agency (EPA), Ireland will breach its annual obligations for GHG emissions under the EU 2020 target in 2017 taking account of the increased emissions from agriculture if the ambitious growth targets in the FH2020 plan are met (Environmental Protection Agency 2012). The implication of this analysis is that Ireland will find it difficult to meet its EU GHG emission commitments unless emissions from agriculture can be reduced.

The research question is whether the growth strategy of FH2020 would be possible in the context of a limit on agricultural GHG emissions and if this is not the case what the consequences for jobs and income would be for the agri-food sector and the wider economy. This chapter addresses this question by combining a partial equilibrium model of Irish agriculture, an environmental model of agricultural GHG emissions for Ireland and a social accounting matrix (SAM) of the Irish economy adapted to examine the agri-food sector in detail. The simulation results from the partial equilibrium sector model are used as exogenous shocks in a SAM multiplier model to analyse the impact of different policies on the Irish agri-food sector and the economy as a whole.

The conflict between a desire to increase agricultural production while reducing GHG emissions from agriculture is not unique to Ireland. It is also an issue for other developed countries that happen to be large net exporters of animal-derived agricultural products. Particularly where animal-based production is dominated by the bovine sector, the associated emissions can be substantial. Bovine-based agriculture is the main output of Irish agriculture, representing close to two-thirds of the value of Irish agricultural output of which close to 90 % is exported. Irish agriculture also benefits from a high level of support, which typically represents about 70 % of the value of agricultural income (Department of Agriculture Fisheries and Food 2006).

The objective of this chapter is to first assess the wider (direct and indirect) economic impact of the achievement of the targets set out in FH2020. In turn, the level of GHG emissions from agriculture under the achievement of FH2020 is determined. Then two alternative scenarios are implemented which involve the imposition of a constraint on GHG emissions from agriculture. The level of agricultural activity and the wider economic and employment impact under these two alternative scenarios are estimated. Comparing the outcomes under the three scenarios makes it possible to assess the economic impacts of constraining emissions from the agriculture sector in Ireland.

The rest of the chapter is organised as follows. Section 12.2 describes the scenarios and modelling framework. The results are discussed in Sects. 12.3 and 12.4 concludes.

12.2 Modelling Approach and Scenarios

Three scenarios are compared in this analysis. The Reference Scenario reflects the achievement of the targets that are set out in the Food Harvest 2020 report. It also includes the agreed series of annual 1 % expansions of the milk quota and its

Table 12.1 Scenario descriptions

Scenario 1	Reference scenario (FH2020) Agricultural output is consistent with the achievement of FH2020 targets Agricultural policy assumptions include: expansion of EU milk quota and elimination in 2015 other CAP policies remain unchanged no World Trade Organization (WTO) agreement
Scenario 2	Agricultural emissions reduction of 20 % by 2020, relative to 2005 Agricultural output is consistent with the achievement of a 20 % reduction in GHG emissions from Irish agriculture Other modelling assumptions are similar to Scenario 1
Scenario 3	Agricultural emissions reduction of 10 % by 2020, relative to 2005 Agricultural output is consistent with the achievement of a 10 % reduction in GHG emissions from Irish agriculture Other modelling assumptions are similar to Scenario 1

eventual elimination in 2015. Trade and CAP policy remain unchanged, as no World Trade Organization (WTO) agreement or CAP reform agreement is assumed to occur. Next, two GHG Emission Reduction scenarios are specified. In these scenarios the reduction in Irish agricultural output required to reduce GHG emissions from the sector by 20 % and by 10 % relative to 2005, respectively, is estimated. The 20 % reduction target is the GHG reduction target for the non-ETS sector as a whole, while the 10 % would represent more favourable treatment of agriculture compared to the other sectors in the non-ETS sector. Estimates of the economic impact of meeting these targets in terms of agricultural output and value added are produced using a partial equilibrium sector model of Irish agriculture. The wider economy impacts of these constraints are then estimated using a SAM model of the Irish economy. The three scenarios are described in Table 12.1.

12.2.1 Model Descriptions and Methodology

In this section the various models that are used in the analysis are briefly described. The manner in which these models are linked together is also detailed.

12.2.1.1 Partial Equilibrium Model and GHG Model

The FAPRI-Ireland partial equilibrium model of Irish agriculture is used to generate projections of agricultural output and income in the three scenarios. The FAPRI-Ireland model (Binfield et al. 2003, 2007, 2008) is a dynamic partial equilibrium model that is integrated within the FAPRI EU Gold model (Hanrahan 2001). The FAPRI approach to the development of agriculture sector models and the conduct of policy analysis is described elsewhere (Meyers et al. 2010; Westhoff and Meyers 2010). The FAPRI-Ireland model has a sub-module which generates projections of GHG and other emissions to air that are associated with agricultural production. Details of the GHG sub-module and earlier policy scenario analysis can be found in Donnellan and Hanrahan (2006). For the purposes of this analysis, it is assumed that the reduction in GHG emissions that is needed to meet the reduction targets does not occur overnight (since this would be an unrealistic means of achieving the target) but that the reduction takes place gradually over the period 2013–2020.

The FAPRI-Ireland model Reference Scenario projections provide an estimate of the distance agriculture would be from achieving particular percentage GHG reduction targets if no policies to address GHG emissions from agriculture were pursued. While it may be possible to reduce GHG emissions per unit of output in the future using abatement technologies currently being developed by scientists, there is likely to be limited scope for these technologies to be effective in the short term, since they must be taken from the laboratory setting into the real world. Farmers will need to be educated in these technologies in order for them to be adopted at the farm level. Therefore, if a substantial GHG emissions reduction target were imposed on Irish agriculture, in the short term, it would need to address this target through a reduction in agricultural activity. If a targeted, least cost, GHG reduction strategy was prescribed for the agricultural sector then, due to its low farm profitability, reduced production of beef from suckler cows would likely be selected as an economically efficient strategy to reduce agricultural GHG emissions. Therefore, it is assumed in this analysis that the reductions in GHG emissions required to achieve each of the emission reduction scenarios (Scenarios 2 and 3) are brought about through a reduction in the number of beef cattle (i.e. suckler cows, and their progeny).

This means that the beef target for FH2020 is not achieved in the two GHG emission reduction scenarios although the other sectoral FH2020 targets for milk, sheep and pigs are still met. Using the FAPRI-Ireland model it would be possible to look at a range of other options by which emissions could be constrained (reductions in the number of dairy cows, sheep, etc.). Each of these options would lead to one or other of the FH2020 sectoral targets not being achieved. While these other options are not considered here, the model could be used to make such assessments. Using our partial equilibrium model of Irish agriculture, integrated with the model of GHG emissions, the required reduction in the suckler herd and the associated decline in beef production is estimated, as well as the reduction in the value of beef output and the impact on aggregate agricultural income. The model produces estimates of the change in agricultural output consistent with the three scenarios and these are then used to develop the shocks implemented in the SAM.

No consideration is given to the mechanism (carbon taxes, carbon quota, etc.) that might achieve the GHG reductions represented by Scenarios 2 and 3. It is assumed that the required reduction in beef herd numbers is achieved purely through the imposition of a quota on animal numbers.

It is assumed that other countries' GHG abatement strategies do not impact on their levels of livestock production and that in turn there is no impact on international commodity prices arising from any agricultural GHG reductions internationally. This means that the possibility that other EU member states might also cut cattle numbers in response to their EU GHG reduction obligations is not considered. If other EU member states did cut cattle numbers, this would drive up EU beef prices, including prices in Ireland. By increasing the value of remaining beef production in Ireland, this effect would offset the reduction in the value of beef output due to the GHG emissions constraints but the overall effect would still be strongly negative.

As beef cattle numbers decline to achieve the emission reduction targets, land would become available for other agricultural purposes (or else production would become increasingly extensive). The impact on GHG emissions of the conversion of land previously used for beef cattle to use for bio-energy crops, dairy, tillage, forestry, land abandonment, etc., would have varying implications for GHG emission levels. Detailed consideration of the impact of these different alternative uses of surplus land has not been made in this analysis, since it would require considerably more work. Instead, the assumption is made that the reduction in beef output occurs through an extensification of stocking rates without implications for the output of other commodities.

12.2.1.2 Social Accounting Matrix

A SAM is an extension of an input–output table. The SAM is a square balanced matrix that portrays the economic flows from one account to another, representing expenditures and receipts of all the economic agents in the economy, with the condition that total expenditure and total receipts of each agent are equal. The 2005 AgriFood-SAM for Ireland is an extension of the Input–Output Table published by CSO (Central Statistics Office 2009). The 2005 Irish Agriculture and Food SAM (Miller et al. 2011) has 180 × 180 accounts, with 75 activities producing 75 commodities, 3 factors of production (labour, capital and land), 11 institutions (9 farm and non-farm households, enterprises and government), 11 tax-related accounts (direct, indirect tax and custom duties, subsidies, etc.), one savings/investments account, one change in stock account and three external accounts (UK, REU and ROW). The agri-food sector is represented by 12 primary agriculture sectors producing 12 agriculture commodities and 10 food processing sectors producing 10 food commodities, providing greater detail about the downstream activities of the food system and its relationship with the agricultural sectors.[1]

The 2005 AgriFood-SAM for Ireland is not a model as such. It can be transformed into an economic model if it is assumed that all relations are of the linear type, prices are fixed and all production activities function under the condition of excess demand, following Pyatt and Round (1979). SAM-based multiplier analysis

[1] The AgriFood-SAM used in this model is based on 2005 data when sugar production was still represented in the agricultural sector in Ireland. This does not significantly influence the results as sugar has little input into any of the four sectors in the model.

assumes that one or more accounts are made exogenous to allow the matrix to be invertible and consequently to calculate the multipliers. The government account, the rest of the world account, the capital account and the account for indirect taxes are regarded as exogenous accounts as these accounts are either politically determined or outside domestic control. The remaining accounts are viewed as endogenous and include the production (activities, commodities), factors of production (value added) and institution (households and enterprises) accounts. Consequently, the model becomes a demand-driven Keynesian model, as supply is assumed to adjust to demand.

The exogenous accounts are aggregated into a single account which records the injections into the system and the leakages from it. The leakages include transfer income sent to the rest of the world, institutional savings, indirect taxes and imports, while the injections include transfers from the rest of the world, government transfers to institutions, government consumption and export demand. The exogenous account can be seen as an independent variable while the endogenous account is the dependent variable. The 2005 AgriFood-SAM and the multiplier model are used to examine the impact of an expansion and/or contraction of different sectors on the wider economy taking account of the direct, indirect and induced effects of these exogenous shocks.

Consider, for example, the effect of a reduction in beef processing output by €1,260 million compared with the base year as assumed in one of the scenarios. The direct impact on output in the other sectors/accounts is derived from the column of inputs reported in the AgriFood-SAM assuming fixed proportions in the production function. Indirect effects arise as the lower production of inputs into the processing industry lead to further reductions in the industries that supply these activities with inputs. The induced effect is captured through the factor incomes account as wages and profits of those working in the beef processing sector and in the other affected sectors decline and so does their spending and demand for consumption goods, housing, financial services, etc.

12.2.1.3 SAM Scenario Shocks

Table 12.2 provides a description of the shocks implemented in the SAM. These FH2020 shocks replicate the required output levels for primary agricultural commodities identified in the FAPRI-Ireland partial equilibrium model. The AgriFood-SAM based multiplier model is given by:

$$y = Ay + x = Mx$$

where y is a column vector of accounts totals in the AgriFood-SAM, x is the exogenous transactions (which include government, capital, and rest of the world accounts), A is a matrix of average expenditure propensities of the AgriFood-SAM and M is the multiplier matrix. In a SAM-based multiplier model the exogenous (shock) variable is a change in final demand; hence the FH2020 shocks are

Table 12.2 The three scenario shocks implemented in the SAM model

Sector	2020 volume change relative to 2005 (%)	Value change (€ millions)	Final demand shock (€ millions)
Scenario 1 (FH2020)			
Milk	50	815	1,394
Cattle	2	50	88
Sheep	−28	−60	−145
Pig	33	102	425
Scenario 2 (FH2020 with 20 % GHG reduction target)			
Cattle	29	−713	−1,260
Scenario 3 (FH2020 with 10 % GHG reduction target)			
Cattle	−17	−420	−742

Source authors' calculations
Note Milk, Sheep and Pig sectors report the same volume changes for each of the three scenarios

translated into changes in final demand in the relevant processing sectors consistent with the output levels in primary agriculture calculated by the FAPRI-Ireland model in the three scenarios (i.e., a change in cattle output is modelled as a change in the final demand for beef products).[2] This reflects an assumption that all of the additional primary production is processed and not exported in raw or live form which is a reasonable assumption in current Irish circumstances.

The FAPRI-Ireland partial equilibrium model simulates the sectoral activity level associated with the FH2020 targets by interpreting the value and volume targets so that volume shocks for four of the main agricultural sectors are defined for this analysis. The shocks used in the multiplier analysis are defined as follows: a 50 % increase in volume in the milk sector, €815 million, requires a final demand shock of €1,394 million transmitted through an increase in dairy processing output; a €50 million increase in cattle output requires a final demand increase of €88 million in beef meat processing; a €60 million decrease in sheep output requires a final demand decrease of €145 million in sheep meat processing; and a €102 million increase in pigs output requires a final demand increase of €425 million in pig meat processing. The only difference between this scenario and the two emissions reductions scenarios is found in the shocks associated with the cattle sector. This is because Scenarios 2 and 3 allow the GHG emissions reduction to be achieved through a decrease in beef output while assuming that the remaining FH2020 targets are met. In Scenario 2, a reduction in cattle output of €713 million is achieved through a negative final demand shock of €1,260 million to the beef processing sector. In Scenario 3, a reduction in cattle output of €420 million is achieved through a negative final demand shock of €742 million to the beef processing sector. The changes to final demand lead to direct, indirect and induced changes in the wider economy which are captured used the SAM multiplier model.

[2] The method used to estimate the final demand shocks is detailed in Miller et al. (2011).

12.2.1.4 Marginal Versus Average Employment Coefficients

Given the scale of the unemployment problem in Ireland arising from the economic crisis, the employment impacts of the three scenarios are calculated in addition to the output and income impacts. Employment multipliers capture the relationship between the change in output in a sector and the resulting change in employment. Assuming the same fixed proportions production function as in the SAM would imply that the marginal employment multiplier is equal to the average employment intensity in each sector (measured as labour per euro of output). There are good grounds to think that this assumption exaggerates the likely employment impact of output changes, particularly for those sectors where output is increasing.

For example, if milk output grows by 50 %, then assuming that the employment multiplier equals the average employment intensity would imply that employment in milk production would also increase by 50 %. The reality will be somewhat different; the future growth in Irish milk production will involve production efficiencies which mean that each unit of milk requires less labour input (a larger number of cows is managed per farm worker). Underemployment, where the labour available at the farm level is less than is required for the actual level of production, must also be considered, since it is a feature of some parts of Irish agriculture. This means that if expansion in the output of the sector takes place it will not necessarily lead to an immediate increase in the numbers employed. The marginal employment impact of the expansion of a sector will tend to be smaller than the average (direct) employment impact.

Thus, marginal employment coefficients are used to calculate the employment impacts of the three scenarios. These are estimated in Miller et al. (2013) using a fixed effect log–log model for 57 out of the 75 SAM activities in the AgriFood-SAM. Table 12.3 provides the direct and marginal employment coefficients for the 22 agricultural and food sectors.[3]

The magnitude of the marginal coefficients for most of the sectors is small so it can be inferred that increased output is associated with productivity improvement or additional capital investment rather than increases in the labour input. For a few of the sectors the marginal coefficients are greater and this could be due to a change in product composition to a range of wider value added products.[4]

---

[3] The complete table with all 57 employment elasticities and marginal employment elasticities can be found in Miller et al. (2013).

[4] The model uses an annual unbalanced panel dataset for the period 1995–2008 measured in constant 2006 prices. Estimates of the marginal employment coefficients would benefit from an updated and/or balanced panel not available at the moment of this exercise.

Table 12.3 Marginal employment coefficients

Employment coefficients (per €1 million output)					
Agriculture sectors	Direct	Marginal	Food processing sectors	Direct	Marginal
Milk	18.075	6.018	Beef meat	3.344	0.086
Cattle	23.122	11.279	Pig meat	3.589	0.619
Sheep	19.071	10.664	Poultry meat	3.624	0.630
Pigs	3.676	4.615	Sheep meat	3.953	0.289
Poultry	5.493	4.653	Fish and other fishing products	9.314	0.259
Horses	6.400	1.204	Fruit and vegetable products	9.724	2.954
Cereals	26.442	1.242	Dairy products	2.572	2.587
Fruit and veg.	1.184	1.729	Animal feed	2.958	1.937
Sugar	26.116	1.938	Other food products	1.408	2.103
Potatoes	8.062	2.917	Beverages	0.664	1.933
Other crops	3.709	3.821			
Fodder crops	1.332	3.870			

Source Miller et al. (2013)
Note direct employment coefficients are calculated for 2005

12.3 Results and Discussion

This section summarises the results from the three scenarios proposed and explores the employment and economic effects of each of these scenarios on the wider economy.

12.3.1 GHG Emissions

The activity levels specified in the FH2020 Scenario can be used to estimate the level of GHG emissions from Irish agriculture to 2020. In Fig. 12.1 historical GHG emissions are presented along with the projected level of GHG emissions under the FH2020 scenario and the two GHG emissions reduction scenarios.

Under the FH2020 scenario, GHG emissions increase in the coming years principally due to the increase in dairy cow numbers and associated dairy emissions required to meet the 50 % milk volume expansion target. This increase in emissions associated with dairy expansion more than offsets the contraction in emissions from a fall in the size of the suckler herd. By 2020 the level of GHG emissions under the FH2020 scenario is almost 20.2 million tonnes CO_2 eq. Agricultural GHG emission projections under the FH2020 scenario are summarised in Table 12.4, along with the calculated reduction targets. The projected level of emissions under the FH2020 scenario does not consider potential emission reductions that might arise through the adoption of abatement technologies.

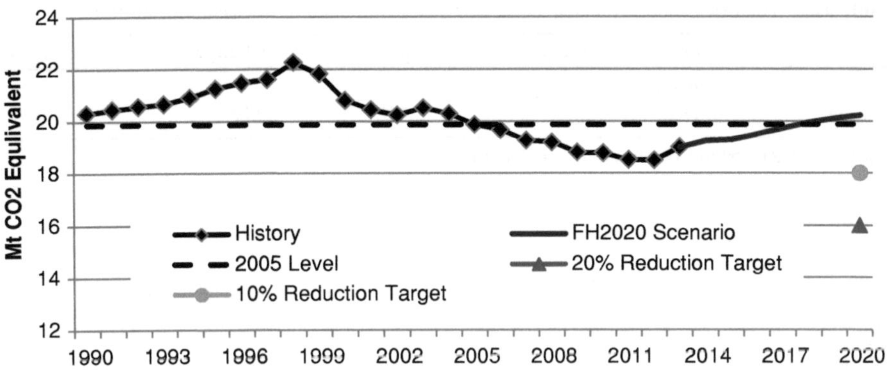

Fig. 12.1 Historical and projected Food Harvest GHG emissions from Irish agriculture. *Source* FAPRI-Ireland (Donnellan and Hanrahan 2012). *Note* excludes emissions due to fuel combustion

Table 12.4 Historical and projected Food Harvest GHG emissions from Irish agriculture and various hypothetical GHG reduction targets

	Mt CO_2 eq.			
	2005	2020	Change	% change
Scenario 1: Food Harvest 2020	20.3	20.23	−0.07	−0.3
Scenario 2: 20 % GHG reduction target		16.23	−4.07	−20.0
Scenario 3: 10 % GHG reduction target		18.26	−2.04	−10.0

Source FAPRI-Ireland (Donnellan and Hanrahan 2012)
Note excludes emissions due to fuel combustion

12.3.2 *Primary Agriculture Under the Reference Scenario (Scenario 1)*

The FAPRI-Ireland model is used to estimate the agricultural income (operating surplus) figure associated with achievement of the FH2020 targets. Achieving the FH2020 targets would lead to an increase in agricultural income of just over €347 million or 16 % relative to the average for the period 2007–2009 (Table 12.5). This mainly reflects the increase in the value of milk output (and milk prices due to the production of higher value added dairy products) associated with the 50 % volume increase in milk production, but it also reflects the fact that beef production with a low level of profitability is being replaced by milk production with a higher level of profitability (Fig. 12.2).

Table 12.5 Historical and projected Food Harvest sector income in Irish agriculture

	2007–2009 average	2020	Change	% change
	Euro million			
Income	2,181	2,528	347	16

Source FAPRI-Ireland (Donnellan and Hanrahan 2012)

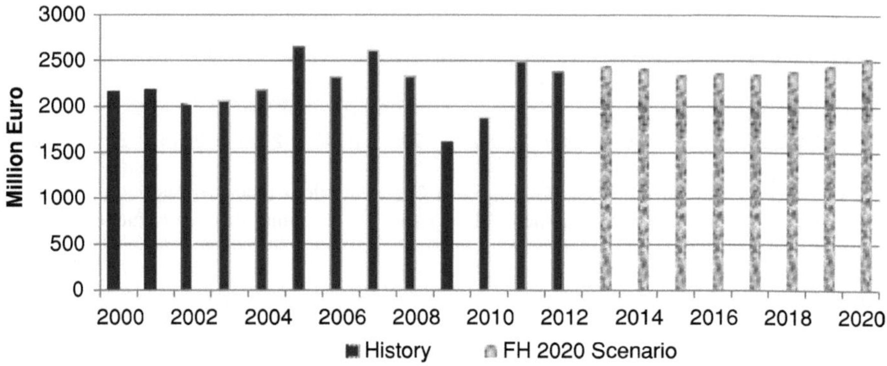

Fig. 12.2 Irish agricultural sector income: historical and Food Harvest projections. *Source* FAPRI-Ireland (Donnellan and Hanrahan 2012)

Table 12.6 Animal numbers, beef production and cattle sector value in 2005 and 2020 under Food Harvest 2020 and under a 20 % GHG reduction target (Scenario 2)

	Base year 2005	FH2020 2020	GHG – 20 % 2020	FH2020 versus 2005	GHG −20 % versus 2005	GHG −20 % versus FH2020
	000 head			*% change*		
Total cattle	6.39	5.98	4.33	−6.4	−32	−28
Dairy cows	1.12	1.26	1.26	12.5	13	0
Suckler cows	1.15	1.02	0.42	−11.3	−63	−59
	Million tonnes			*% change*		
Beef production	0.545	0.558	0.389	2.4	−29	−30
	Euro millions			*% change*		
Cattle sector value	1,413	1,921	1,284	36.0	−9	−33

Source FAPRI-Ireland (Donnellan and Hanrahan 2012)

12.3.3 Implications for Primary Agriculture of Reduction in GHG Emissions (Scenarios 2 and 3)

Table 12.6 illustrates that to reach the specified 20 % GHG reduction target by 2020 requires that cattle numbers are reduced to 4.33 million head by 2020. Total cattle numbers were 6.39 million head in 2005 and are projected to be 5.98 million in 2020 under FH2020. Suckler cow numbers are reduced to just over 0.42 million head by 2020 to achieve the 20 % target. Sucker cows were at 1.15 million head in 2005 and are projected to be 1.02 million in 2020 under FH2020. Accordingly, by 2020 Irish beef production would be reduced to 0.389 mt to achieve the 20 % reduction target. Beef production was 0.545 mt in 2005 and is projected to be 0.558 mt in 2020 under FH2020.

Beef prices in 2020 are projected to be higher than in 2005. Rising beef prices partially offset the impact of the reduction in the quantity of beef produced in 2020 in both the FH2020 and the 20 % GHG reduction scenarios. This means that the percentage decline in beef output value is smaller than the percentage quantity reduction in both cases. By 2020 the value of the cattle sector under the 20 % reduction target will be €1,284 million. In 2005 the value of cattle output was just over €1,413 million and is projected to be €1,921 million in 2020 under FH2020.

A 20 % GHG reduction strategy targeted at the Irish suckler beef herd imposes a large percentage cut on activity in that sector especially when measured against the possible value of that sector under the FH2020 initiative. Under FH2020 the sector would grow in output value terms by 36 % by 2020. To facilitate a 20 % reduction in GHG emissions would require that all of that expansion in the output value of the sector would be abandoned. Since beef production is just one GHG emission source within agriculture, the reduction in its activity level will need to be larger than 20 % to offset the expansion in other sectors. Under Scenario 2 the volume of beef produced would contract by 29 % relative to 2005. This is a significant reduction in the volume of beef available for trade on the EU market of the order of 100,000 tonnes. This is consistent with the results in Fellmann et al. (2012) that a 20 % emission reduction from EU agriculture generates a decrease in the EU beef meat production of between 11.5 and 23 % depending on the mitigation scenario adopted. Given that the EU is moving to the point where it is becoming a net importer of beef, achieving a GHG reduction through this means would increase the EU's requirement to import beef from third countries with potentially a larger increase in GHG emissions than the emissions saved in Ireland (O'Mara 2011). It is therefore useful to examine a more pragmatic scenario which limits the reduction in the volume of beef produced in Ireland.

As an alternative to a 20 % reduction in agricultural GHG emissions by 2020 relative to 2005, the implications of a smaller reduction in agricultural GHG emissions of 10 % by 2020 relative to 2005 is investigated. All of this reduction is again assumed to be achieved through a cut in the suckler herd. As illustrated in Table 12.7, to reach the specified 2020 10 % GHG reduction target requires that cattle numbers are reduced to 5.13 million head by 2020. Suckler cow numbers are reduced to just over 0.73 million head by 2020 to achieve the 10 % target. By 2020, Irish beef production would be reduced to 0.453 mt in order to meet this target. By 2020 the value of the cattle sector under the 10 % reduction target will be €1,612 million.

The impact of the scenarios on agricultural income is summarised in Table 12.8. The first point to note is that the impact under all of the scenarios is somewhat similar, which is in marked contrast to the impact of the scenarios on

Table 12.7 Animal numbers, beef production and cattle sector value in 2005 and 2020 under Food Harvest 2020 and under a 10 % GHG reduction target (Scenario 3)

	Base year (2005)	FH2020 (2020)	GHG minus 10 % (2020)	FH2020 versus 2005	GHG minus 10 % versus 2005	GHG minus 10 % versus FH2020
	000 head			*% change*		
Total cattle	6.39	5.98	5.13	−6.4	−20	−14
Dairy cows	1.12	1.26	1.26	12.5	13	0
Suckler cows	1.15	1.02	0.73	−11.3	−37	−28
	Million tonnes			*% change*		
Beef production	0.545	0.558	0.453	2.4	−17	−19
	Euro billions			*% change*		
Cattle sector value	1.413	1.921	1.612	36.0	14	−16

Source FAPRI-Ireland (Donnellan and Hanrahan 2012)

Table 12.8 Aggregate income in Irish agriculture (operating surplus) under the three scenarios examined

	2005 Euro Millions	2020	% change
Scenario 1 (FH2020)	2.650	2.528	−4.6
Scenario 2 (20 % GHG reduction)		2.372	−10.5
Scenario 3 (10 % GHG reduction)		2.475	−6.6

Source FAPRI-Ireland (Donnellan and Hanrahan 2012)

agricultural output shown in Tables 12.6 and 12.7. The main reason is that agricultural subsidies make up a large share of agricultural income in Ireland and particularly in beef production where they account for more than 100 % of income (Department of Agriculture Fisheries and Food 2006). These subsidy payments are decoupled which means that their value is not affected by the reduction in beef sector activity. Therefore, the impact of aggregate agricultural income arising from contraction in the beef sector is far smaller than the decline in the output of beef.

Thus far the analysis has only considered the impact of the GHG reduction scenarios in primary agriculture. While it would appear that the different scenarios impact significantly on beef output, the impact on agricultural income is less significant for the reasons outlined. However, policy makers are also interested in the extent of the impact on the wider economy in terms of jobs and incomes. To examine this question requires a different model in the form of a SAM.

12.3.4 Wider Economy (SAM) Implications of the Scenarios

The three scenarios are simulated using different assumptions to see how household income and employment respond to the changes in output. The GDP and

Table 12.9 Income effect of the Food Harvest 2020 (Scenario1) targets and GHG emission reduction (Scenario 2 and Scenario 3)

	Scenarios 1, 2 and 3			Scenario 1	Scenario 2	Scenario 3
	50 % Milk	−28 % Sheep	33 % Pig	(2 %) Cattle	(−29 %) Cattle	(−17 %) Cattle
	Euro Millions					
Output	2,552	−268	715	246	−3,532	−2,081
Labour	602	−60	113	64	−920	−542
Land	143	−15	6	14	−207	−122
Capital	465	−69	114	48	−694	−409
GDP	1,210	−144	233	127	−1,821	−1,073
Households						
Urban	493	−52	96	52	−751	−442
Rural non-farm	164	−18	32	17	−250	−147
Rural dairy farm	44	−5	4	5	−65	−38
Rural dairy and other farm	28	−3	2	3	−41	−24
Rural cattle rearing farm	44	−5	3	5	−65	−38
Rural cattle and other farm	30	−3	2	3	−44	−26
Rural mainly sheep farm	27	−3	2	3	−39	−23
Rural tillage farm	12	−1	1	1	−17	−10
Rural other farm	3	0	0	0	−5	−3
Total household income	845	−89	143	89	−1,276	−752

Source Authors' calculations using a SAM multiplier model

household income multiplier effects are shown in Table 12.9. The focus is on the differences in GDP and household income arising from the differences in levels of beef processing activity and exports corresponding to the different levels of cattle output in the three scenarios. For comparison, the output, GDP and household income effects of meeting the FH2020 targets for milk, sheep and pig production (which are common across all scenarios) are also shown. The output figures are the result of summing the changes in output in individual sectors across the whole economy.

Under Scenario 1 (the reference scenario), an €88 million increase in beef exports (reflecting achievement of the FH2020 growth target) leads to a €246 million increase in the total gross output in the economy. Because there is double-counting when the output changes in individual sectors are summed, looking at the changes in GDP gives a better indication of the effect on national welfare, although this is still an imperfect indicator given that GDP does not account for net foreign payments. Under Scenario 1, the overall increase in output of €246 million leads to an increase of €127 million in factor income distributed between labour, land and capital. This reference scenario outcome is then contrasted with the outcome under Scenarios 2 and 3 which involve constraints on the level of cattle output in order to meet GHG reductions targets as set out in Table 12.1. The results presented in Table 12.9 provide an indication of the overall economic impact of the implementation of the GHG constraints, decomposing the effects in terms of land labour and capital and in terms of the impact on household income.

Looking first at the impact on GDP, under Scenario 2, where the requirement is to meet the 20 % GHG emissions reduction target, this would result in an overall decrease in output of €3.532 billion and a fall in GDP of €1.821 billion. Under Scenario 3, where the requirement is a 10 % reduction in GHG emissions, the impacts are not as drastic as under Scenario 2, but would still lead to a fall of €1.073 billion in GDP.

Moving to the impact on household income, under Scenario 2, a reduction of €713 million in the value of cattle output would lead to a total decrease in household income of €1,276 million assuming households do not switch in producing another type of output. Under Scenario 3 a €420 million reduction in the value of cattle output gives rise to a €752 million reduction in household income.

These changes in factor income have different consequences for households depending on the sources of their income. In Ireland, urban households represent 70 % of the total number of households, while farm households represent almost 5 %. In aggregate absolute terms the fall in income is felt mainly by urban households as urban households are the largest household groups in terms of numbers. The next affected cohort is rural non-farm households. Given the widespread cattle ownership among Irish farmers, most farm-household types are also adversely affected.

While the aggregate losses are largest for the urban household cohort, on average each urban household incurs a loss in income of only about 2 %. By contrast, for a farm household the loss varies between 10 and 18 % depending on the farm enterprise type of the household. The largest share of 18 % is in the dairy and other enterprise household, while the cattle enterprise household registers a loss of 11 %.

Under Scenario 2, a €713 million reduction in the cattle output will cost each cattle enterprise farm-household €3,511 on average. Each urban and rural non-farm household will be worse off by approximately €746 and €686 per annum, respectively. The highest reductions will be felt in the dairy and other enterprise farm households with €5,485 per annum, per household.

The employment effects are calculated in each scenario using marginal employment multipliers for 57 sectors and zero employment multipliers for the remaining sectors in the economy.[5] As it is assumed that the reduction in GHG emissions required to achieve each of the emission reduction scenarios is brought about through a reduction in the numbers of beef cattle, the differences between scenarios is due solely to differences in the level of beef processing final demand associated with the different levels of cattle output in each scenario.

Table 12.10 reports the results for the three scenarios using these employment multiplier assumptions. The increase in milk production associated a 50 % increase in milk output which will generate 4,903 jobs in the primary agricultural milk

---

[5] For those sectors where marginal employment coefficients could not be calculated, the employment changes associated with output changes are assumed to be zero.

Table 12.10 Employment effect of the Food Harvest 2020 targets and GHG emission reduction, using marginal employment coefficients

Sectors	Scenario 1, 2 and 3			Scenario 1		Scenario 2		Scenario 3	
	50 %	−28 %	33 %	(2 %)		(−29 %)		(−17 %)	
	Milk	Sheep	Pig	Cattle	Total	Cattle	Total	Cattle	Total
Milk	4,903	−5	10	36	4,943	−523	4,384	−308	4,599
Cattle	56	−6	13	560	623	−8,040	−7,977	−4,736	−4,673
Sheep	10	−641	2	1	−629	−13	−643	−8	−637
Pig	5	−1	471	1	476	−8	469	−5	471
Beef meat	5	0	1	56	61	−800	−795	−471	−466
Pig meat	2	0	123	0	125	−3	122	−2	123
Sheep meat	5	−429	1	1	−422	−8	−430	−5	−427
Dairy products	2,976	−5	9	4	2,984	−64	2,916	−38	2,942
Other sectors	2,373	−313	1,166	272	3,496	−3,906	−681	−2,301	923
Total	10,335	−1,399	1,795	931	11,658	−13,365	−2,636	−7,873	2,855

Source Authors' calculations using a SAM multiplier model

sector, 2,976 jobs in the dairy processing sector and another 2,373 in other sectors of the economy. This amounts to a total of 10,335 jobs created in the economy if the FH2020 milk output target is met. If all the four FH2020 targets are met as assumed, then the indirect and induced effects of the changes in final demand in the other three sectors would marginally increase the numbers engaged in milk production at the farm level to 4,943 workers, the numbers of workers in the dairy processing sector to 2,984 workers plus 3,496 workers in the other sectors of the economy (column 6, Table 12.10). Summing the results across all four sectors, employment would increase by 11,658 jobs if the FH2020 targets are met in full.

The 29 % drop in cattle output assumed to meet the 20 % GHG emissions reduction target in Scenario 2 leads to a loss of 8,040 jobs in cattle production at the farm level and a loss of 800 jobs in the beef processing industry. The total job loss in the economy (combining the direct, indirect and induced effects) from the reduction in cattle output is 13,365 compared to an increase of 931 jobs under Scenario 1. The employment impact of Scenario 2 on the economy is shown in column 8 of Table 12.10. In total, there would be a decrease of 2,636 jobs if the FH2020 targets with the exception of the beef target are met under a 20 % GHG emission reduction constraint. The total employment impact of the 20 % constraint is given by the difference between the net increase in jobs in Scenario 1 and the net decrease in jobs in Scenario 2, i.e. 11,658 − (−2,636) = 14,294 jobs. The final column on the right in Table 12.10 shows the employment outcome if the GHG emissions reduction constraint is relaxed to 10 %. In this scenario, the reduction in total employment due to the fall in cattle output is of 7,873 jobs. The achievement of the FH2020 targets with the exception of the cattle target under a 10 % GHG emissions reduction target would result in an overall employment increase in the economy of 2,855 jobs. The overall employment cost of the 10 % reduction target is thus 11,658 − 2,855 = 8,803 jobs.

The employment results are sensitive to the values chosen for the employment multiplier. For example, using average employment intensities per unit of output, which imply that employment would change proportionally with any change in output, would yield a larger employment increase if the FH2020 targets were achieved and, conversely, would imply a larger net cost in terms of jobs arising from climate policy constraints. For the reasons discussed earlier, marginal employment coefficients are used as a more realistic alternative as it takes into account that increased output would largely be driven by increased use of capital and increased productivity rather than increased employment.

12.4 Conclusions

This chapter addresses the trade-off between conflicting objectives in the case of Irish agriculture in an economy recovering from a severe economic downturn and with high unemployment. The government has set ambitious expansion targets for the agri-food sector in the Food Harvest 2020 strategy and expects that meeting these targets can make an important contribution to job creation and lowering unemployment. On the other hand, it is required by commitments entered into under the EU's Effort Sharing Directive to lower GHG emissions by 20 % in 2020 over 2005 levels for sectors not covered by the EU's emissions trading scheme. Agriculture accounts for 40 % of non-ETS emissions, and if expansion occurs as projected in the FH2020 strategy, then Ireland is expected to be in breach of its emission reduction commitments as early as 2017.

Agriculture could therefore be required to make some contribution to reduced emissions. In this chapter, it is assumed that this contribution would take the form of a reduction in cattle output given that a high proportion of agriculture's emissions take the form of methane produced as a result of enteric fermentation by bovine animals, and given that technical GHG abatement technologies in agriculture are assumed to be either unavailable or un-adopted at the farm level. The objective of the chapter is to quantify the employment effects of meeting the FH2020 targets and also the employment effects if these targets are constrained by climate change policy.

This chapter uses a partial equilibrium agricultural sector model with an associated GHG emissions module to calculate the required reduction in cattle output to meet two a priori targets: a 20 % reduction in GHG emissions by 2020 compared to 2005 (thus proportional to the overall Irish target for the non-ETS sector), and a less demanding 10 % reduction target. Reductions in cattle output will mean reductions in throughput in the beef processing sector. The GDP, household income and employment effects of these enforced reductions are then calculated using a SAM model of the Irish economy with a detailed representation of the agri-food sectors.

The results demonstrate that a substantial constraint on GHG emissions in agriculture via a 20 % reduction requirement require a 29 % reduction in cattle output, but the impact on income in the agriculture sector is quite limited, given the low profitability of the sector and its high dependence on decoupled farm payments. However, the implied reduction in the throughput of the beef processing sector would have adverse consequences for the rest of the economy which policy makers also need to consider. Achieving the FH2020 targets could contribute a further 11,650 jobs to the Irish economy. If agriculture is required to meet a 20 % GHG emissions reduction target and this is achieved solely through a reduction in the output of cattle, then there would be a net loss of 2,630 jobs or a gross cost of 14,290 jobs across the economy. Even a more relaxed 10 % emissions reduction target would reduce the job gain under the FH2020 strategy to 2,630 jobs, implying a gross job cost of this constraint of 8,800 jobs across the economy.

These results are influenced by a number of assumptions necessary in the analysis. An important assumption is that a reduction in beef production takes place through an extensification of stocking densities so that the land is not diverted to the production of other commodities. The principal alternative land use to suckler cows is either sheep or dairy production where increased output would also imply increased GHG emissions contrary to the assumptions of the analysis. Taking account of potential production responses at the farm level to the introduction of a GHG emissions constraint would improve the accuracy of the GDP, household income and employment change figures. The other parameter with an important influence on the employment outcome is the value chosen for the employment multipliers. This chapter has used marginal employment multipliers derived econometrically from the recent relationship between output and employment changes in individual sectors. The assumption is made that these marginal employment multipliers are valid for both increases and decreases in output. If, instead, average employment intensities were used to calculate the likely employment changes, then the employment effects of climate policy constraints would be significantly higher.

Ultimately, policy makers may need to adopt a pragmatic approach to GHG emissions reductions which place a smaller emissions reductions burden on the primary agriculture sector. Given the limited potential for technical emissions abatement strategies to reduce emissions per unit of output, in the short term at least, a modest GHG emissions reduction target for agriculture would allow the agri-food sector to expand, would limit the adverse economic impact associated with a more onerous constraint, and would have a reduced impact on the volume of beef Ireland ships to international markets. Of course, relaxing the constraint on agricultural emissions implies that the constraints on the transport and household sectors must be further increased, which may also have associated GDP and employment costs.

References

Binfield, J., Donnellan, T., Hanrahan, K., & Westhoff, P. (2003). *The luxembourg cap reform agreement: implications for EU and irish agriculture*. Ireland: The Luxembourg CAP Agreement, Teagasc.

Binfield, J., Donnellan, T., Hanrahan, K., & Westhoff, P. (2007). *CAP health check analysis: impact on EU milk quota expansion*. Carlow: Teagasc.

Binfield, J., Donnellan, T., Hanrahan, K., & Westhoff, P. (2008). *Issues in examining the impact of WTO reform on the Beef and Dairy Sectors in the European Union*. Beijing: International Association of Agricultural Economists.

Breen, J., Donnellan, T., & Westhoff, P. (2010). Food for thought: EU agriculture presents new challenges for EU agriculture. *EuroChoices, 9*(3), 24–29.

Central Statistics Office 2009. Supply, Use and Input-Output Tables for Ireland, 2005.

Department of Agriculture Fisheries and Food 2006. Annual Review and Outlook 2005/2006.

Department of Agriculture Fisheries and Food 2010. Food Harvest 2020. A vision for Irish agri-food and fisheries.

Donnellan, T., & Hanrahan, K. (2006). It's the environment, stupid! To what extent can WTO trade reform lead to environmental rather than economic impacts? In *3rd World Congress of Environmental and Resource Economists,* Kyoto, Japan.

Donnellan, T., & Hanrahan, K. (2012). *Greenhouse Gas Emissions by Irish Agriculture: Consequences arising from the Food Harvest Targets*. Agricultural Economics: Teagasc.

Environmental Protection Agency 2012. Ireland's Greenhouse Gas Emissions Projections 2011–2020.

Environmental Protection Agency 2013. Ireland National Inventory Report 2012.

European Parliament 2009. Decision No 406/2009/EC of the European Parliament and of the Council of 23 April 2009 on the effort of Member States to reduce their greenhouse gas emissions to meet the Community's greenhouse gas emission reduction commitments up to 2020.

Fellmann, T., Pérez Domínguez, I., Witzke, H.-P., & Oudendag, D. (2012). Mitigating GHG emissions from EU agriculture—What difference does the policy make? In *International Association of Agricultural Economists (IAAE) Triennial Conference,* Foz do Iguaçu, Brazil.

Hanrahan, K. (2001). *The EU GOLD model 2.1: An introductory manual*. Dublin: Mimeo, Teagasc.

Meyers, W., Westhoff, P., Fabiosa, J., & Hayes, D. J. (2010). The FAPRI global modelling system and outlook. *Journal of International Agricultural Trade and Development, 6*(1), 1–20.

Miller, A. C., Matthews, A., Donnellan, T., & O'Donoghue, C. (2011). A 2005 Agricultural-Food SAM (AgriFood-SAM) for Ireland. Trinity College Dublin, IIIS Discussion Paper No. 372.

Miller, A. C., Matthews, A., Donnellan, T., & O'Donoghue, C. (2013). The employment effects of Food Harvest 2020 in Ireland. In *87th Annual Conference of the Agricultural Economics Society*. University of Warwick, United Kingdom.

O'Mara, F. (2011). The significance of livestock as a contributor to global greenhouse gas emissions today and in the near future. *Animal Feed Science and Technology, 166–167*, 7–15.

Pyatt, G., & Round, J. I. (1979). Accounting and fixed price multipliers in a social accounting matrix framework. *Economic Journal, 89*(356), 850–873.

Westhoff, P., & Meyers, W. (2010). The FAPRI approach: A few key principles. *Journal of International Agricultural Trade and Development, 6*(1), 133–135.

Chapter 13
Development and Application of Economic and Environmental Models of Greenhouse Gas Emissions from Agriculture: Some Difficult Choices for Policy Makers

Trevor Donnellan, Kevin Hanrahan and James P. Breen

Abstract This chapter describes how economic models designed to examine agricultural policy can be adapted to explore environmental applications such as the greenhouse gas (GHG) emissions from agriculture to a 2050 time horizon. The tensions between environmental policy aimed at reducing GHG emissions, and policies promoting agricultural production to increase food security are explored. Ireland is a major net exporter of beef and milk products, with agriculture representing a high share of non-Emissions Trading Scheme (non-ETS) GHG emissions. Ireland is used to illustrate an issue which has wide-scale global implications. The feasibility of achieving emission reductions is examined in the absence of technical abatement measures. Instead, to reduce emissions the size of the suckler herd is limited. However, it is found that even eliminating the suckler herd would leave emissions well short of achieving a 20 % reduction target. Even a 10 % GHG emissions reduction, while possible under this approach, is likely to be politically unfeasible. The tension between environmental and food security is likely to be replicated at a global level, given the significant contribution of agricultural production to anthropogenic climate change. The chapter highlights the importance of detailed modelling of future emissions in advance of setting feasible emissions reduction targets.

T. Donnellan (✉) · K. Hanrahan
Rural Economy Development Programme, Teagasc, Athenry, Galway, Ireland
e-mail: Trevor.Donnellan@teagasc.ie

J. P. Breen
School of Agriculture, University College Dublin, Dublin, Ireland

C. Zopounidis et al. (eds.), *Agricultural Cooperative Management and Policy*,
Cooperative Management, DOI: 10.1007/978-3-319-06635-6_13,

13.1 Introduction

Projections suggest that the world's population will rise from 7 billion in 2012 to over 9 billion by 2050 (FAO 2009a). Over that period real income levels will also increase, particularly in developing countries. Both these factors will greatly increase global food requirements by 2050. The increase in global population is being accompanied by economic growth which is leading to a change in diet, with meat and other livestock products becoming more prominent in the diet as incomes increase. The growth in global food demand has created twin concerns about the ability to limit the growth in greenhouse gas (GHG) emissions related to future food production and at the same time ensuring future global food security.

Achieving a rapid increase in food production to cope with a growing global population is not a new global challenge. Food production has increased dramatically over the last 50 years in line with the increase in the world's human population. This increase in food production was achieved by bringing additional land into agriculture and by using technology developed in research to substantially increase crop yields and animal performance. Today agriculture is a significant source of human-induced global GHG emissions and at present up to 30 % of human-induced GHG emissions globally are estimated to arise due to food production. This encompasses emissions from agriculture and the production of associated inputs, as well as emissions from food processing, packaging and distribution (Vermeulen et al. 2012).

Compared with the past, the challenge faced over the next 30 years is more complex in that there is limited additional land available for food production and the pace of yield growth is likely to be lower, in the short-term at least, due to reduced investment in agricultural research in recent decades. A further complication faced in producing more food for the growing global population is the need to constrain the impact of agriculture on the environment, including its contribution to global GHG emissions.

Internationally, ambitious medium and long-term targets are being set for GHG emissions reductions and these are likely to include emissions reduction targets for agriculture. Taking Ireland, a major exporter of meat and dairy products, as an example, this chapter uses a modelling approach to assess the level of emissions that would be generated by agriculture in Ireland under two scenarios associated with the future level of agricultural activity in the period to 2050. The modelling approach relies on linking a partial equilibrium economic model for agriculture to a model of greenhouse gas emissions from agriculture. This allows the level of GHG emissions associated with the scenarios to be determined and the cost of emissions abatement, in these scenarios, to be assessed.

A partial equilibrium model of the Irish agricultural sector known as the FAPRI-Ireland model was first developed in 1998 (Binfield et al. 2003). Initially, the motivation for this model was to look exclusively at issues associated with the economics of agriculture in Ireland, e.g. production, consumption, trade and income levels, and the main applications of the model have been ex ante analysis of agricultural policy change (Binfield et al. 2003, 2007, 2008). The structure of the economic model was adapted over time and a submodule, which provides projections of agricultural GHG and other emissions to air, has been added (Donnellan and Hanrahan 2006). Up until now this model projection activity has maintained a time horizon of 10 years. This is a reasonable time horizon in the context of economic projections since it facilitates business planning, but in the context of objectives such as addressing climate change (and food security) longer term projections are desirable.

Making long-term projections for agriculture in the European Union (EU),where policy is a significant determining factor of production, is a challenging undertaking. This chapter presents projections to 2050 of the level of agricultural activity and it is acknowledged at the outset that different price and policy scenarios would give rise to an alternative set of projections. The chapter also presents projections of the associated GHG emissions produced by agriculture to 2050, reflecting the projected level of activity.

The chapter is relevant in a number of aspects. Projections of GHG emissions to 2050 provide a baseline indicator for policy makers of future GHG emissions,in the absence of GHG abatement measures. Such projections provide guidance to policy makers as to the feasibility of setting particular GHG emissions abatement targets and the potential impact associated with these targets in terms of income and agricultural output. In turn, this provides a rationale for investment in research to address GHG emissions in agriculture.

The remainder of this chapter is structured as follows. In the next section, we discuss the issue of food security, the role of agriculture in the Irish economy and the need to limit GHG emissions from the agriculture sector in Ireland. This is followed by a discussion of the strategies that could be used to deal with the problem of agricultural GHG emissions. That is followed by a description of the modelling approach and the scenarios to be examined. The results of the analysis are then presented and the final section provides a discussion and conclusion.

13.2 Background

This section summarises the challenge presented in meeting global food security objectives and the dilemma involved in finding ways to increase agricultural production while at the same time reducing the adverse impact of agriculture on the environment.

13.2.1 Food Security

Food security is defined by the FAO as:

> when all people, at all times, have physical, social and economic access to sufficient, safe and nutritious food that meets their dietary needs and food preferences for an active and healthy life. Household food security is the application of this concept to the family level, with individuals within households as the focus of concern (FAO 2009b).

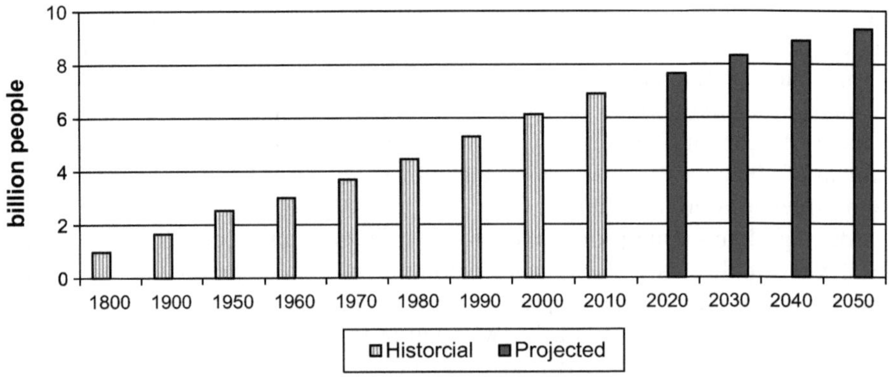

Fig. 13.1 World Population estimates from 1800, with projections to 2050 (FAOSTAT 2013)

The rate of world population growth is slowing down but the absolute increase in population over the next 30 years will still be of the order of 2 billion as shown in Fig. 13.1. Accordingly, food security is an important objective in global politics and has received much greater attention in the period since the food price spike of 2008 and the global recession beginning in 2009 (FAO 2009b). That food crisis has been identified as an event which exacerbated food insecurity. While food prices declined in the aftermath of this commodity price spike, it was followed by the wider global economic crisis. The economic crisis has manifested itself in slower economic growth globally and greater volatility in exchange rates, with adverse consequences for the desire to achieve food security. Overall, food prices remain elevated relative to the average levels of the previous decade. While the issue of food security is an everyday problem in some parts of the developing world, it has also risen in priority within the EU in recent years. This is largely as a reaction to the food price spike of 2007/2008, reflecting the substantial increase in the number of people globally falling into the category of the undernourished, as shown in Fig. 13.2.

Analysis of global stock data for agricultural commodities such as grains shows that the balance between annual production and consumption of many commodities is now tight. Globally, commodity stockholding activity, which traditionally acted as a buffer in times of production shortfalls or surpluses, is now less prevalent due to changes in government policy and limited availability of surplus commodities on the world market in recent years due to a series of adverse weather-related shocks.

Improvements in global transportation and the liberalisation of trade policies have increased the overall volume of commodities being traded internationally. This has also increased the number of people benefitting from/dependent on global markets, where historically some would have relied on local production systems with limited or no connection to the world market. This means that international

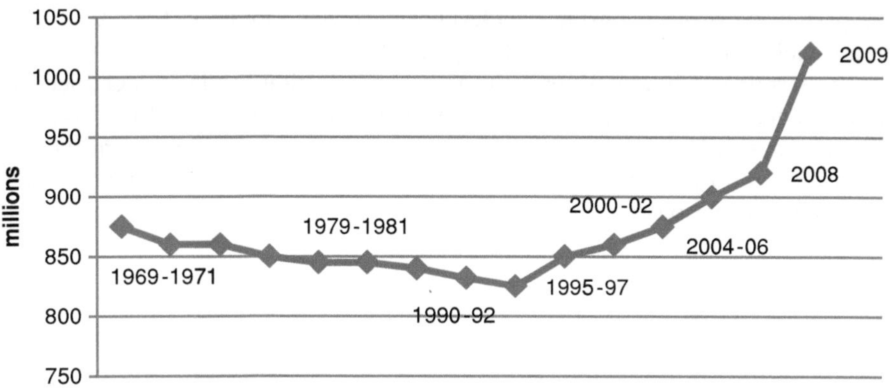

Fig. 13.2 Number of undernourished in the world (FAO 2009b)

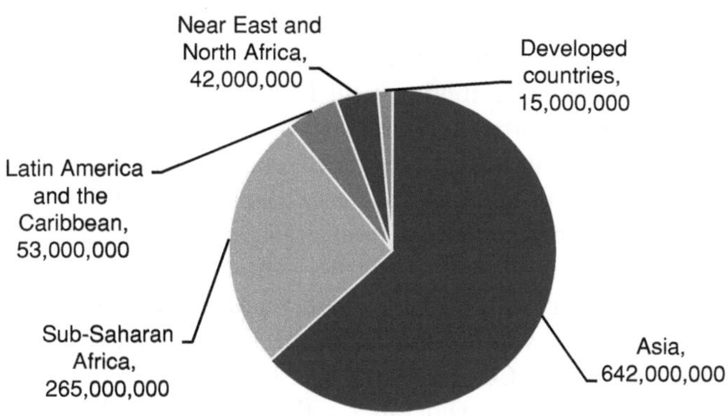

Fig. 13.3 Number of people undernourished by region, 2009 (FAO 2009b)

factors such as world agricultural commodity prices, macroeconomic growth, energy prices, exchange rates, agricultural, trade and environmental policies are all of increased importance to people in developing countries.

While it is true that the proportion of the world's population that is undernourished is on the decrease, the rate of reduction is unlikely to be sufficient to meet the hunger reduction targets set out in the Millennium Development Goals (MDG). The objective of MDG 1 is to halve, between 1990 and 2015, the proportion of population that is undernourished. Outside of Latin America and the Caribbean the number of hungry people has been on the increase (FAO 2009b). A regional breakdown in the numbers of undernourished is shown in Fig. 13.3.

Internationally, the growth in meat production itself places greater pressure on resources, since it will require the production of additional grains as animal feed. Annual meat consumption per capita ranges from as low as 10 kg in sub-Saharan Africa to about 90 kg in the industrialised world. Rising incomes mean that over the past 20 years meat consumption has been growing rapidly in East Asia, home to as much as one-fifth of the world's population.

The farming of livestock for meat or milk production has been identified as a major source of methane, a potent greenhouse gas emanating from livestock digestive systems. Some have therefore advocated a switch away from livestock production towards forms of agriculture (or food consumption) which generate lower GHG emissions.

13.2.2 Agriculture and the Irish Economy

For a country of just 4.5 million people, Ireland has a large agri-food sector based mainly on milk and beef production, most of which is exported. This production is facilitated by favourable climatic conditions for the growing of grass, which, in contrast with much of the rest of the world, is the main item in the diet of Irish cattle. The extent of milk and beef production in Ireland gives agriculture prominence as a source of Irish GHG emissions (Breen et al. 2010).

In terms of its economic contribution to the Irish economy, primary agriculture has receded significantly in importance, particularly over the past 20 years. While the size of the agricultural sector in Ireland has remained relatively stable, the non-agricultural economy grew rapidly over much of this period, so that the contribution of agriculture to Irish GDP declined. The share of total employment represented by agriculture in Ireland also declined considerably over this period, as shown in Fig. 13.4.

The more recent history of the Irish economy saw a sharp contraction following a banking crisis in 2009, which resulted in the intervention of the International Monetary Fund, European Central Bank and European Commission. Domestic demand has since contracted considerably with a rapid rise in both unemployment and emigration. Government policy has focused on job creation and in rural areas this has focused on the future expansion of the agri-food sector.

In 2009, the Irish Government appointed a Committee of Irish agri-food industry experts to develop a "…draft strategy for the medium-term development of the agri-food (including drinks) fisheries and forestry sector for the period to 2020. The strategy will outline the key actions needed to ensure that the sector contributes to the maximum possible extent to our export-led economic recovery…" (DAFF 2010). The Committee's report, which has become known as the Food Harvest 2020 (FH2020) report, was published in July 2010 and subsequently adopted as Irish Government policy.

The FH2020 report sets ambitious targets for output growth from the different subsectors of Irish agriculture. By 2020, total output from agriculture, forestry and fisheries is to increase by €1.5 billion. Within agriculture, specific targets were set for growth in milk output and for growth in the value of output from the beef, sheep and pig subsectors. The report envisages that by 2020 the volume of milk produced in Ireland will have grown by 50 % when compared with a base period of 2007–2009. The target growth in the value of output from the beef and sheep

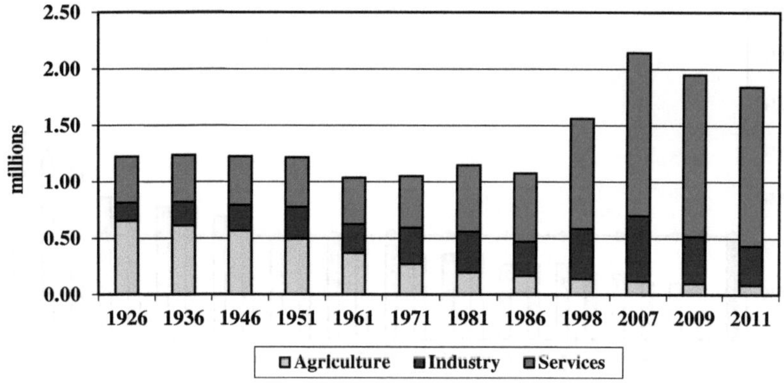

Fig. 13.4 Employment by broad economic sector in Ireland (CSO 2013; Kennedy et al. 1988)

subsectors is 20 % by 2020 and the target for growth in pig sector output value by 2020 is 50 %. In the case of beef, dairy products and sheep meat, exports represent approximately 90, 80 and 70 %, respectively of total annual production of these commodities. Therefore, Ireland's contribution to world trade in these commodities is much larger than an examination of its total production figures might suggest.

Given the outlook for continued growth in world population and the expectation, that with increasing affluence, global diets will increasingly include more meat and dairy products, the outlook for Irish agricultural commodity markets is relatively positive from a producer perspective (OECD-FAO 2012; FAPRI-ISU 2012). The market outlook, when combined with agreed changes in EU agricultural policy (the ending of the milk quota system in 2015), will see perceived impediments to growth in milk production being removed. This has led to increased optimism concerning the potential contribution of Irish agriculture to Irish economic recovery.

13.2.3 Climate Change Policy

In 2010, about 30 % of Ireland's GHG emissions came from the agriculture sector (EEA 2013) whereas the corresponding EU average in 2010 was just over 10 %, making the Irish agricultural sector relatively unique in the EU, as shown in Fig. 13.5 (EPA 2012).

Unlike some other sectors of the economy, agriculture is not part of the EU Emissions Trading Scheme (ETS) and thus there are no policy measures in place at present that would lead to a reduction in agricultural GHG emissions. Apart from agriculture, the other non-ETS GHG emissions sources include transport, households, services, smaller industrial installations and waste. Agriculture represented 41 % of emissions in the non-ETS sector in Ireland in 2010 (EEA 2013). With this

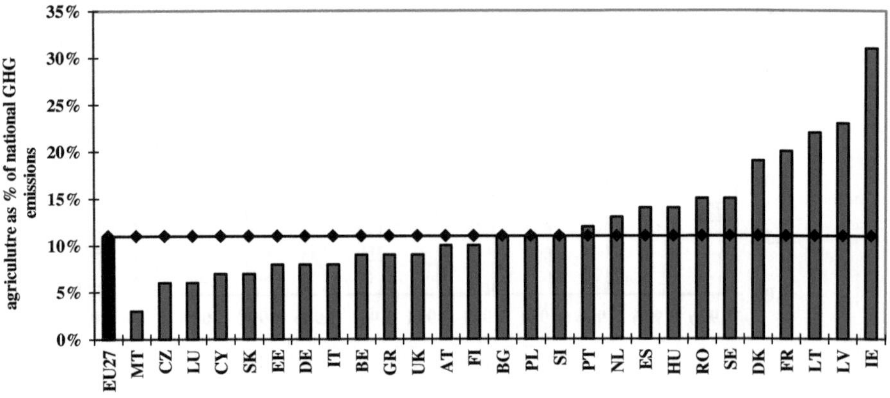

Fig. 13.5 Proportion of GHG emissions from agriculture in EU Member States in 2010 (EEA 2013)

large share of non-ETS emissions, insulating agriculture from any GHG emissions reduction requirement would be controversial, as it would require other non-ETS sectors to make greater emissions reductions.

Measures which could reduce the GHG emissions from agriculture already exist and further measures are in development. Examples include alterations to farm management practices such as reduced fertiliser usage and extending the grazing season for livestock. These technologies will need to be taken from the laboratory and deployed by farmers if they are to be effective and this is a process which takes time to achieve results. Over the short-term, the global capacity to reduce GHG emissions from agriculture remains limited and the necessary growth in global food production will likely mean that global emissions from agriculture continue to rise.

The Irish Government policy objective of facilitating and encouraging strong growth in agricultural production (FH2020) contrasts with policy relating to climate change. Ireland's public policy in relation to climate change is framed within the EU climate change policy framework. Ireland, as an EU member state, is committed to reducing GHG emissions through the EU Effort Sharing Agreement (European Parliament 2009). Under the agreement, Ireland's GHG emissions are to be reduced by 20 % relative to the level in 2005 by 2020. In the event that an international agreement on climate change is reached, Ireland's reduction commitment under the EU agreement increases to 30 %. Recent reports prepared for the Irish Government by the secretariat of the National Economic and Social Council (NESC 2012a, b) set out strategies for decarbonising Irish society by 2050. Within these documents the "vision" for the agriculture sector (and its decarbonisation) is framed within the context of the Intergovernmental Panel on Climate Change (IPCC) conventions relating to agriculture, land use and land use change and forestry (LULUCF). While current and proposed national policy (DECLG 2013) does not yet specify any sectoral allocation, a reduction in emissions of 20 % by 2020 will have to be achieved by the non-ETS, since emission reductions by industries covered by the ETS are governed by the operation of the EU ETS. Any special or differential treatment for agriculture within the context of the national emissions reduction target would necessarily imply significantly higher reductions in the transport and residential sectors that account for the vast majority of the non-agricultural elements of non-ETS GHG emissions in Ireland.

The conflict between GHG emissions reduction commitments and associated environmental policies and agricultural policies that seek to encourage agricultural production are particularly acute for large agricultural exporters. In an EU context Ireland is unique in both its export orientation and the ruminant animal basis of its agri-food industry. Milk and meat output from ruminant animals accounts for the majority of Irish agricultural output (61 %) and the majority of that output is exported. Increasing the production of food while, at the same time, reducing GHG emissions is a dilemma faced by Ireland and by the wider global community. Agriculture accounts for a very large share of GHG emissions globally and increasing agricultural production, while simultaneously reducing the contribution of agriculture to climate change, is listed as one of the major challenges facing the international food system in the Beddington Report (GO-Science 2011).

Given the agreements on climate change policy entered into by Ireland, Irish policy makers face the task of implementing policies that will allow Ireland to meet its GHG reduction commitments. As McCarthy and Scott (2008) note, for any emissions target, such as the 20 % and potential 30 % reductions in the EU Effort Sharing Agreement, there are competing menus of policy actions which could deliver the desired reduction. Of interest to policy makers is that some GHG abatement policy measures have much higher economic costs per tonne of emissions reduction than other measures. Given that a one-tonne reduction in CO_2 equivalent emissions from source x has the same beneficial effect on climate change as a one-tonne reduction from source y, society and public policy on climate change should seek to identify the least-cost abatement strategy. Where the costs associated with reductions from various sectors differ, it follows, based on a least-cost objective, that the share of the reduction sought from each sector should not be uniform.

In the context of reducing agriculture's GHG emissions, those agricultural activities that contribute to GHG emissions and that are currently marginally economic or uneconomic (in the sense of their profitability) should be the first focus of policy makers in their search for the least-cost abatement policy. If an activity, such as steel production or cattle production is unprofitable, then from a societal perspective, the costs of reducing the GHG emissions associated with such activities are likely to be negative (i.e. it will actually be economically beneficial). This means that such activities will be close to the top of most policy makers' climate change policy menus. This climate change policy "arithmetic" explains the focus in the remainder of this chapter on the contribution of the dairy and beef subsectors to Irish agriculture's GHG emissions and the changes that are likely to be necessary to resolve the tension between government policy relating to climate change and agricultural production.

13.3 Strategies to Abate Emissions in Agriculture

While science and technology holds out the promise of a more carbon-efficient agricultural sector, it should be understood that there are limits to what can be achieved in the short term. The contribution to GHG abatement of the technologies that flow from agricultural production research programmes will first have to be accepted by the IPCC. Farmers will then have to adopt the technologies proposed. Experience suggests that neither of these processes is either rapid or guaranteed.

The process whereby farmers adopt new technology will only begin, and subsequently accelerate, if farmers see an economic incentive to change their behaviour (i.e. adopt more carbon-efficient production practices). Creating these economic incentives involves the internalisation of the external costs of GHG emissions, through the imposition of measures such as a carbon tax, GHG quota or an emissions permit trading scheme (Clark 2008).

The question which arises is whether the scale of the reductions in Ireland's agricultural GHG emissions that would need to be made, could be achieved from technological solutions alone. A complicating factor is that the extent to which the different GHG abatement technologies under development can be considered additive in terms of their contribution to GHG abatement is also unclear. Thus, if agriculture were to make a proportionate contribution to the achievement of the GHG reduction targets set in EU and national climate change policies, a range of other agricultural and environmental policy options will have to be considered.

By increasing the costs of production,the application of a carbon tax in Ireland would reduce the international competitiveness of the traded goods sector,including agriculture.Reduced incomes in the agricultural sector following from the imposition of a carbon tax would reduce the production of agricultural output in Ireland and thereby reduce emissions of GHG in Ireland. However, since global emissions of GHG are driven by global demand for agricultural and food products, lower agricultural output in Ireland would almost certainly be offset completely by increased agricultural production elsewhere with next to no change in global emissions of GHG (so-called carbon leakage).

Taxing the consumption of beef and dairy products in Ireland would face similar difficulties to those associated with the imposition of a carbon tax on agricultural production. Consumption taxes could change consumer choices by raising the price of these foods; however, the Irish Government can only levy such taxes in Ireland. Consumers in the rest of the EU, where most Irish food production is consumed, would not be subject to such taxes and would continue to demand Irish beef and dairy commodities. The incentives, i.e. commodity prices, faced by Irish farmers, would not change very much, and consequently the reduction in GHG emissions from Irish agriculture through an Irish consumption tax is likely to be minimal.

A command and control approach to addressing the problem of achieving GHG reduction targets in agriculture could involve the allocation of a non-tradable GHG quota to each farmer based on their agricultural activity in a base period. Over the period to 2020, the level of this quota would then be reduced so that a national GHG reduction target for agriculture would be achieved.

There are a number of problems with the command and control approach. Some involve the calculation of the initial GHG emissions quota at the farm level. Actual emissions vary from farm to farm, due to scale, the intensity of production (e.g. yield of milk per cow) and according to production practices (e.g. application of artificial fertilisers). Establishing what these variables were for each farm in some reference period would be a costly exercise. Assessing, on an ongoing basis, farmers' compliance with their GHG emissions quota would likely present an even more formidable challenge in terms of designing and implementing monitoring and enforcement mechanisms.

The introduction of a GHG emissions quota would ensure that a least-cost abatement solution to the achievement of Irish agriculture's GHG emissions targets would not be achieved, since all farmers would have to achieve the same percentage reduction (Baumol et al. 1988). From an economic perspective, those farms with lower abatement costs should reduce their GHG emissions more than those with higher abatement costs, since otherwise the total cost of achieving the reduced GHG emissions level will be higher than necessary.

If GHG quotas were tradable amongst farmers then the overall costs of achieving a given reduction would be reduced compared with a fixed GHG quota regime. In the GHG context such a policy regime is known as a "cap and trade" system. The costs of carbon permits traded by agricultural processors and input suppliers will be reflected in the output and input prices faced by farmers.

From the perspective of the wider economy and with the objective of minimising the total economy costs of GHG abatement, it could make more sense for emissions permits to be tradable outside of agriculture. However, this could lead to a flow of GHG permits out of agriculture to the non-agricultural economy, with negative consequences for the level of agricultural and food production in the EU. A compromise might involve confining the tradability of 'agricultural' GHG quota to the agricultural sector. This would prevent the flow of GHG quota out of agriculture and most likely reduce the price at which such quota would trade between farmers or other economic agents involved in an agricultural ETS.

In the analysis in the next section, the consequence of the imposition of a binding GHG constraint on Irish agricultural production is investigated. In this analysis the least profitable sub-sector of Irish agriculture (beef) adjusts and thereby facilitates the expansion in production of the most profitable subsector (dairy). The change in production that occurs is likely to be akin to that which would arise if an agricultural sector cap and trade scheme were used to achieve a given reduction in GHG emissions. In the presence of a cap and trade scheme those farms expanding their level of agricultural activity would purchase GHG quota from those reducing their level of agricultural activity. This trade would not affect total agricultural sector income since it would involve transfers from one part of the sector to another. However, it would alter the costs of expanding production and the benefits of curtailing or ceasing loss-making production activities. These expansion and contraction costs and benefits and the operation of a carbon permit market have not been incorporated in this analysis. Breen (2008) found, using a linear programming model, that the economic costs of reductions in agricultural greenhouse gas emissions in Ireland are smaller using a tradable emissions permit approach rather than a command and control approach.

13.4 Model Description and Methodology

Future GHG emission levels from agriculture will be the product of emissions factors and the future level of agricultural activity. Considerable work has been done to provide GHG emissions factors which are specific to Ireland, notably the work by O'Mara et al. (2006). The other element of the future GHG emissions equation is the projected future level of agricultural activity. We use the FAPRI-Ireland partial equilibrium model of the Irish agriculture sector to generate these activity projections.

The FAPRI-Ireland model (Binfield et al. 2003, 2007, 2008) is a dynamic partial equilibrium model that is integrated within the FAPRI EU GOLD model (Hanrahan 2001). The GOLD model in turn can form a component of the FAPRI world modelling system for world agriculture. In this way the model for Ireland can incorporate the consequences of changes in international trade policy as they relate to agriculture. The model has an agricultural commodity coverage that extends to markets for grains (wheat, barley and oats), other field crops (potatoes, sugar beet and vegetables), livestock (cattle, pigs, poultry and sheep) and milk and dairy products (cheese, butter, whole milk powder and skim milk powder). Many of the equations in the model are estimated using annual data from the period 1973–2011 or over shorter periods in cases where data are not available or where, for policy reasons, longer estimation periods would not be meaningful. The FAPRI approach to the development of agriculture sector models and the conduct of policy analysis is described in Meyers et al. (2010) and Westhoff and Meyer (2010). The FAPRI-Ireland model has a submodule which generates projections of GHG and other emissions to air, which are associated with agricultural production. Details of the GHG submodule and earlier policy scenario analysis can be found in Donnellan and Hanrahan (2006).

The primary purpose of the FAPRI-Ireland model is to analyse the effect of policy changes on economic indicators such as the supply and use of agricultural products, agricultural input expenditure and sector income. In doing so the model produces future projections of animal numbers, input usage volumes (e.g. fertiliser, feed, fuel, energy) and other indicators. These data can be incorporated into the satellite GHG models to enable the provision of base data and projections relating environmental indicators, such as GHG emissions, fertiliser usage and ammonia emissions. The projections of commodity outputs and input usage from the FAPRI-Ireland model can be converted into projections of GHG emissions using the default Intergovernmental Panel on Climate Change (IPCC) conversion coefficients, modified, where possible, with specific coefficients for Ireland.

GHGs in the form of methane and nitrous oxide emissions from each agricultural subsector *i* are a function of the number of animals, crop areas harvested and nitrogen application. Since the global warming potential of CH_4 and N_2O differ, for the purpose of their addition these are brought to a common base of CO_2 equivalents using standard weighting systems. CH_4 produced in each agricultural sector can be represented as:

$$CH_{4,i,t} = f(q_{i,t}, \alpha_i) \tag{1}$$

where $CH_{4,i,t}$ is the total amount of CH_4 produced by sector i in year t, q is the quantity of animal or crop category i in year t and α is the methane conversion coefficient associated with the animal or crop category i. Similarly, N_2O produced in each agricultural sector can be represented as:

$$N_2O_{j,t} = f(q_{j,t}, \beta_j) \tag{2}$$

where $N_2O_{j,t}$ is the total amount of N_2O produced by sector j in year t, q is the quantity of animal or crop category j in year t and β is the nitrous oxide conversion coefficient associated with the animal or crop category j.

Finally, total GHG emissions in the common base of CO_2 equivalents can be expressed as:

$$\text{EquivCO}_{2t} = \delta \sum_{i=1}^{n} CH_{4,i,t} + \gamma \sum_{j=1}^{m} N_2O_{j,t} \tag{3}$$

where EquivCO_2 is CO_2 equivalent, while $\delta = 21$ and $\gamma = 310$ are the global warming potentials of methane and nitrous oxide, respectively.

The next section provides a brief review of the projected level for some of the agricultural variables used in the generation of GHG emissions. The consequent scenario projections of GHG emissions from Irish agriculture are then presented.

13.4.1 Scenario Development

The assessment of the economic impact of a future constraint on agricultural activity needs first to establish the future level of agricultural activity under specific policy assumptions. Accordingly, a *Reference Scenario*, based on a specific set of future policy assumptions, is set out. This reference scenario reflects the achievement of the targets that are set out in the FH2020 for the period to 2020. Creating a set of projections over the longer term horizon of 2050 is difficult in the absence of the required exogenous data, in particular projections of future agricultural commodity prices. Therefore, assumptions were made about the change in output and input prices for the period 2021–2050. Under these assumptions the real price of agricultural commodities remains relatively fixed. In this reference scenario CAP policy remains unchanged, so the value of agricultural support payments declines in real terms over time.

Using the model, we arrive at a set of projections which provides an estimate of the distance agriculture would be from achieving a particular percentage GHG reduction target if no policies to address GHG emissions from Irish agriculture were pursued. This Reference Scenario also provides projections of the future value of agricultural output, input expenditure and agricultural income out to 2050.

Next a *GHG Emission Reduction* scenario is specified. At present there are no indications as to what policy might prevail in the period to 2050 to constrain agricultural GHG emissions, so an assumption must be made about the percentage reduction in agricultural GHG emissions that would need to be achieved by 2050 and a relevant base year against which such reductions would be measured. In the GHG Emission Reduction scenario the reduction in Irish agricultural output required to reduce GHG emissions from the sector by 10 % (relative to their level in 2005) by 2050 was estimated. Initially, it was planned to look at a scenario with a 20 % reduction, but as will become clear in the next section, this was not feasible. Estimates of the economic impact of meeting the target in terms of value added in Irish agriculture are produced. This economic impact assessment is carried out for primary agriculture only and does not extend to include the impact on the wider economy.

13.5 Results and Discussion

The model is used to produce the reference scenario projections of agricultural activity and associated GHG emissions to a 2050 time horizon. As shown in Fig. 13.6, GHG emissions decline in the first couple of years of the projection period relative to current levels, but then increase once milk quotas are removed in 2015 and the milk sector begins to grow. The total cattle population remains relatively stable until 2025, but there is a rise in the number of dairy cows and a fall in the number of suckler cows (and their progeny) as illustrated in Fig. 13.7. Over the period 2030–2050, the suckler cow population continues to contract and the dairy cow population stabilises, leading to an overall decline in the bovine breeding herd and a contraction in the number of bovines, which decline to 4.6 million head by 2050. The consequences for GHG emissions over the projection period are that emissions rise in the short- to medium-term, reaching about 22 Mt CO_2 eq by 2030. Emissions continue to rise beyond 2030 but at a much lower rate. The principal drivers for the increase in emissions are the growing dairy cow, pig and poultry populations and an associated increase in fertiliser emissions and

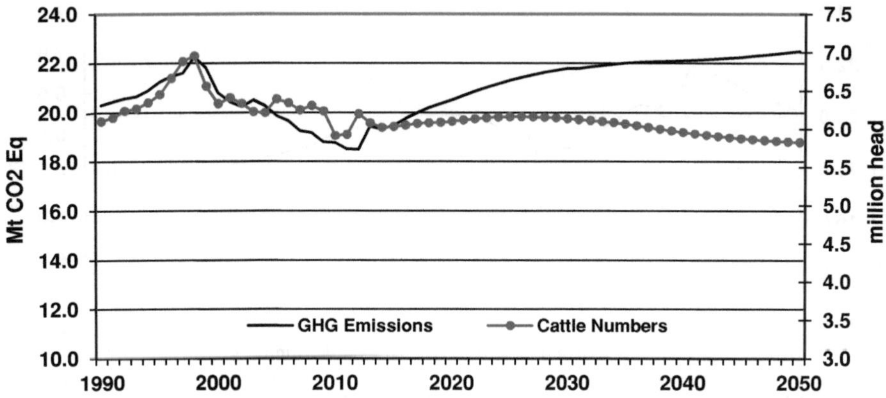

Fig. 13.6 GHG emissions and cattle population under the reference scenario

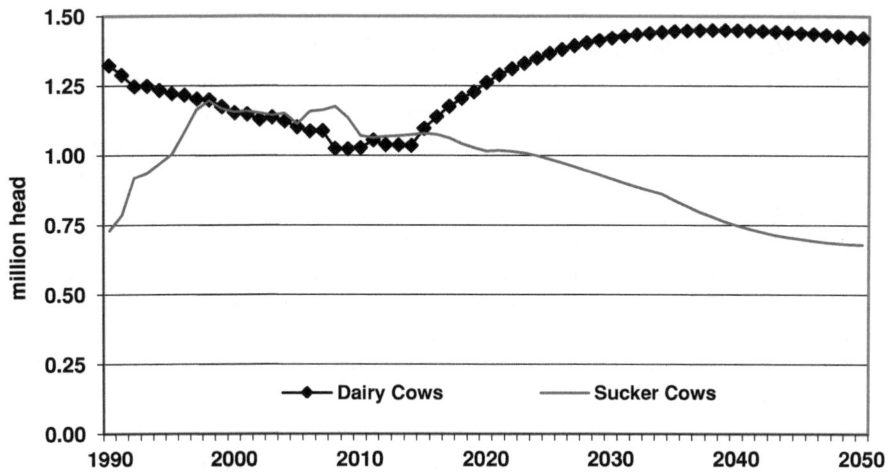

Fig. 13.7 Dairy cow and sucker population under the reference scenario

emissions associated with animal waste. This overall growth in emissions is only partially offset by a decline in the number of beef animals and emissions associated with this category of activity.

This outcome reflects the fact that dairy production is a much more profitable enterprise than beef production in Ireland. Beef production is heavily supported by subsidy payments and the value of these payments is held fixed in nominal terms out to 2050, which means that the real value of these payments declines substantially. While these payments are decoupled from production, they are assumed to have some supply inducing impact, so a fall in their real value over the long term has adverse implications for the size of the Irish suckler cow herd.

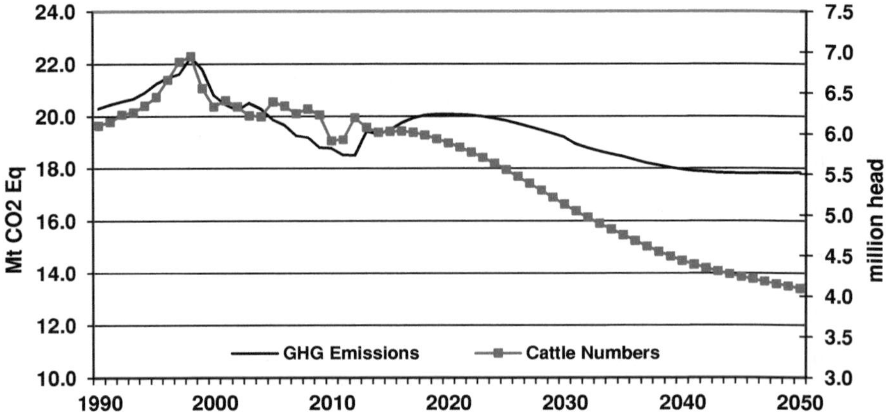

Fig. 13.8 GHG emissions and cattle population under the 10 % GHG reduction scenario

Next the model is run with a GHG constraint to determine the outcome under the GHG Reduction Scenario, where agricultural GHG emissions must be reduced so that they are 10 % below the 2005 level by 2050. For the purposes of this scenario we assume that the approach taken is to limit GHG emissions from the suckler herd (and associated progeny), while other areas of agricultural activity continue in an unconstrained fashion. The basis for this approach is the low profitability of the beef sector and the large contribution that it makes to total agricultural GHG emissions in Ireland. In the absence of low-cost technical GHG abatement technologies, reducing the size of the suckler herd would represent a least-cost abatement strategy in meeting an agricultural GHG reduction target. The model is therefore used to find a cattle population and production intensity that is consistent with the imposed 10 % GHG reduction by 2050. As shown in Fig. 13.8 the extent of the decrease in the cattle population required to achieve the GHG reduction target of 10 % below 2005 levels is dramatic.

Given that the GHG reduction strategy is targeted at the specialist bovine sector, the decrease that is required in the suckler cow herd, as shown in Fig. 13.9, is even more pronounced than the overall reduction in the cattle population. By 2050, the suckler herd is reduced to the point where it almost disappears. It should be clear at this point that placing a quota on the suckler herd would not deliver a larger agricultural GHG reduction (such as a reduction of 20 %) by 2050, since the sucker herd is almost completely eliminated, even under a 10 % GHG reduction scenario.

To reach the specified 10 % GHG reduction target by 2050 requires that cattle numbers be reduced to 4.09 million head by 2050. Total cattle numbers were 6.34 million head in 2005 and are projected to be 5.82 million in 2050 under the Reference Scenario. Suckler cow numbers are reduced to 0.08 million head by 2050 to achieve the 10 % GHG reduction target. For comparison, suckler cow numbers were 1.11 million head in 2005 and are projected to be 0.68 million in 2050 under the Reference Scenario. The dramatic decline in suckler cow numbers under the

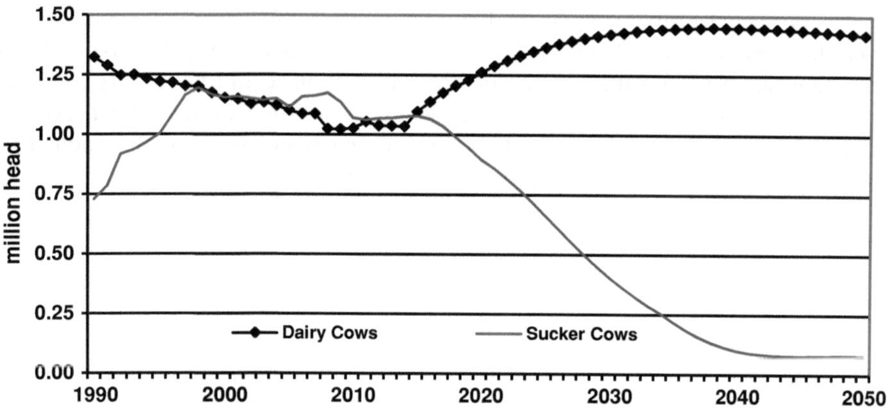

Fig. 13.9 Dairy cow and sucker population under the 10 % GHG reduction scenario

Table 13.1 Animal numbers, beef production and cattle sector value under a reference 2050 scenario and under a 10 % GHG reduction target

		Reference scenario	GHG −10 %	2050 reference versus 2005	GHG −10 % versus 2005	GHG −10 % versus reference 2050
	2005	2050	2050			
	Million head			*Percent change (%)*		
Total cattle	6.34	5.82	4.09	−8	−35	−30
Dairy cows	1.10	1.43	1.43	30	30	0
Suckler cows	1.11	0.68	0.08	−39	−93	−88

Source FAPRI-Ireland GHG Model

Table 13.2 Beef production and cattle sector value under a reference 2050 scenario and under a 10 % GHG reduction target

		Reference scenario	GHG −10 %	2050 reference versus 2005	GHG −10 % versus 2005	GHG −10 % versus reference 2050
	2005	2050	2050			
	Million tonnes			*Percent change (%)*		
Beef Production	0.545	0.521	0.197	−4	−64	−62
	Euro millions			*Percent change (%)*		
Cattle Output Value	1,413	3,585	1,430	154	1	−60

Source FAPRI-Ireland GHG Model

GHG Emissions Reduction Scenario is reflected in the reduced production of beef in Ireland. By 2050 Irish beef production would decrease to 0.197 mt to achieve the 10 % reduction target. For comparison, beef production was 0.545 mt in 2005 and is projected to be 0.521 mt in 2050 under the Reference Scenario.

It is projected in these scenarios that beef consumption in Ireland will decline over the period to 2050; however, the impact which meeting the 10 % GHG reduction target has on the volume of beef production would be largely mirrored by a broadly similar percentage reduction in the volume of Irish beef exports. Table 13.1 summarises the impact on the bovine population, while Table 13.2 summarises the impact on beef production and the value of cattle output.

13.6 Conclusion

It is important to note that in these scenarios no attempt has been made to incorporate the possible impact of GHG abatement technologies. As well as reducing GHG emissions per unit of output, these technologies may also have implications for the costs of production. Anticipating the uptake of these technologies over a long timescale is complicated by our limited knowledge of which technologies will exist in the future and how much they will cost. In addition, the uptake of any such technologies is likely to be influenced by the extent of the GHG emissions constraint that is faced by agriculture.

Some abatement technologies may be prohibitively expensive and hence uneconomic. Other abatement technologies may be cost neutral and there are even some abatement technologies which are said to be cost negative, i.e. these technologies when adopted actually improve farm productivity. The difficulty with negative cost abatement technologies is that even though they may reduce emissions on a per unit of output basis, they also improve farm profitability. Other things being equal, measures which improve farm profitability would also lead to increased production and GHG emissions, which may then counteract the beneficial impact of the abatement technology.

If progress with these GHG abatement technologies is limited, or if these technologies remain prohibitively expensive, then what are the alternatives? In these circumstances either the volume of agricultural activity would need to be reduced (with adverse consequences for the exportable surplus, implications for international food prices and food security) or the target level of agricultural GHG emissions reduction needs to be set at a relatively moderate level.

The incoherence in Irish Government policy between agricultural and climate change policies can be generalised to the international stage. The results presented in this chapter illustrate the potential difficulty that agricultural net exporters will face in the context of future GHG reduction commitments which impact on agriculture. Increasingly policy makers and the general public in Ireland and internationally will need to consider the implications which international agreements designed to tackle climate change have on global food production. An argument can be made that the current system of agreement for the monitoring and reduction of GHG emissions is inappropriate and in conflict with the desire to produce affordable food for the global population.

Under existing global agreements the GHG emissions created in the production of food are associated with the food exporter rather than the food importer. There is already a precedent for an alternative treatment of particular sectors. For example, fossil fuel emissions are associated with the fuel-consuming country rather than the fuel-producing country.

Do alternative mechanisms to constrain GHG emissions from agriculture, such as an approach focused on the intensity of GHG emissions per unit of food output, deserve consideration? Given the scale of the increase in global food production required over the coming decades, and the desire to produce food at affordable price levels, GHG emissions abatement in agriculture may require an approach that is not in conflict with countries' desire to exploit their comparative advantage in the production of particular food commodities. This argument has even greater merit in the case of food net exporting countries, such as Ireland and New Zealand, where agricultural production has a low GHG emission intensity (Leip et al. 2010).

Ultimately, this chapter illustrates the limitations of setting EU emissions targets at the EU member state level without first modelling the potential impact of such targets at a sectoral level. Future target setting will need to take account of the sectoral consequences of imposing particular emissions reduction targets for elements of the economy. While least-cost economic solutions may indicate a particular course of action, ultimately this will be balanced by concern on the part of policy makers to find a solution which will not be seen to place a disproportionate burden of adjustment on certain economic sectors. While a strategy for emissions reductions targeted at certain economic sectors or on certain regions within an economy may be shown to be economically efficient, the reality is that such a strategy is likely to be politically unfeasible.

References

Baumol, W. J., & Oates, W. E. (1988). *The theory of environmental policy* (2nd ed.). Cambridge: Cambridge University Press.

Binfield, J., Donnellan, T., Hanrahan, K., & Westhoff, P. (2003). The Luxembourg CAP reform agreement: Implications for EU and Irish agriculture. *The Luxembourg CAP Agreement*, October 2003. Ireland: Teagasc.

Binfield, J., Donnellan, T., Hanrahan, K., & Westhoff, P. (2007). CAP health check analysis: Impact on EU milk quota expansion, Teagasc.

Binfield, J., Donnellan, T., Hanrahan, K., & Westhoff, P. (2008). *Issues in examining the impact of WTO reform on the beef and dairy sectors in the European Union*. Contributed paper, International Association of Agricultural Economists 2009 Conference, August 16–22, 2009, Beijing, China. Retrieved March 22, 2013, from http://ageconsearch.umn.edu/bitstream/51752/2/51752.pdf.

Breen, J. P. (2008). *Controlling greenhouse gas emissions by means of tradable emissions permits and the implications for Irish farmers*. Contributed paper European Association of Agricultural Economists (EAAE), 107th Seminar, January 30-February 1, 2008, Sevilla, Spain.

Breen, J., Donnellan, T., & Westhoff, P. (2010). Food for thought: EU agriculture presents new challenges for EU agriculture. *EuroChoices, 9*(3), 24–29.

Central Statistics Office CSO (2013). *Quarterly National Household Survey*. Retrieved March 22, 2013, from http://www.cso.ie/en/qnhs/.

Clark, H. (2008). New Zealand's agricultural GHG emissions: policies and approaches. Teagasc GHG Seminar. Retrieved March 22, 2013, from www.teagasc.ie/publications/2008/20080328/index.asp.

DAFF (Department of Agriculture, Forestry and Fisheries). (2010). Food harvest 2020. Dublin, Ireland. Retrieved March 22, 2013, from http://www.agriculture.gov.ie/agri-foodindustry/foodharvest2020/.

DECLG (Department of Environment, Community and Local Government). (2013). Draft heads of climate action and low carbon development bill 2013. Retrieved March 22, 2013, from http://www.environ.ie/en/Publications/Environment/ClimateChange/FileDownLoad,32468,en.pdf.

Donnellan, T., & Hanrahan, K. (2006). It's the environment, stupid! To what extent can WTO trade reform lead to environmental rather than economic impacts? 3rd World Congress of Environmental and Resource Economists, Kyoto, Japan. Retrieved March 22, 2013, from http://www.webmeets.com/files/papers/ERE/WC3/352/donnellanhanrahan.pdf.

European Environment Agency. (2013). EEA data viewer. Retrieved March 22, 2013, from http://www.eea.europa.eu/data-and-maps/.

Environmental Protection Agency. (2012). Ireland National Inventory Report 2012. Wexford, Ireland. Retrieved March 22, 2013, from http://coe.epa.ie/ghg/nirdownloads.jsp.

European Parliament. (2009). Decision No 406/2009/EC of the European Parliament and of the Council of 23 April 2009 on the effort of Member States to reduce their greenhouse gas emissions to meet the Community's greenhouse gas emission reduction commitments up to 2020.

FAO Food and Agriculture Organisation of the United Nations. (2009a). How to feed the world in 2050. Retrieved March 11, 2013, from http://www.fao.org/fileadmin/templates/wsfs/docs/expert_paper/How_to_Feed_the_World_in_2050.pdf.

FAO Food and Agriculture Organisation of the United Nations. (2009b). The state of food insecurity in the world: Economic crises—impacts and lessons learned. Retrieved March 11, 2013, from http://www.fao.org/docrep/016/i3027e/i3027e.pdf.

FAOStat. (2013). Statistical Database of the Food and Agricultural Organisation of the United Nations. Retrieved March 22, 2013, from http://faostat.fao.org/.

FAPRI-ISU. (2012). 2012 World Agricultural Outlook. Iowa State University, Ames IA. Retrieved March 22, 2013, from http://www.fapri.iastate.edu/outlook/2012/.

GO-Science. (2011). *Foresight*. The Future of Food and Farming. The Government Office for Science, London, UK.

Hanrahan, K. (2001). The EU GOLD Model 2.1: An introductory manual. Mimeo, Teagasc, Dublin.

Kennedy, K., Giblin, T., & McHugh, D. (1988). *Economic development in Ireland in the 20th century*. Dublin: Routledge.

Leip, A., Weiss, F., Wassenaar, T., Perez, I., Fellmann, T., Loudjani, P., Tubiello, F., Grandgirard, D., Monni, S., & Biala, K. (2010). Evaluation of the livestock sector's contribution to the EU greenhouse gas emissions (GGELS)—final report. European Commission, Joint Research Centre. Retrieved March 22, 2013, from http://ec.europa.eu/agriculture/analysis/external/livestock-gas/index_en.htm.

NESC. (2012a). *Towards a new national climate policy: Interim report of the NESC secretariat*. Report to the Department of Environment, Community and Local Government. June 2012. Retrieved March 22, 2013, from http://www.environ.ie/en/Publications/Environment/ClimateChange/FileDownLoad,31202,en.pdf.

NESC. (2012b). *Ireland and the climate change challenge: Connecting 'How Much' with 'How To'*. Final Report of the NESC Secretariat to the Department of Environment, Community and Local Government. December 2012. Retrieved March 22, 2013, from http://www.environ.ie/en/Publications/Environment/ClimateChange/FileDownLoad,32467,en.pdf.

McCarthy, C., & Scott, S. (2008). *Controlling the cost of controlling the climate: The Irish government's climate change strategy*. ESRI Working Paper No. 229.

Meyers, W., Westhoff, P., Fabiosa, J., & Hayes, D. J. (2010). The FAPRI global modelling system and outlook. *Journal of International Agricultural Trade and Development, 6*(1), 1–20.

OECD-FAO. (2012). Agricultural outlook 2012–2021. Organisation for Economic Cooperation and Development (OECD).

O'Mara, F., Ryan, M., Connolly, J., O'Toole, P., Carton, O., Lenehan, J. J., et al. (2006). *Climate change—estimation of emissions of greenhouse gases from agriculture and strategies for their reduction*. Wexford: Environmental Protection Agency.

Vermeulen, S. J., Campbell, B. M., & Ingram, J. S. I. (2012). Climate change and food systems. *Annual Review of Environmental Resources, 37*, 195–222.

Westhoff, P., & Meyers, W. (2010). The FAPRI approach: A few key principles. *Journal of International Agricultural Trade and Development, 6*(1), 133–135.

Part V
Regulatory Changes and Management of Emissions

Chapter 14
Economic Incentives and Alternative Nitrogen Regulation Schemes: A Spatial Sector Economic Modelling Approach

Jørgen D. Jensen and Jens E. Ørum

Abstract The objective of this chapter is to investigate economic incentives associated with changes in nitrogen regulation, including the distribution between farm types and geographically. The analysis is carried out on a partial equilibrium simulation model of the Danish agricultural sector—ESMERALDA. The model is based on farm-level production and economic data for all Danish farms, which allow the analysis of spatial aspects related to the alternative regulations, in terms of environmental (in terms of nitrogen use) and economic effects. Results of the model analyses suggest that replacing the current flat-rate quota on nitrogen input on all farms with a more differentiated quota on nitrogen leaching will in particular be binding for crop and pig farms in environmentally sensitive areas.

14.1 Introduction

Regulating agricultural use of nitrogen is a crucial issue in agri-environmental policy in most European countries. The overall regulation of nitrogen pollution in Europe is regulated by the Water Framework Directive (European Commission 2000), which establishes a framework for national regulations and their implementation. The issue of nitrogen regulation in agriculture has been the subject of a large number of studies all over the world, e.g. Schou et al. (2000), Anselin et al. (2004).

The current regulation of nitrogen use in Danish agriculture was established in the 1990s and is based on farm-level quotas. Quotas are determined on the basis of a calculated economically optimal nitrogen application for each of the respective crops in the farm's production plan. Optimal application is based on partial crop

J. D. Jensen (✉) · J. E. Ørum
Department of Food and Resource Economics, University of Copenhagen, Copenhagen, Denmark
e-mail: jorgen@ifro.ku.dk

C. Zopounidis et al. (eds.), *Agricultural Cooperative Management and Policy*, Cooperative Management, DOI: 10.1007/978-3-319-06635-6_14,

yield functions derived from generally recognised field experiments (Videncentret for Landbrug 2010). For the production year 2012/2013, the farm-level quota was determined as approximately 85 % of the economically optimal amount of nitrogen for each crop, summed over the farm's crop areas. Nitrogen from the farm's animal manure production is incorporated in the farm's quota in terms of a required utilisation rate for this nitrogen. The farmer is allowed to buy an amount of nitrogen mineral fertilizer corresponding to the difference between the total farm quota and the assumed utilisation of nitrogen from the farm's own production of animal manure. Fertilizer purchases beyond this quota are subject to heavy taxation.

One problem with the current regulation is that it is not cost-effective as a tool to regulate leaching of nitrogen from agriculture to aquatic environments, because the relationship between nitrogen application and leaching is complicated and depends on the crop, soil, geography, climatic conditions, etc. For this reason, a revision of the regulation is under consideration. One approach to a more cost-effective regulation would be to impose quotas (or taxes) directly on the amount of nitrogen leached—a regulation that would put high requirements on data and modelling of nitrogen run-off on all locations. An alternative would be to revise the quota on application with a more differentiated account of the crops' leaching potential and local soil and vulnerability characteristics.

The objective of this chapter is to evaluate the economic incentives involved in a suggested new scheme for regulating agricultural nitrogen use in Denmark. Furthermore, the chapter addresses economic perspectives related to the extent of transferability, in particular by addressing the differences in cost-effectiveness with transferability within water catchments versus quota transferability across the entire country.

The next section outlines the economic model and data framework, as well as the considered approaches to nitrogen regulation, and the following two sections show some results from the model analysis and discuss these results and their perspectives. The final section concludes the chapter.

14.2 Methodology and Data

The analysis is carried out on a partial equilibrium simulation model of the Danish agricultural sector—ESMERALDA. The model is based on econometrically estimated parameters and farm-level production and economic data for all Danish farms, which allow the analysis of spatial aspects related to the alternative regulations, in terms of environmental (in terms of nitrogen use) and economic effects.

14.2.1 Data

The point of departure for establishing a data set for Danish farms is a combined database comprising data from the Danish General Agricultural Register (GLR) and the Central Livestock Register (CHR) for the population of Danish farms (Kristensen and Rasmussen 2002; Kristensen and Kristensen 2004; Kristensen et al. 2005; Børgesen et al. 2009; GLR/CHR 2007). The GLR/CHR database is an integrated system of registers comprising all Danish farms, and it includes structural information such as land use and livestock density, obtained from the administration of various subsidy and control schemes. The database has been established to fulfil the Danish obligations with respect to EU regulation 3508/92 (November 27, 1992) regarding monitoring of various EU support schemes, as well as some Danish regulations requiring monitoring on a regular basis. Since the 1980s, substantial collections of geo-coded data have been established, with the aim of administration and specific tasks related to land use. Regarding livestock, the registers comprise information about animal stocks, collected for administrative and veterinary purposes, e.g. for administration of cattle payments and for tracing outbreaks of animal diseases, such as foot-and-mouth disease, BSE, salmonella, etc. To ensure regular updating, farmers are informed about the current information in the registers once every year, so they have the opportunity to check if the recorded information in the register is correct.

Economic information is not included in these registers. Therefore, the register data have been supplemented with micro-simulated economic variables, based on a sample of individual farm accounts (FADN) from Statistics Denmark. The Danish FADN data are collected from approximately 1,900 farms on an annual basis. For each farm, data are collected concerning structural variables such as land use and animal numbers, as well as economic variables such as the value of assets at the beginning and end of the year, various cost items, gross yields for different commodities and economic performance, such as gross and net farm profits, labour, capital endowments, etc. (Institute of Food and Resource Economics 2007; Pedersen 2006).

Several of the structural variables in the FADN dataset are also present in the GLR/CHR registers. Assuming that the structural variables are important determinants for the economic variables, and that farms which are structurally similar (regarding size, crop composition, stocking density, animal composition) should also be expected to possess fairly similar economic characteristics, the structural variables can be used as a key to estimate economic variables for the farms in the registers. Hence, a methodology for estimating economic variables on register farms as a weighted average of corresponding variables from FADN farms, with weights reflecting the degree of similarity (Jensen and Kristensen 2013), has been developed and applied. Thus, a complete data set of observed farm-level data for structural variables (land use, livestock, etc.) and estimated economic variables (output value, costs, income, etc.) was established for the entire population of Danish farms for the year 2006.

14.2.2 The ESMERALDA Model

ESMERALDA is an econometric sector model describing production, input use, allocation of agricultural land, livestock activity, etc. in Danish agriculture, distributed on 36 lines of production, including 25 land uses (spring barley, winter barley, wheat, rye, oats, triticale, mixed grain, other grain, pulses, rapeseed, grass seeds for sowing, clover seeds for sowing, potatoes for consumption, starch potatoes, sugar beets, silage cereals, maize, grass in rotation, permanent grass, fodder beets, non-food and fallow, plus three optional 'new' crops, depending on the scenario) and 11 livestock types (dairy cows, suckling cows, heifers, male calves, sows, fattening pigs, horses, sheep, hens, chicken and furred animals). The model can evaluate changes in these variables as a consequence of, e.g. changed prices or subsidies, environmental regulations, quotas, cross-compliance requirements, etc. The model is based on farm-level data from the above database, which are aggregated into a number of representative farms, as convenient in relation to the specific topic purpose of analysis (e.g. farm-type groups, farm size groups, geographic groups, or combinations thereof), and hence obtain results for each of these farm groups (Jensen and Kristensen 2013; Jensen 2001). For the present purpose, farms have been aggregated to 21 types, cf. below. The model calculates farm-level optimal adaptations of production, input use, land use, livestock, etc. to changes in the farms' economic framework conditions, e.g. prices, subsidies or restrictions on production or input use. Adaptations on the respective farm types can subsequently be distributed on the underlying individual farms in the GLR/CHR data material, and hence provide a spatial description of the sector's adaptation to the changed framework conditions.

The production technology for each representative farm is modeled in nested CES production functions for each individual line of production, using the input nesting structure displayed in Fig. 14.1. At each stage of the nesting structure, input substitution between two 'branches' is given by an elasticity of substitution.

For example, the composition of mineral phosphorus (P) and mineral potassium (K) is assumed to be influenced by the relative prices of the two nutrients, and a change in the relative prices would lead to a change in this composition depending on the elasticity of substitution between the two inputs—the larger the elasticity, the more the composition will change. Similarly, in the next step, the composition of nitrogen fertilizer (N) and the PK-aggregate is determined by the relative prices of nitrogen and the PK-aggregate, and the elasticity of substitution between N and PK, and so forth.

Some of the elasticities of substitution (associated with energy, labour, capital and pesticide aggregates) in the individual lines of production were estimated econometrically on the basis of FADN data from 2000 to 2008, whereas others (related to nutrients, individual pesticides and land) were estimated on the basis of experimental data from field trials (Jacobsen et al. 2004; Ørum 1999). A few examples of these elasticities are displayed in Table 14.1.

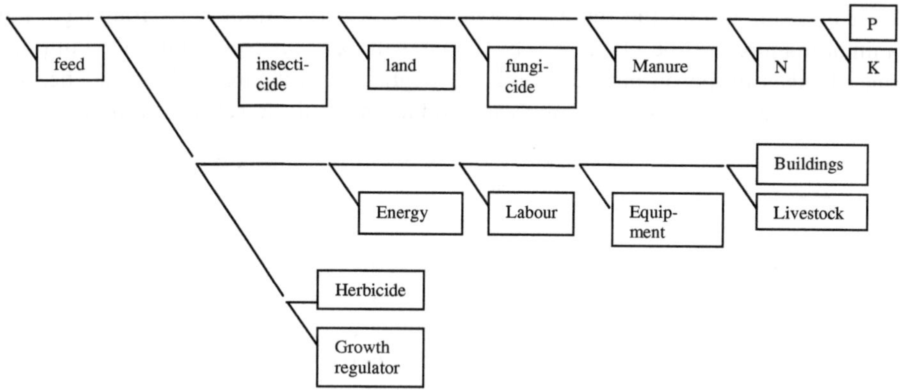

Fig. 14.1 Input nesting structure in ESMERALDA lines of production

Table 14.1 Examples of elasticities of substitution

	Wheat		Potatoes		Grass in rotation	
	Clay	Sand	Clay	Sand	Clay	Sand
Land versus (fungicides, nutrients)	0.100	0.100	0.150	0.150	0.010	0.010
Fungicide versus nutrients	0.648	0.833	0.306	0.437	0.336	0.369
Mineral fertilizer versus manure	0.100	0.100	0.100	0.100	0.100	0.100
Nitrogen versus (phosphorus, potassium)	0.164	0.219	0.031	0.044	0.057	0.062
Herbicide etc. versus (energy, labour, capital)	0.095	0.161	0.057	0.081	0.072	0.080
Energy versus (labour, capital)	0.939	0.939	2.744	2.744	0.378	0.378
Labour versus capital	0.193	0.193	0.250	0.250	0.450	0.450

For each line of agricultural production, the model is calibrated based on average 2006 data for individual lines of production (Institute of Food and Resource Economics 2006), adapted to farm-level data for each farm type.

At the representative farm level, different lines of production are linked to each other via their use of agricultural land, the use of on-farm produced roughage and the production and use of animal manure. With respect to land allocation, farm *f*'s use of land in the respective lines of production (a_{cf}) is restricted by the total on-farm availability of agricultural land (a_f), i.e.

$$\sum_{c \in crops} a_{cf} = a_f \tag{14.1}$$

A dual variable (λ_{af}) associated with this land availability restriction reflects the bindingness of the restriction. Hence, if the farm's total demand for agricultural land increases due to, e.g. a higher crop price, this dual variable will also increase, thus affecting the economically optimal allocation of the available land on the farm, and substitution between land and other inputs.

It is assumed that the representative farms' external trade with roughage and animal manure will be unaffected by changes in external conditions. This implies that changes in the on-farm demand for roughage or manure will be matched by corresponding changes in the on-farm production of these respective inputs, i.e.

$$\sum_{l \in livestock} n_{lf} \cdot x_{ilf} - \sum_{l \in livestock} n_{lf}^0 \cdot x_{ilf}^0 = \sum_{c \in crops} a_{cf} \cdot x_{icf} - \sum_{c \in crops} a_{cf}^0 \cdot x_{icf}^0 \qquad (14.2)$$

where n_{lf} is the number of animal type l on farm type f. Like the land allocation, this internal balance for animal manure and roughage is regulated in the model by farm-type specific dual variables, reflecting internal opportunity cost of the manure or internally produced roughage. Increased demand will increase the dual value, affecting the demand and supply, as well as substitution with other inputs. All other inputs are assumed to be unconstrained at the farm level, and hence the dual values associated with these inputs are equal to zero.

A quota regulation on nitrogen use (as the current regulation) also implies a shadow price on nitrogen quota, which is taken into account in the farmers' economic optimization.

In the farmers' economic optimization problem, the internal price of an input is equal to the sum of the market price and the dual value associated with this input. Within the CES framework, the composition of inputs i and j in production line c is determined by the expression

$$\left(\dot{x}_{icf} - \dot{x}_{jcf}\right) = \sigma_{ijcf} \cdot \left(\frac{w_{jc} + \lambda_{jcf}}{w_{jc}^0 + \lambda_{jcf}^0} - \frac{w_{ic} + \lambda_{icf}}{w_{ic}^0 + \lambda_{icf}^0}\right) \qquad (14.3)$$

where w's are market prices of inputs, λ's are dual values associated with the same inputs, σ is the partial elasticity of substitution between the two inputs, and $\dot{x}$ is the relative change in quantity variable x.

Farmers are assumed to exhibit profit maximizing behavior, implying equality between price and marginal cost for each individual line of production—and thus implying equality between changes in price and marginal cost (including internal opportunity cost) in case of changed framework conditions, i.e.

$$\sum_i \frac{\left(w_{ic}^0 + \lambda_{icf}^0\right) \cdot x_{icf}^0}{\sum_j \left(w_{jc}^0 + \lambda_{jcf}^0\right) \cdot x_{jcf}^0} \cdot \left(\frac{w_{ic} + \lambda_{icf}}{w_{ic}^0 + \lambda_{icf}^0} - 1\right) = \frac{p_c}{p_c^0} - 1 \qquad (14.4)$$

The first ratio in this expression represents input i's share of total internally perceived cost in production line c, the ratio in parenthesis represents the relative price change on input i, and the expression on the right-hand side represents the relative output price change.

In the present application, the model is extended with crop-specific nitrogen leaching functions, which depend on the input of nitrogen N—and its composition of nitrogen from animal manure N_m and mineral fertilizers N_f—and retention rate. In particular, for crop c on site s with retention rate r_s, nitrogen leaching L_{cs} can be described by the leaching function

$$L_{cs} = (1 - r_s) \cdot e^{((\beta_{1c} + 1{,}2 \cdot \beta_{0c} + N_{fc} + 2 \cdot N_{mc}) / \beta_{2c})} \tag{14.5}$$

where β's are parameters, which depend on crop and soil quality. The leaching functions are parameterized on the basis of field experiment data (Vinter 2011a, b, c; Ørum 2012).

14.2.3 Scenario Design

We consider a quota regulation, where the current horizontal regulation (with quotas determined as 15 % below the economic optimum) is replaced by a differentiated quota, where grass and pulses are allocated a quota corresponding to the economic optimum, beets and seed grasses are given a quota 10 % below the economic optimum and grains and maize get a quota 20 % below economic optimum. In addition to this crop differentiation, the quota is also differentiated according to the vulnerability and retention in the considered catchment, with larger quota reductions in areas with low retention.

In the model analysis, farm types are defined according to

- four types of farming (crop, cattle, pig or mixed farming, dependent on the composition of Standard Gross Margin—SGM)
- two farm size categories—large (SGM > 100.000 DKK) or small (SGM < 100.000 DKK) (1DKK ~ 0.13€)
- three water catchment types (robust—light, semi-vulnerable—medium, or vulnerable—dark), cf. Fig. 14.2

In total, Danish agriculture is thus divided into 21 farm types. The land use structure on the 21 farm types is illustrated in Fig. 14.3. For each of the farm types, a nitrogen quota is determined based on the farm types' crop composition, combined with economically optimal nitrogen norms in the baseline, and the above-mentioned principles for crop-specific quota reductions.

The extent to which the nitrogen quota is perceived as binding on the farms' optimization is represented by the dual value associated with the quota (shadow price). The shadow price indicates how much the farmer's potential profit could increase in the case of a marginal increase in the quota. Hence, the shadow price could also be interpreted as a measure of the producer's marginal willingness to pay for additional quota.

It is however assumed that the authorities cannot monitor the farm's application of nitrogen at the field level. Hence, the farmers may allocate their use of nitrogen differently than the quota. In economic optimum, each farm type will allocate its

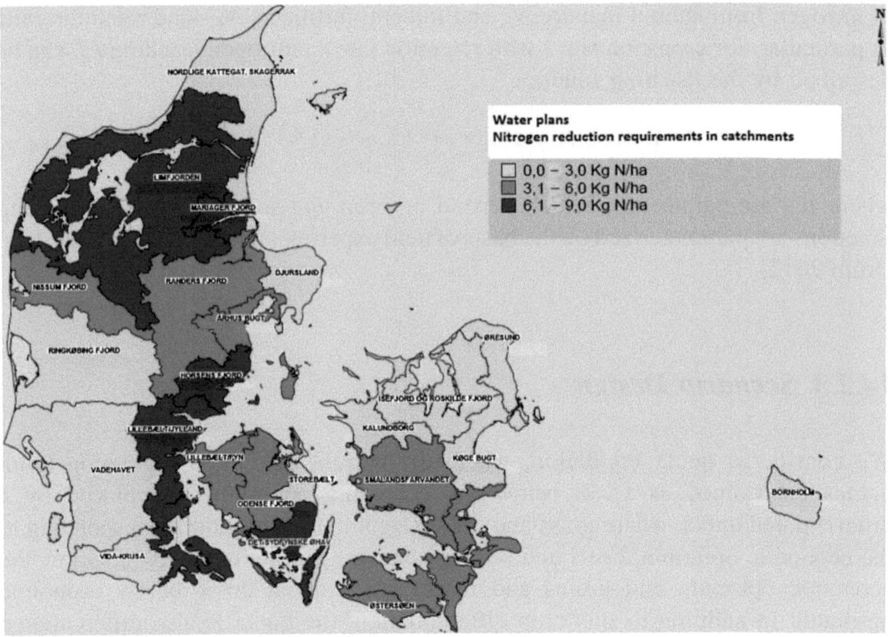

Fig. 14.2 Categorization of water catchments according to vulnerability (Danmarks Miljøundersøgelser 2008)

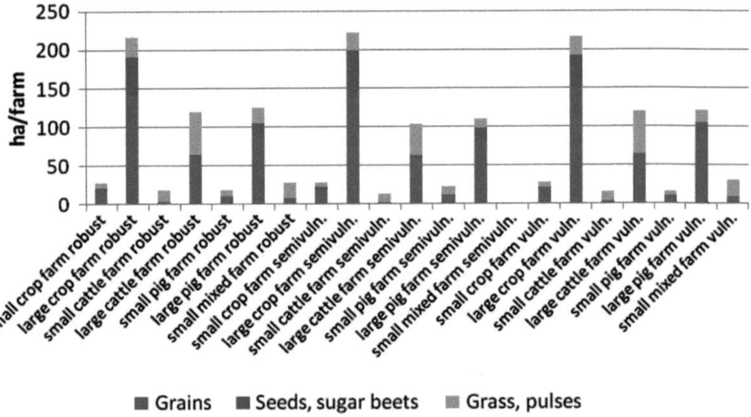

Fig. 14.3 Land use on 21 farm types

nitrogen application such that the dual value of the quota restriction is the same in all crops.

Based on agriculture's farm and production structure in a given base year (2006), ESMERALDA calculates, static-comparatively, how changes in framework conditions affect the farms' land use, livestock, output composition, and input use in agriculture's respective lines of production. But in this chapter, we will focus on the quota revision's effects on the shadow prices of nitrogen quota, with a view to investigating the changes in farmers' incentives introduced by the quota revision, including, e.g. the incentives to trade quota or incentives to circumvent the regulation.

14.3 Results

With regard to nitrogen quota, the market price of nitrogen is considered as exogenous, and the farm's total nitrogen application is determined by the quota (whereas nitrogen within the individual lines of production is determined endogenously). Due to the quota, the model determines the shadow price of the nitrogen quota, and an equilibrium condition states that this shadow price should be equal across lines of production within the farm. Farmers are hence assumed to allocate the nitrogen quota internally on the farm, in order to make marginal returns to the quota equal across lines of production. In a geographical area, inter-farm differences in the shadow price suggest a potential for inter-farm trade with nitrogen quotas, if these quotas are tradable.

In the current horizontal regulation, the quota is particularly binding on farms located on relatively robust soils with high crop yield potential, and less binding in areas with low-yielding (and possibly environmentally vulnerable) soils. Introducing differentiated quota schemes, where quotas are differentiated according to crop type and retention, will change this pattern. Table 14.2 displays changes in calculated shadow prices of nitrogen quotas, measured relative to the market price of nitrogen fertilizer for the considered typology of farms.

As the calculations show, especially crop and pig farms in vulnerable catchments will consider the considered quota model as binding, compared with the baseline situation. The estimated shadow price of nitrogen quota indicates that crop farms in highly vulnerable catchments will be willing to pay an amount corresponding to half of the nitrogen price to be able to expand their use of nitrogen. This follows naturally from the fact that the considered quota model is most binding in these areas. But also in semi-vulnerable areas, crop and pig farms will perceive the quotas as binding. It should also be noted that large farms tend to perceive the quota as more binding than smaller farms.

The geographic localization of farms with high shadow price on nitrogen quota after implementing the considered regulation is illustrated in Fig. 14.4. The figure describes the average change in nitrogen shadow price in 4 × 4 km squares (where the shadow prices within each square are weighted with the area of the involved farms). As the figure reveals, farms with an increase in quota shadow prices tend (not surprisingly) to be located in areas with high vulnerability, cf. Fig. 14.2. Based on this picture, there seems to be a relatively large propensity to trade nitrogen quota, if such trade is possible, even within individual catchments.

Table 14.2 ESMERALDA calculation of the potential for trade with nitrogen quota

	Catchments with …		
	… low vulnerability (%)	… medium vulnerability (%)	… high vulnerability (%)
	Percent of nitrogen fertiliser price		
Small crop farms	0	13	47
Large crop farms	0	19	55
Small cattle farms	0	0	0
Large cattle farms	0	1	0
Small pig farms	0	0	23
Large pig farms	0	14	53
Mixed farms	0	0	0

Source ESMERALDA calculation

And the dark spots in Fig. 14.4 also indicate where the potential economic gains of circumventing the quota, e.g. by illegal trade, would be most likely to occur. Such indications could perhaps serve as an input in designing risk-based control schemes in the monitoring and enforcement of the quota regulation.

A further aspect related to nitrogen quotas at the farm level is to what extent the farms will be prone to re-allocate their nitrogen use within the farm, e.g. by reducing the application of nitrogen in some crops in order to maintain the application of nitrogen in other crops. We attempt to illustrate this in Fig. 14.5, where the calculated reduction in the use of mineral nitrogen fertilizer per hectare is shown for four crops on the 21 farm types: wheat, barley, sugar beets and grass in rotation. The figure ignores farms in the most robust areas, as the quota is assumed to have only a small effect in these areas.

For many of the farm types, the reduction of nitrogen use is almost proportional in the four crops (to the extent that these crops are cultivated on the considered farms). There are however also exceptions to this overall picture, in particular among smaller cattle farms, where reduction in nitrogen application on barley fields seem to be stronger than in these farms' wheat production.

14.4 Discussion

The above calculation is an example of some of the types of analyses that ESMERALDA is capable of doing in relation to effective assessment of nitrogen quotas. It should be mentioned that the shown calculations mainly serve to illustrate the potential uses of the model, rather than as a concrete policy evaluation.

As mentioned in the introduction, an alternative to such differentiated application quota could be quotas on the farms' leaching of nitrogen. The concept of

Fig. 14.4 Geographical localization of farms with different shadow prices on nitrogen quota

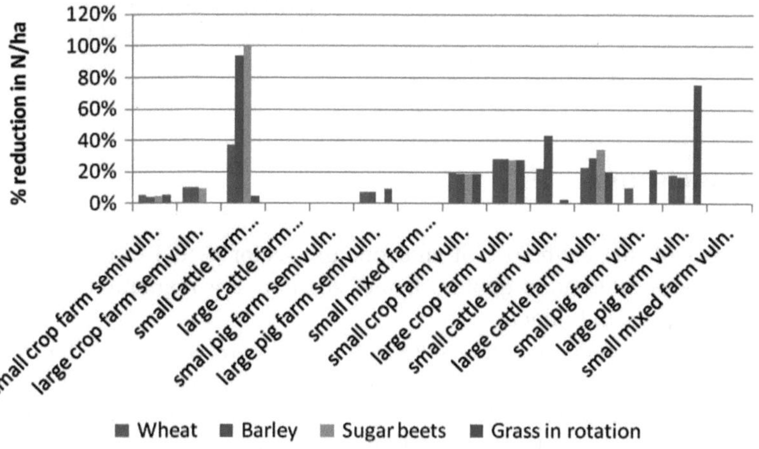

Fig. 14.5 Relative change in nitrogen use per hectare in wheat, barley, sugar beets and grass on different farm types

regulating nitrogen use on the basis of leaching puts some serious knowledge requirements regarding the relationship between nitrogen use and leaching at an aggregate level as well as at the very specific level, such as specific fields on specific locations. Extensive research efforts during the last couple of decades have led to the generation of much important data, models and insights, which are considered useful in this respect. However, there is still some way to go before such data and model frameworks are fully operational for administrative and legislative purposes.

The considerable between-farm differences in the shadow price of nitrogen quota suggests a potential for trade in these quotas. From a purely economic point of view, such trade might bring about the potential to raise economic efficiency of the regulation, and hence be welfare-improving. However, in this regard, some precautions should be underlined. First, there is the conclusion that full flexibility in the allocation of leaching quotas ignores the spatial aspects in the environmental problems related to nitrogen use and leaching, spanning from eutrophication and pollution of surface waters to the protection of ground water resources for drinking purposes. Many of these environmental concerns are highly local, for example the protection of rare and valuable species, or particularly vulnerable aquatic environments, such as lakes, fjords, streams, etc. Hence, the model results illustrate a potential trade-off between concerns for locally specific environmental problems related to nitrogen use on the one hand, and overall cost-effectiveness on the other hand. The applied analytical framework might be developed further to incorporate such concerns, e.g. by modelling site-specific restrictions on local nitrogen leaching in particularly vulnerable locations, and a more detailed modelling of the farm structure in these locations. Although the data framework for modelling agricultural production and economy is in place, such more locally oriented analyses would however pose relatively high demands on data regarding the vulnerability of different locations.

14.5 Conclusion

This chapter illustrates the use of an economic sector model to estimate the cost-effectiveness of alternative quota schemes for nitrogen regulation in agriculture, with a view to spatial effects.

Results of the model analyses suggest that replacing the current flat-rate quota on nitrogen input on all farms with a differentiated quota on nitrogen use (with differentiations based on average leaching properties) implies a redistribution of farmers' economic incentives in relation to the use of nitrogen—geographically and between farm types. Compared with the current flat-rate regulation, this poses new challenges in the monitoring of compliance with the regulation and the potential for trade with quota, which should be held up against the challenges related to the existing quota scheme. The applied model and data framework constitutes a useful tool for analysing such effects, for example as an input to risk-based monitoring. Cost effectiveness in nitrogen leaching reduction may be enhanced further, the more transferable these leaching quotas are—if cost-effectiveness is measured per unit of leaching—but cost-effectiveness measured in relation to the specific environmental effects on the recipients may modify this conclusion.

References

Anselin, L., Bongiovanni, R., & Lowenberg-DeBoer, J. (2004). A spatial econometric approach to the economics of site-specific nitrogen management in corn production. *American Journal of Agricultural Economics, 86*(3), 675–687.

Børgesen, C. D., Kristensen, I. T., & Grant, R. (2009). Landbrugsregisterdata anvendt i regionale og landsdækkende beregninger af N og P tab (Agricultural register data applied in regional and national analysis of N- and P-losses). In C. D. Børgesen, J. Waagepetersen, T. M. Iversen, R. Grant, B. H. Jacobsen & S. Elmholdt (Eds.), Midtvejsevaluering af vandmiljøplan III: Hoved- og baggrundsnotater. red., University of Aarhus, Faculty of Agricultural Sciences, 2009, pp. 77–96 (DJF Rapport Markbrug; 142). http://pure.au.dk/portal/files/2841678/djfma142.pdf.pdf.

Danmarks Miljøundersøgelser. (2008). Nitratklassekortværk. Opdatering af nitratklassifikationen i Danmark. 2. generation. Fagligt notat af 23. oktober, 2008. Danmarks Miljøundersøgelser, Aarhus Universitet.

European Commission. (2000). Directive 2000/60/EC of the European Parliament and of the Council establishing a framework for the Community action in the field of water policy.

GLR/CHR. (2007). www.glr-chr.dk.

Institute of Food and Resource Economics. (2006). Economy of Agricultural Enterprises 2005 (Fødevareøkonomisk Institut, Økonomien i landbrugets driftsgrene 2005).

Institute of Food and Resource Economics. (2007). Agricultural Accounts Statistics 2006 (Fødevareøkonomisk Institut, Landbrugsregnskabsstatistik 2006).

Jacobsen, B. H., Abildtrup, J., Andersen, M., Christensen, T., Hasler, B., Hussain, Z. B., Huusom, H., Jensen, J. D., Schou, J. S., & Ørum, J. E. (2004). Omkostninger ved reduction af landbrugets næringsstoftab til vandmiljøet.

Jensen, J. D. (2001). A regional Econometric Sector Model for Danish Agriculture—A Documentation of the Regionalized ESMERALDA Model, FOI rapport nr. 129.

Jensen, J. D., & Kristensen, I. T. (2013). Geographical patterns in the agricultural economy—The Case of Denmark, Danish. *Journal of Geography* (forthcoming).

Kristensen, I. T., & Kristensen, I. S. (2004) Farm types as an alternative to detailed models in evaluation of agricultural practise in a certain area. In C.A. Brebbia (Ed.) *Management Information Systems 2004: Incorporating GIS and Remote Sensing*. 1-85312-728-0. Udg, pp. 241–250. WIT Press, 2004. http://library.witpress.com/pages/PaperInfo.asp?PaperID=14201.

Kristensen, I. T., Rasmussen, B. M. (2002). Determining the geographical variation in Danish farming. In *Proceedings of Third International Conference on Management Information Systems Incorporating GIS & Remote Sensing,* pp. 35–42. WIT Press, 2002. http://library.witpress.com/pages/PaperInfo.asp?PaperID=70.

Kristensen, I. T., Rasmussen, B. M., Staunstrup, J. K., Mouritsen, A. K. M., Bøcher, P. K., Hansen, J. F., & Kragelund, A. (2005). Den geografiske variation i nøgleparametre for dansk landbrug : Metoder og hovedresultater, (The geographical variation in key parameters for Danish agriculture: Methods and main results) 128 pp. (DJF Intern rapport; 222). http://djfgeodata.dk/datasaml/Litteratur/intern_rapport.pdf.

Ørum, J. E. (1999). Driftsøkonomiske konsekvenser af en pesticidudfasning—optimal pesticid- og arealanvendelse for ti bedriftstyper i udvalgte scenarier, Statens Jordbrugs- og Fiskeriøkonomiske Institut, rapport nr. 107.

Ørum, J. E. (2012). Udbytte og udvaskning for vinterhvede og vårbyg med og uden normkorrektion, Notat til kvælstofudvalget 8. maj 2012, Fødevareøkonomisk Institut, Københavns Universitet.

Pedersen, H. B. (2006). Landbrugsregnskabsstatistik 2005—udvælgelse, beregninger og definitioner, Fødevareøkonomisk Institut. www.foi.life.ku.dk.

Schou, J. S., Skop, E., & Jensen, J. D. (2000). Integrated agri-environmental modelling: a cost-effectiveness analysis of two nitrogen tax instruments in the Vejle Fjord watershed, Denmark. *Journal of Environmental Management, 58*(3), 199–212.

Videncentret for Landbrug. (2010). Oversigt over Landsforsøgene 2010. https://www.landbrugsinfo.dk/Planteavl/Landsforsoeg-og-resultater/Oversigten-og-tabelbilaget/Sider/pl_Oversigten_2010_Afsnit_N_Goedskning.pdf?download=true.

Vinter, F. P. (2011a). Mulighed for rangordning af afgrøder efter deres udvasknings-karaketeristika. Institut for Agroøkologi, Aarhus Universitet.

Vinter, F. P. (2011b). Notat om marginaludvaskning med N-LES3. Institut for Agroøkologi, Aarhus Universitet.

Vinter, F. P. (2011c). Udvaskning fra sædskifter beregnet med FarmN(N-LES3) og FASSET. Institut for Agroøkologi, Aarhus Universitet.

Author Biographies

Jørgen Dejgård Jensen is an Associate Professor at the Department of Food and Resource Economics at the University of Copenhagen. He is an economist by training and has many years' experience in agricultural sector modelling.

Jens Erik Ørum is a senior advisor at the Department of Food and Resource Economics at the University of Copenhagen. He is educated as agricultural economists and he is a specialist in the economics of crop production and crop protection.

Chapter 15
Conservation Agriculture as a Driving Force to Accumulate Carbon in Soils: An Analysis of RDP in Lombardy

Stefano Corsi, Stefano Pareglio, Marco Acutis, Andrea Tosini, Alessia Perego and Andrea Giussani

Abstract Carbon Dioxide (CO_2) emission credits and C-sequestration are measures that are largely applied to limit the rising concentration of CO_2 in the Earth's atmosphere. In this context an increasing role is played by conservation agriculture (CA). This chapter aims to present the policies pursued in Lombardy and to calculate, with the soil CN-cycle model ARMOSA, the potential of C-storage in soils with the adoption of CA measures for 20 years. The analysis is performed on 600 farms (24,550 ha), and it is implemented here taking into account the economic incentive provided by the 2007–2013 Rural Development Program (RDP) of Lombardy. The results show that C-accumulation in soils by CA can contribute to achieve Kyoto targets, but it needs a significant economic effort. Suggestions for policy-makers are here briefly outlined in relation to similar policies applied at the international level.

15.1 Introduction

Agriculture and forestry play a key role in producing public goods, notably environmental such as landscapes, farmland biodiversity, climate stability, and greater resilience to natural disasters, such as flooding, drought, and fire. At the same time, many farming practices have the potential to put pressure on the

S. Corsi · A. Tosini (✉)
Department of Economics, Management and Quantitative Methods, University of Milan, Milan, Italy
e-mail: andrea.tosini@gmail.com

S. Pareglio
Department of Mathematics and Physics, Catholic University of Sacred Heart, Milan, Italy

M. Acutis · A. Perego · A. Giussani
Department of Agricultural and Environmental Sciences, University of Milan, Milan, Italy

C. Zopounidis et al. (eds.), *Agricultural Cooperative Management and Policy*, Cooperative Management, DOI: 10.1007/978-3-319-06635-6_15,

environment, leading to soil depletion, water shortages and pollution, loss of wildlife habitats and biodiversity (COM(2010) 672/5). The increase in the atmospheric concentration of CO_2 and in CO_2 equivalent emissions are now considered a global concern and indicators of the ongoing global warming. It has been estimated that agriculture accounts for 14 % of all anthropogenic emissions into the atmosphere (Beach et al. 2008).

Farmers have a significant impact on soil carbon sink and altering their management practices can reduce losses and increase the absorption of carbon by protecting soils from erosion. It has been estimated that European soils of the cropland could sequester between 50 and 100 million tons of carbon annually, by adopting agricultural practices to reduce the loss of organic carbon from the soil and the use of machinery (EU-Comm 2009).

In compliance with the Kyoto Protocol (UN 1998), Europe has developed policies to support greenhouse gas emissions reduction. Directive 2003/87/EC established a scheme for greenhouse gas emission allowance trading within the Community (Emissions Trading Scheme (ETS)) in order to promote reductions of greenhouse gas emissions in a cost-effective and economically efficient manner (Article 1), and it relates mainly to the energy and industrial sectors. The climate and energy package (20-20-20) in 2008 and Decision No. 406/2009/EC of the European Parliament and of the Council of 23 April 2009 provide for the evaluation and implementation of a more rigorous commitment of the Community in the field of emission reductions, aiming to ensure the European Union meets its ambitious climate and energy targets for 2020. Other European Community policies have been prepared for the containment of greenhouse gasses such as energy efficiency and use of renewable sources. The so-called "Effort Sharing Decision" (406/2009/EC) sets binding annual targets in terms of emissions of greenhouse gasses for every Member State for the period 2013–2020 related to areas not included in the EU ETS (Emissions EU-ETS), such as transport, buildings, agriculture, and waste. The total share fixed at the European level of abatement of emissions from these sectors for 2020 is equal to 10 % compared to 2005. This reduction in emissions, added to the dimension reduction coming from the ETS sectors, should allow to achieve the objectives of 20-20-20 (EU-Climate Energy 2007).

European, national, and regional policies support directly or indirectly the agricultural and forestry practices for GHG emissions reduction: agricultural and forestry practices that affect carbon sequestration are no tillage, minimum tillage (Lal 2004), energy crops, and permanent crops, such as meadows and pastures, the incorporation of crop residues into the soil, conversions from land or abandoned land to forest and energy crops, and the adoption of organic farming techniques (Freibauer et al. 2004). The importance of soils in climate change mitigation is emphasized both in the implementation of the commitments of the Kyoto Protocol and in the priority areas of the EU Common Agricultural Policy (CAP) as well as in the document that directs the choices for the future CAP (2014–2020): it is known that it ' *... is important to further unlock the agricultural sector's potential to mitigate, adapt and make a positive contribution through GHG emission* *reduction, production efficiency measures including improvements in energy efficiency, biomass and renewable energy production, carbon sequestration and protection of carbon in soils based on innovation.*' (EU-Comm 2010). Agricultural practices could help to sequester between 50 and 100 million tons of carbon annually in European soils (COM, 2006, IP/06/1241).

The requirement for the agricultural sector to intensify efforts to reduce GHG emissions in the framework of EU strategy on climate change is also mentioned in the Regulations 74/2009/CE on support for rural development (Health Check), which requires the adoption of specific measures for the reduction of GHG emissions addressed from 2010.

Conservation agriculture and in particular minimum tillage, no tillage, and sod seeding allow us to reduce CO_2 emissions from the soil due to a decrease in soil layer mixing. Moreover, cover crops and crop residues allow lower losses of soil due to the reduced impact of rain drops and of erosive action produced by the wind; the top cover also improves infiltration and retention of soil moisture and mitigates the temperature variations on and in the soil.

Conservation Agriculture is characterized by three linked principles: continuous minimum mechanical soil disturbance, permanent organic soil cover, and diversification of crop species grown in sequences and/or associations (FAO 2013).

The advantages that FAO (FAO 2013) identifies in conservation agriculture relate primarily to the reduced loss of soil and consequently a lower loss of pollutants in water also due to the coverage, air protection, thanks to a lower level of emissions of CO_2, CH_4, and N_2O, and factors related to manure injection, and not overturning of the layers maintaining and rebuilding soil architecture. Indirect effects are the accumulation of carbon in soils, an improving presence of soil fauna and biological activity through the recycling of organic matter and plant nutrients, and therefore a greater biodiversity. Furthermore, conservation agriculture techniques enable the reduction of production costs, e.g., fuel, machinery operating costs, and maintenance. The disadvantages, mostly concentrated in the transition from a conventionally tilled system to conservative agriculture, are mainly related to the initial investment in specialized machinery and the use of appropriate/improved seeds adapted to local conditions.

At the beginning of the tillage conversion, farmers need greater amounts of herbicides (Pisante 2007) being careful not to create negative conditions for soil organisms (microbes and soil fauna); runoff is one of the primary mechanisms of contamination of surface waters by herbicides (Krutz et al. 2005). However, the techniques of conservation agriculture, which allow the reduction of runoff, can break down these losses. Subsequently, the weed control can also be managed through rotations and crop residues as well as different seeding time. A central role is played by technical support and training to farmers, compared to conventional till farming, which needs a radical change in approach and management in particular as regards the control of weeds.

The advantages and disadvantages reveal the divergence between the social desirability of conservation agriculture and its potential attractiveness to individual farmers. While many of the costs associated with the exchange of tilling techniques fall at the farm level, most of the benefits relate to the production of public environmental goods (Knowler et al. 2007). Without policies and funding to farmers the adoption of conservation techniques will be a function of perceived profitability at the farm scale.

The European Rural Development Programme (RDP) aims to protect and enhance the rural environment and contribute to the development of a competitive and sustainable farm, improving quality of life of rural communities; it is split into three main areas linked to Farming and Food (also known as Axis 1), Environment and Countryside (also known as Axis 2) and Rural Life (also known as Axis 3). The Lombardy Region supports conservative agriculture through a specific measure of the RDP (measure 214 action M).

The work aims to estimate the amount of carbon stored in the soil due to a change in the management of agricultural techniques from conventional to conservation and to assess the amount of funding necessary to achieve predetermined objectives of storage. Analysis of the contribution will also be compared with similar policies at the international level.

The work is part of "AgriCO_2ltura," a research project coordinated by ERSAF (regional entity for services to agriculture and forests) funded by the Direzione Generale Agricultura ed Ambiente of the Lombardy Region, which has the purpose to verify, under different soil and climatic conditions of the plain, if conservation agriculture techniques allow accumulation of carbon in soils, reduce CO_2 emissions in the atmosphere, and promote the conservation of soil biodiversity as compared to conventional tillage. The project follows several issues; from the study of storage and emission of carbon in cultivated soils as a function of the different soil and climatic conditions and farming techniques, to the identification of regional deposits of carbon in agricultural soils, to the possibility of using methodologies or techniques to achieve carbon (CO_2) balance, and greenhouse gas intensity of Lombard agricultural systems. The research project compares the analysis outcome with the impact of EU and regional policies related to carbon storage in soils.

15.2 Case Study

From 2010 the Lombardy Region has been funding the activity of farmers who decided to introduce and manage all or part of their land through conservation agriculture. The measure has the main objective to increase the amount of carbon in soils by counteracting the adverse side effects resulting from the simplification of cropping systems and the intensive management of the soil due to deep plowing with inversion of the soil layers and repeated periods of bare soil and tillage, such as CO_2 emissions, high energy consumption, reduction of biodiversity, and soil fertility (organic matter reduction, increased erosion, in particular solids transposed in the plains, compaction, or sealing). The policy is applied to arable land of the region as a result of the conditions shown in Table 15.1 for at least 5 continuous years. For farms to join the financing, they must guarantee a minimum area

Table 15.1 The amount of subsidies paid to farmers who have responded to calls of measure 214 action M of RDP for the use of conservation techniques for the area concerned

Subsidies (€/ha/year)	Techniques
208.00	Direct seeding (SD)
290.00	Direct seeding + cover crop
278.00	Direct seeding + direct injection of sewage farming
360.00	Direct seeding + cover crop + direct injection of liquid manure
190.00	Minimum tillage
272.00	Minimum tillage + cover crop
260.00	Minimum tillage + direct injection of sewage farming
342.00	Minimum tillage + cover crop + direct injection of liquid manure

of 1 ha and in any case not less than 10 % of the total area of the holding. The payments are disbursed according to the areas covered by conservation practices and contribution is described in Table 15.1.

The development of integrated policies for the promotion of the agricultural sector as part of an action to combat climate change requires extensive research, and data concentration of carbon in soils subjected to different forms of management and cultivation in the presence of soil and climatic conditions, and in turn, different evolutionary dynamics and trends in the regional context. It is also necessary to show at the field scale that the conservative techniques are able to be actually effective in influencing climate change, certifying variations in the carbon content in soils and the reduction of emissions, so that the environmental benefits and credits become potentially recognizable.

Data of the participation of farmers on conservative agriculture policy in the Lombardy Region were collected with the help of the regional offices and the presidency that provided the agricultural information system (Agricultural Information System of the Lombardy Region (SIARL)), a database that collects the company files describing agricultural enterprises falling in the region, which prepared for the call for financing of the CAP and RDP. Data on farmers' participation in 2011 and 2012 were required to perform the analysis. The cropland involved in the tillage conversion was about 1 % of the regional farmers and about 3 % of the agricultural regional area (UAA) (8,306 ha in 2011 and 24,450 ha in 2012, while the UAA TOT of the Lombardy Region is 986,853 ha (ISTAT 2013)). The amount of the payments was 2,039,522.25 € for the first year (2011) and 5,721,607.44 € for the second (2012). Figure 15.1 displays the distribution of funding over the Lombard plain.

In order to estimate the organic carbon sequestration in arable soil, we first identified the main cropping systems adopted under the pedoclimatic conditions of the Lombardy plain. The land use data at the cadastral scale were derived through linking of the regional databases. Data were then aggregated at agrarian regions, which is meant as a territorial subdivision consisting of a few neighboring municipalities being homogeneous in terms of land use and pedoclimatic conditions.

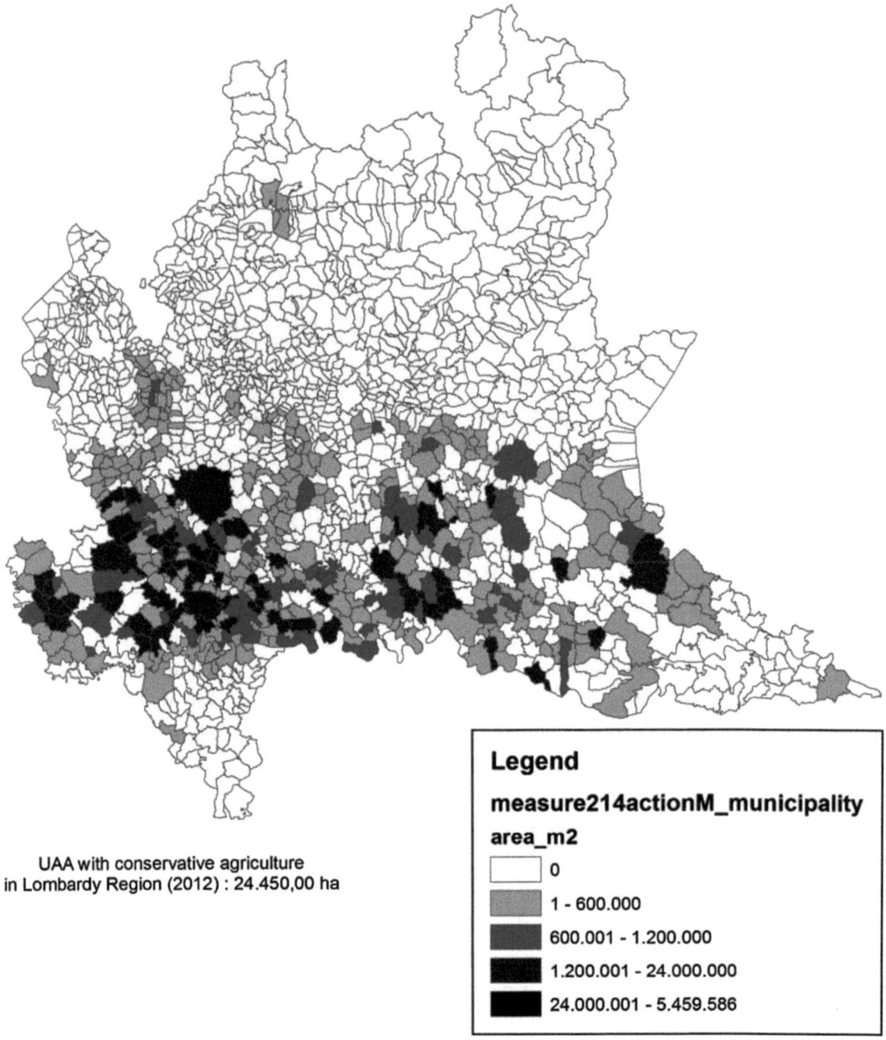

Fig. 15.1 Municipalities using conservation agriculture in 2012. The black municipalities receive more funding and have more surfaces affected by conservative techniques

We identified 214 types of land use, which were aggregated in 17 groups taking into account the possible double cropping system. For each agrarian region a number of 1–9 herbaceous crops was therefore obtained through the analysis, being representative of about 85 % of the Utilized Agricultural Area (UAA) as reported in Table 15.2; such information was used to derive the type of the cropping system adopted in the studied area (Fig. 15.2).

The carbon balance in the soil was calculated on the basis of the output variables simulated by the crop simulation model (Acutis et al. 2007; Perego et al. 2011,

Table 15.2 Representative cropping systems of the Utilized Agricultural Area (UAA cropland and permanent grassland) of the agrarian regions in Lombardy plain

	ID	UAA (ha)	
Permanent grassland and grazing	PP	5,232	
Grassland in rotation with maize	PVM	57,908	
Grain maize (possible cover-crop)	MG	192,995	
Silage maize (possible cover-crop)	MF	55,286	
Wheat (possible cover-crop)	F	26,048	
Maize–Wheat	M_F	92,134	
Grain maize–Soybean	MG_L	14,089	
Alfalfa–Maize–Wheat	Med	70,648	
Rice	R	102,656	
Fruit trees (*not simulated*)	A	16,193	
Open-field vegetables (*not simulated*)	O	7,880	
Total UAA		641,068	
Total UAA related to the simulated systems		514,338	80 %

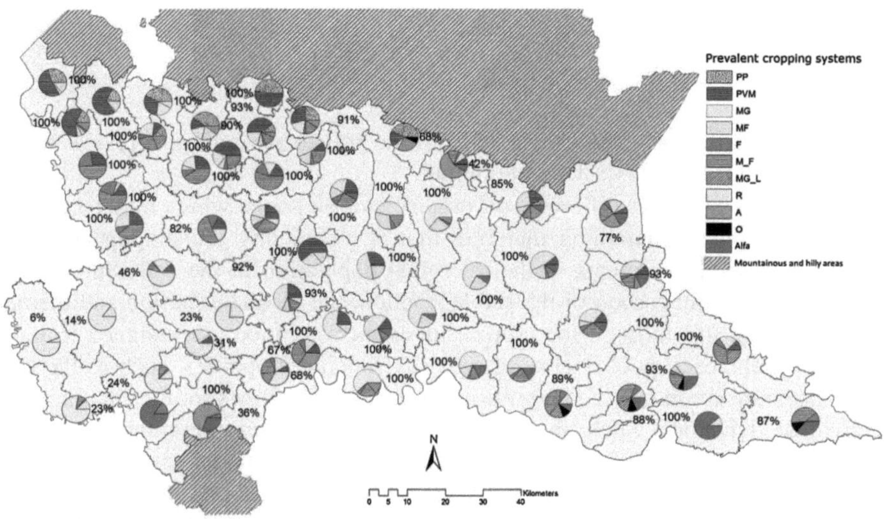

Fig. 15.2 Representative cropping systems in the agrarian regions in Lombardy plain. The current Utilized Agricultural Area (UAA %) related to the simulated systems is displayed in each agrarian region

2013) applied under the cropping systems, which were previously identified, comparing the two techniques of agronomic management under examination, conventional, and conservative.

ARMOSA simulates the crop growth and development together with water, carbon, and nitrogen dynamics in the soil-plant atmosphere continuum at daily

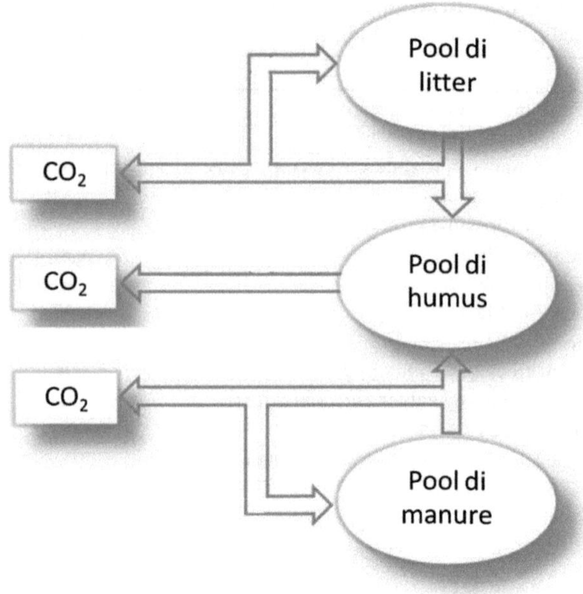

Fig. 15.3 The three types of organic carbon pool implemented in the ARMOSA crop simulation model

time-steps. Particularly the CO_2 loss from the soil to the atmosphere via mineralization is calculated as a variable output.

The model simulates different soil organic carbon pools, which may be defined as a compartment containing material, that is chemically indistinguishable and equally accessible by plants or the microbial population in the soil (Smith et al. 2002). The organic carbon of the stable fraction, the fertilizer-carbon and the one applied to the soil through the crop residues incorporation are assigned to independent pools; each pool is characterized by specific properties, such as the C:N ratio, decomposition rate, and humification rate, which allows for the simulation of individual dynamics. Such an approach leads to a better representation of the actual conditions occurring in the soil. The ARMOSA model implements three type of pools of which two are characterized by a quicker rate of decomposition (30 up to 400 days), named "litter" (L) and "manure-derived-feces" (M), which represent the crop residues and the fertilizer contribution, respectively. As a consequence, the organic carbon of any fertilizer application and crop residues incorporation is assigned to an independent pool of the M or L type. In particular, the decomposition rate of the plant residues is a function of the crop type and organ (i.e., stem, leaves, root, and storage). The third type of organic carbon pool, "humus" (H) is the one characterized by a slower decomposition rate, being the stable fraction of the organic matter in soil (Fig. 15.3).

The carbon balance is computed as the difference between the input data and the simulated output of the carbon-related variables. The input data are: (i) the atmospheric CO_2-C fixed via photosynthesis, (ii) the addition of C through manuring, and (iii) the amount of C contained in the crop residues. The output data

Table 15.3 Estimated contribution to be allocated to agrarian regions for the portion of land cultivated according to conservative agriculture if it were financed by the action M measure 214 for 20 years (conv: conventional tillage, cons: conservative tilling)

Agrarian region	Prevailing cluster	Soil carbon T_0 OC (ton)	UAA (ha)	Soil carbon T_0 OC (t/ha)	% OC from conv to cons	5 % UAA (ha)	Payments T_{20} (190 €/ha/y)	Carbon stored T_{20} OC (ton)
12-03	6	83,324.12	1,058.25	78.74	0.47 %	52.91	201,067.58	388.48
12-04	6	138,277.23	1,757.88	78.66	0.59 %	87.89	333,996.74	817.07
12-05	6	108,517.13	1,374.38	78.96	0.61 %	68.72	261,132.20	659.88
12-06	6	200,815.35	2,541.14	79.03	0.54 %	127.06	482,817.46	1,078.79
13-09	6	198,733.51	2,531.76	78.50	0.41 %	126.59	481,034.93	810.82
13-10	6	143,082.04	1,839.69	77.78	0.25 %	91.98	349,541.04	358.17
13-13	6	216,889.58	2,794.40	77.62	0.42 %	139.72	530,935.96	917.51

are: (i) the C content of the harvested biomass and (ii) the C mineralized by the microbial biomass. Moreover, the difference in C content of the M, L, and H pools are items of the balance. The C balance allows for the estimation of the increase or decrease of the soil C content over the years of simulation.

To determine funding relative to each simulated scenario, the entities of the annual compensation of the measure 214 action M (conservative agriculture) have been included. Two simulations were carried out with different levels of contribution, one with a contribution of 190 €/ha, linked to minimum tillage, the second with 260 €/ha, relating to conservative agriculture with minimum tillage, and direct injection of effluent farming or manure burying. The model considers rotations for most of the agricultural regions and the possibility of the manure incorporation into the soil. These two scenarios of compensation were simulated by the ARMOSA model. This model is not meant to simulate the sod seeding system, so that the minimum tillage scenarios were chosen.

ARMOSA was parameterized for simulating the two different tillage systems. For the conservation scenario, the depth of tillage was limited to a depth of 15 cm without crop residual soil incorporation, equal to the minimum tillage (M measure), which determines a least soil disturbance and leaves the maximum of crop residue on the soil surface. For the conventional tillage scenario the model was defined with the plowing depth for each crop and at the kind of crop residues to optimize incorporation; the tillage depth varied from 0 cm for meadow grass to 30 cm for maize. The ARMOSA parameterization was made according to Oorts et al. (2007): the rate of C decomposition of organic carbon was reduced by 30 % under the conservative scenario with respect to traditional tillage. For the two tillage systems the crop management was defined with details of fertilization and manure (time and amount), planting, and harvest dates, tillage depths for conventional tillage, and crop residue management.

The model results showed a significant improvement of soil carbon amount ($p < 0.01$) from TA to CA for all crop rotations. The carbon sequestration potential ranges from 0.1 to 0.48 t C/ha/year according to rotation and soil type.

A lot of experiments confirm these results; Freibauer et al. (2004) showed a potential carbon sequestration: for CA the range of year sequestration potential is between 0.1 and 0.5 t C/ha. From these simulations an example of the simulations and the results obtained with the two types of financing is reported in Table 15.3.

For each scenario it was possible to estimate the contribution that farmers should receive for 20 years of conservative agriculture keeping unchanged the values of the financing of the 2012 deadline call for action measure 214 M. From these simulations amounts were deducted for annual funding related to each percentage of land cultivated with conservative agriculture and the effectiveness in the storage of C.

15.3 Results

Applying the ARMOSA model under territorial analysis it was possible to calculate the carbon balance in the different scenarios identified comparing the two techniques of agronomic management, conventional, and conservative.

Table 15.3 reports the agrarian regions involved in the action measure 214 M with the simulated amount of carbon stored under conservative tillage over 20 years. Table 15.4 shows the estimated value of the storage of carbon (C) in the soil for 20 years under conservative techniques in the UAA currently involved in measure 214 M in 2012.

Table 15.5 shows the model outcome under the first scenario (190 €/ha) considering the increasing area involved in the measure 214 M, which corresponds to a different amount of contribution. Table 15.5 reports the analysis outcome under the scenario of conservative techniques used in up to a maximum of 15 % of the UAA. The analysis took into account only a minimum tillage with a subsidy of 190 €/ha.

Through the modeling analysis it was possible to estimate the amount of carbon stored in the soil for which the grant is meant to be constant. In particular funding of 535.23 € would be paid for 1 ton of carbon stored, which corresponds to 145.84 € for a ton of CO_2 not emitted.

On the basis of the UAA (ha) cultivated according to the practices of conservative tillage the total fund for 20 years can be determined multiplying the compensation for the chosen type of invitation extended to 2012, and 20 years for the simulated hectares. The carbon stored in soils can be put in relation with the CO_2 emission (rate of conversion from C to CO_2; 3.67).

In Table 15.5, the columns €/t C incorporated and €/t CO_2 not emitted were included to show the possible financing granted for any unit of C stored per unit or per unit of CO_2 emission by the European policies. The last column shows the annual funding that should be provided for the amount of land concerned by conservative techniques.

Table 15.4 Utilized Agricultural Areas (UAA) where conservative tillage is adopted and carbon is stored in soils for 20 years

Agrarian region	Prevailing cluster	Soil carbon T_0 CO (ton)	UAA (ha)	Soil carbon T_0 CO (t/ha)	% UAA under 214 M	% C from conv to cons	Carbon stored T_{20} C (ton)
12-02	6	98,055	1,245	78.74	0.36 %	0.00 %	3.60
12-03	6	83,324	1,058	78.74	0.0 %	0.00 %	0.52
12-04	6	138,277	1,758	78.66	0.0 %	0.00 %	0.00
12-05	6	108,517	1,374	78.96	3.5 %	0.42 %	455.46
12-06	6	200,815	2,541	79.03	2.7 %	0.28 %	571.75
13-01	6	406,351	5,177	78.50	0.01 %	0.18 %	722.62
13-09	6	198,734	2,532	78.50	5.7 %	0.47 %	931.67
13-10	6	143,082	1,840	77.78	0.2 %	0.01 %	16.36
13-13	6	216,890	2,794	77.62	9.1 %	0.77 %	1,664.02
14-05	6	410,236	5,285	77.62	0.01 %	0.18 %	729.53
14-06	6	726,323	9,358	77.62	0.04 %	0.18 %	1,291.64
15-01	6	122,990	1,587	77.48	1.3 %	0.17 %	206.82
15-02	6	259,920	3,284	79.14	2.9 %	0.26 %	676.22
15-03	6	201,115	2,591	77.62	2.5 %	0.29 %	586.05
15-04	6	450,429	5,757	78.24	2.3 %	0.24 %	1,068.21
15-05	3	461,048	7,444	61.94	6.9 %	0.75 %	3,478.93
15-06	3	250,927	4,268	58.79	13.0 %	0.95 %	2,384.59
15-07	3	1,014,380	16,786	60.43	6.0 %	0.56 %	5,648.03
15-08	3	1,231,057	20,924	58.84	14.1 %	0.80 %	9,900.57
15-09	4	41,737	645	64.68	2.6 %	0.22 %	93.48
16-06	6	129,323	1,639	78.88	0.0 %	0.00 %	0.00
16-07	6	160,855	1,995	80.61	0.0 %	0.00 %	0.00
16-08	6	165,325	2,178	75.92	0.0 %	0.00 %	0.00
16-09	3	969,130	16,415	59.04	0.7 %	0.06 %	567.60
16-10	3	765,541	12,297	62.25	1.6 %	0.16 %	1,188.71
17-10	1	430,461	5,792	74.32	2.2 %	0.14 %	582.32
17-11	1	806,549	11,121	72.52	1.0 %	0.05 %	404.63
17-12	3	1,384,142	22,063	62.74	3.3 %	0.35 %	4,828.81
17-13	4	1,866,536	26,215	71.20	2.5 %	0.38 %	7,142.18
17-14	1	2,692,925	35,315	76.26	1.4 %	0.13 %	3,388.53
18-02	1	949,705	13,568	70.00	0.3 %	0.00 %	12.25
18-04	1	2,107,495	29,773	70.79	7.4 %	0.08 %	1,770.35
18-05	2	1,781,732	28,231	63.11	3.9 %	0.06 %	1,046.09
18-06	2	283,459	4,454	63.64	12.6 %	0.41 %	1,173.16
18-07	3	1,058,832	18,276	57.93	16.1 %	0.54 %	5,704.82
18-08	1	776,398	10,733	72.34	8.3 %	0.30 %	2,363.91
18-09	1	662,802	9,172	72.26	0.6 %	0.02 %	137.79
18-10	1	1,120,894	15,337	73.08	2.1 %	0.24 %	2,699.25
18-11	4	677,864	10,036	67.54	9.3 %	0.97 %	6,565.73

(continued)

Table 15.4 (continued)

Agrarian region	Prevailing cluster	Soil carbon T_0 CO (ton)	UAA (ha)	Soil carbon T_0 CO (t/ha)	% UAA under 214 M	% C from conv to cons	Carbon stored T_{20} C (ton)
19-01	3	489,215	8,191	59.73	1.7 %	0.19 %	950.25
19-02	4	1,467,548	20,931	70.11	3.3 %	0.55 %	8,067.38
19-03	4	536,898	7,658	70.10	2.1 %	0.32 %	1,691.29
19-04	4	1,387,355	18,987	73.07	6.0 %	1.09 %	15,085.34
19-05	4	1,425,413	19,985	71.32	4.3 %	0.67 %	9,592.69
19-06	4	1,340,323	18,573	72.17	1.7 %	0.29 %	3,912.54
19-07	1	1,423,385	18,779	75.80	2.5 %	0.23 %	3,224.42
20-01	4	698,182	10,079	69.27	4.3 %	0.58 %	4,053.24
20-02	1	1,952,352	25,408	76.84	0.8 %	0.09 %	1,662.12
20-03	4	1,418,170	20,272	69.96	0.7 %	0.10 %	1,378.70
20-04	1	1,430,076	18,873	75.77	1.0 %	0.09 %	1,276.79
20-05	4	1,290,810	18,467	69.90	0.0 %	0.00 %	0.00
20-06	1	1,168,950	15,786	74.05	0.0 %	0.00 %	0.00
20-07	1	1,351,872	18,266	74.01	0.0 %	0.00 %	43.67
20-09	4	216,471	3,420	63.29	3.2 %	0.14 %	298.87
97-04	6	28,352	353	80.27	0.0 %	0.00 %	0.00
97-05	6	294,598	3,783	77.87	0.0 %	0.01 %	15.82
98-01	3	1,059,516	17,109	61.93	5.9 %	0.61 %	6,486.67
98-02	4	1,196,428	16,354	73.16	4.5 %	0.94 %	11,212.56
98-03	4	734,093	9,888	74.24	5.5 %	1.09 %	7,998.59
						TOT (ton)	146,957.16

Table 15.6 shows the model outcome related to the financing for the minimum tillage coupled with the manure and crop residues incorporation into the soil, which corresponds to a payment of 260 €/ha.

The estimated subsidies appeared to be pretty high, if we take into account the public financial resources allocated to the Rural Development Program for the Lombardy Region for the whole period 2007–2013, which amounted to 1.025.193.491 € (Mid-term evaluation of the RDP 2010). Such an amount includes 503.958.147 € for axis 2 and 273.797.954 € to the measure 214 (Mid-term evaluation of the RDP 2010). Considering that the funding program lasts 7 years, and that the whole budget should be divided on 4 axes, it is reasonable to think that between 6 and 12 million euro (estimated from previous levels of funding and hypotheses A and B) can be allocated to measure 214 M. In the Lombardy Region Sub-Measure 214 action M is one of the ten in which the measure 214 RDP is split.

If the funding amount of the next rural development program (2013–2020) remains similar to the current, as suggested by the press of the European Commission, it is conceivable to allocate 5–10 % of the Lombardy UAA land to conservative agriculture for hypothesis A (only minimum tillage, 190 €/ha) or 5 % of Lombardy UAA for hypothesis B (260 €/ha).

The new European agricultural policy (CAP and RDP) is under discussion and it is unclear as to what might be the amount of contributions to farmers. However,

Table 15.5 Hypothetical grant (hypothesis A) of measure 214 action M for surfaces cultivated with conservative agriculture—Minimum tillage (190 €/ha)

% UAA (55 agrarian region)	UAA ha (55 agrarian region)	Funding € for 20 years (RDP)	C stored (t)	Ton CO_2 eq.	€/t OC stored	€/t CO_2 not emitted	€/year
5.00 %	32,147.83	122,161,736.42	228,242.83	837,651.18	535.23	145.84	6,108,086.82
10.00 %	64,295.65	244,323,472.85	456,485.66	1,675,302.36	535.23	145.84	12,216,173.64
15.00 %	96,443.48	366,485,209.27	684,728.49	2,512,953.54	535.23	145.84	18,324,260.46

Table 15.6 Hypothetical grant (hypothesis B) of measure 214 action M for surfaces cultivated with conservative agriculture—Minimum tillage coupled with the burying of manure (260 €/ha)

% UAA (55 agrarian region)	UAA ha (55 agrarian region)	Funding € for 20 years (RDP)	OC stored (t)	Ton CO_2 eq.	€/t C stored	€/t CO_2 not emitted	€/y
5.00 %	32,147.83	167,168,691.95	228,242.83	837,651.18	732.42	199.57	8,358,434.60
10.00 %	64,295.65	334,337,383.89	456,485.66	1,675,302.36	732.42	199.57	16,716,869.19
15.00 %	96,443.48	501,506,075.84	684,728.49	2,512,953.54	732.42	199.57	25,075,303.79

the outstanding tendency is to promote the agricultural practices, which are multi-functional and environmental respectfully so that the action measure 214 will probably account for the same amount of the 2007–2013 programming period.

15.4 Conclusion

In conclusion from the allocation of RDP funding and the amount paid in 2012 for the "action M of the measurement 214" it is conceivable to believe in an investment of 5–10 % of the territory of the region of Lombardy with conservative agriculture.

The percentage impact of C stock is quite limited compared to the CO_2 annual emissions of the Lombardy Region (the levels of 2010 totaled 83 Mt CO_2 equivalent INEMAR, 2010), which is about 1 %. But if the emissions reported from the same study for the agricultural sector are taken into account, the percentage becomes significant: the C stored and not emitted would represent 10.7 % of all emissions recorded in 2010 (7.8 Mt CO_2 equivalent).

A further comparison was made with the quantities of domestic emissions with the objectives outlined in the Kyoto Protocol, and in 2010 the issue of Italy stood at 501.317 Mt CO2 equivalent, while the objective of the Kyoto Protocol was 485.7 Mt CO_2 for 2012 (EEA 2012). The difference between the two quantities was compared with those of equivalent CO_2 not emitted as reported in Tables 15.5 and 15.6. As a percentage this represents 5.31 %, which is significant considering the limited area affected by the techniques and that the land falls only in the

Lombardy Region. At the national level, it may be charged for the land under conservation agriculture for all Italian regions. This could be very significant for the inventory of climate-altering gasses and should be counted in the land use and land use change and forestry (LULUCF, art 3.3 and 3.4).

Payments to farmers to store a ton of C (€ 535.23/t C) turn out to be one of the largest in comparison to other programs or policies developed in international contexts.

A pilot program introduced in Canada, Canada's Pilot Emission Removals, Reductions, and Learning (PERRL), enabled farmers to receive $11.08/t (1€ = 1.33$ Canada) CO_2 stored. Tons of CO_2 stored were counted through the coefficients of carbon sequestration. Farmers had to respect the conservative techniques such as no-till, not using a plow and not burning the stubble. The PERRL in 2007 was abandoned after $15 million funding (AAFC 2012).

In Australia, the Australian Soil Carbon Accreditation Scheme (ASCAS), a system of carbon credits, pays $90/t/year (1€ = 1.33$ AUD) retrospectively for the increase of carbon in soil for the first 3 years for the real carbon storage in soil measured each year and compared with the initial stock of departure (McKenzie et al. 2000). For each increase of 0.15 % of carbon in the considered soil sample (110 cm), it has an equivalent increase of 23.1 t/ha soil in carbon stored.

Conservation agriculture has not only the purpose of incorporating soil C, but it sets objectives to reduce erosion and nutrient losses to water, to increase biodiversity and reduce the emission of greenhouse gasses from soils. In addition, the measure tied to the conservative techniques (RDP measure 214 M) does not provide funds for the purchase of agricultural machinery specialists. In Australia, a study related to the increase of carbon sequestration by land use and costs for farmers who carry out management changes to increase the storage (Kragt et al. 2012) highlights the burden on farmers in achieving the national target reduction of CO_2 emissions. In a scenario it has been estimated that an increase of 10 kg/ha/year of carbon sequestration (compared to C-storage under the profit-maximizing mix rotation) would cost the farmer about $87 (1€ = 1.33$ AUD) per t CO_2-e. The next steps of the project will be to assess the costs of transition from conventional to conservation agriculture and to determine the higher/lower costs required for conservation agriculture compared with the yields.

It is however a very high funding as estimated in the two scenarios proposed for the amount of carbon stored in soils and therefore an increasing participation of farmers and the areas concerned can be expected in future years.

The model ARMOSA sets the conditions for the accounting of organic carbon stored in soil subject to conservative techniques. The regional measure 214 action M allows to obtain a dual result, namely to keep unchanged the level of yield and food production and at the same time to increase the pool of organic carbon in the soil; payments for farmers are considerably greater than policies present at the international level. The amount of carbon storable in Italian soil is high and could be included in the inventory of LULUCF to reach the standards set by Kyoto and post-Kyoto.

References

AAFC Agriculture and Agri-Food Canada. (2012). http://www.agr.gc.ca/eng/programs-and-services/my-aafc-account/?id=1210684614941.

Acutis, M., Brenna, S., Pastori, M., Basile, A., De Mascellis, R., & Bonfante, A. et al. (2007). Modellizzazione della dinamica dell'acqua e dell'azoto nei suoli agricoli lombardi—Progetto ARMOSA. Quaderno della ricerca. Regione Lombardia, Vol. 65, p. 128. http://www.ersaf.lombardia.it/upload/ersaf/gestionedocumentale/QdRArmosa_784_4537.pdf.

Beach, R. H., DeAngelo, B. J., Rose, S., Li, C., Salas, W., & DelGrosso, S. J. (2008). Mitigation potential and costs for global agricultural greenhouse gas emissions. *Agricultural Economics, 38*(2), 109–115.

EU-Climate Energy. (2007). Comunicazione della commissione al consiglio europeo e al parlamento europeo "Una politica energetica per l'europa". Bruxelles 10.1.2007 COM(2007). http://eur-lex.europa.eu/LexUriServ/LexUriServ.do?uri=COM:2007:0002:FIN:IT:PDF.

EU-Comm. (2009). Cambiamenti climatici: La Commissione mette in luce l'importanza del suolo. Rapporto commissione europea, IP/09/353, 2009. http://europa.eu/rapid/press-release_IP-09-353_en.htm?locale=en.

EU-Comm. (2010). La PAC verso il 2020: Rispondere alle future sfide dell'alimentazione, delle risorse naturali e del territorio. Bruxelles, COM(2010) 672/5. http://ec.europa.eu/agriculture/cap-post-2013/communication/com2010-672_it.pdf.

European Environment Agency (EEA). (2012). Greenhouse gas emission trends and projections in Europe 2012—Tracking progress towards Kyoto and 2020 targets. European Environment Agency report N° 6/2012. http://www.eea.europa.eu/publications/ghg-trends-and-projections-2012.

FAO. (2013). Site of conservative agricolture, 2013. http://www.fao.org/ag/ca/.

Freibauer, A., Rounsevell, M., Smith, P., & Verhagen, J. (2004). Carbon sequestration in the agricultural soils of Europe. *Ed. Geoderma, 122*(1), 1–23.

ISTAT. (2013). 6° Censimento generale dell'agricoltura in Lombardia, Risultati definitivi. http://www.istat.it/it/archivio/83482.

Kragt, M. E., Pannell, D. J., Robertson, M. J., & Thamo, T. (2012). Assessing costs of soil carbon sequestration by crop-livestock farmers in Western Australia. *Agricultural Economics, 112*, 27–37.

Knowler, D., & Bradshaw, B. (2007). Farmers' adoption of conservation agriculture: A review and synthesis of recent research. *Ed. Food Policy, 32*, 25–48.

Krunz, L. J., Senseman, S. A., ZablotowiczR, M., & Matocha, M. A. (2005). Reducing herbicide runoff from agricultural fields with vegetative filter strips: A review. *Weed Science, 53*, 353–367.

Lal, R. (2004). Soil carbon sequestration to mitigate climate change. *Ed. Geoderma, 123*, 1–22.

McKenzie, N. J., Jacquier, D. W., Ashton L. J., & Cresswell, H. P. (2000). Estimation of Soil Properties Using the Atlas of Australian Soils. CSIRO Land and Water, Canberra ACT—Technical Report 11/00. http://www.clw.csiro.au/aclep/documents/tr11-00.pdf.

Oorts, K., Garnier, P., Findeling, A., Mary, B., Richard, G., & Nicolardot, B. (2007). Modeling soil carbon and nitrogen dynamics in no-till and conventional tillage using PASTIS model. *Soil Science Society of America Journal, 7*, 336–346.

Perego, A., Acutis, M., Brenna, S., & Carozzi, M. (2011). L'azoto nei suoli agricoli: Interazioni con acqua e atmosfera—Progetto ARMOSA. Quaderno della ricerca n. 130. Regione Lombardia, p. 40. http://www.ersaf.lombardia.it/upload/ersaf/pubblicazioni/QdRn130-Armosa%202_13383_107.pdf.

Perego, A., Giussani, A., Sanna, M., Fumagalli, M., Carozzi, M., & Alfieri, L. et al. (2013). The ARMOSA simulation crop model: Overall features, calibration and validation results. *Italian Journal of Agrometeorology, 3*, 23–38.

Pisante, M. (2007). *Agricoltura Blu—la via italiana dell'agricoltura conservative*. Ed. Edagricole.

Smith, J. U., Smith, P., Monaghan, R., & MacDonald, A. J. (2002). When is a measured soil organic matter fractione quivalent to a model pool? *European Journal of Soil Science, 53*, 405–416.

UN. (1998). Kyoto protocol to the united nations framework convention on climate change. United Nations. http://unfccc.int/resource/docs/convkp/kpeng.pdf.

Part VI
Assessing Differences in Policy Implementation Across Countries and Sectors

Chapter 16
Cross-Atlantic Differences in Biotechnology and GMOs: A Media Content Analysis

Lena Galata, Kostas Karantininis and Sebastian Hess

Abstract Different regulations about the permission and approval rate of biotechnology and genetically modified organisms (GMOs) between the USA and Europe have been controversial for decades. While there is a wide scientific coverage of what may be the cause of this divergence, little is known about the role that popular media play in the related political discourse. We analyzed the media coverage of biotechnology topics in both the USA and UK from 2011-2013 by examining two opinion-leading newspapers. We test the hypothesis that the respective media content reflects differences in transatlantic policies towards biotechnology. The two newspapers differed in reporting intensity but were alike in their content about GMOs: with the central actors being scientists and NGOs, arguing mostly in the field of the agricultural sector, the debate seems to be locked in a stalemate of potential risks re-iterated against potential benefits, with none of the two positions clearly dominating the discourse.

16.1 Introduction

It has been 40 years already since biotechnology was born, and more than 30 years since it was widely introduced into our lives; and despite that since its emergence, biotechnology has been "a bone of contention," this technology remains highly controversial; launching various debates across almost every aspect of contemporary life, from organic farming and agriculture to regulation and public participation in science and technology. Throughout these debates, arguments and

L. Galata (✉)
Mediterranean Agronomic Institute of Chania (MAICh), Crete 73100, Greece
e-mail: lenaglt@gmail.com

K. Karantininis · S. Hess
Swedish University of Agricultural Sciences (SLU), Uppsala, Sweden

C. Zopounidis et al. (eds.), *Agricultural Cooperative Management and Policy*, Cooperative Management, DOI: 10.1007/978-3-319-06635-6_16,

actors have appeared and vanished; conflicts and theories have boomed and diminished. This makes profound the idea that the movement of biotechnology, as it has often been named (Bauer and Gaskell 2002), is a social process which changes over time and space, and as such, abiding reconsideration is needed, especially since experience and afterthought are due.

Genetically modified organisms (GMOs) on the right side of the Atlantic have faced a relatively protracted public debate and have yet to achieve a transnational consensus. This should not raise any queries, since European regulators, from its incurrence, saw biotechnology as a novel process requiring novel regulatory provisions, while at the same time, a complex series of national and European initiatives have until today embraced a wide range of known and unknown risks (including risks to the environment). The EU is focusing on process rather than product-oriented analysis (Carlarne 2007) and the primary components of the EU framework regulating GM products are "pre-marketing safety assessments; and a single 'one-stop' authorization procedure to achieve the internal market" (Tsioumani 2004). All GMOs, along with irradiated food, are considered "new food" and subject to extensive, case-by-case, science-based food evaluation by the European Food Safety Authority (EFSA). Furthermore, the regulations concerning the import and sale of GMOs for human and animal consumption grown outside the EU involve providing freedom of choice to the farmers and consumers (European Commission 2007). Accordingly, all food, including processed food or feed which contains greater than 0.9 % of approved GMOs, must be labeled.

In contrast with Europe, America did not see biotechnology as posing special risks, so regulation was contained within the existing laws, addressing known physical risks of new products. Furthermore, transgenic crops move through standard grain supply channels and have been substituted for traditional crops in the production of a wide variety of food products and animal feeds. The key to the US approach to regulation of GMOs is the principle of minimal oversight of food products that are generally regarded as safe (GRAS). Conventional food products are considered GRAS, and this is the standard by which GM foods are being judged in the US. As a result, the concept of substantial equivalence has been developed as part of the process of evaluating the safety of GM foods (WHO/FAO 2000). The objective of such an approach is not to establish absolute safety, but to consider whether a GM food (ingredient) is as safe as its conventional counterpart; the latter has been developed as part of the process of evaluating the safety of GM (Sheldon 2002).

Unquestionably, the case of the GMOs is not troubling only Europe or the USA. On the contrary, most of the countries of the world seem to take some position, hence acting in such an antithetical matter on the subject of GMOs. Many surveys reveal that in developing countries—as in developed countries—opinions expressed about GMOs seem to vary from country to country and from social group to social group. Most countries have handed regulatory responsibility for agricultural biotechnology to multiple agencies that deal with agriculture, the environment, and food safety, and agencies have then typically grafted regulations concerning agricultural biotechnology onto existing regulations relating to the release of new varieties, use of pesticides, and marketing of food products. For example, they appear relatively more favorable in China, and in many countries views are influenced by gender, income level, age, and socio-professional status (Reid and Canadian Medical Association 2002). Due to the lack of international coordination, countries are being separated into one key attitude on GMOs (precautionary principle), or the other (substantial equivalence).

16.1.1 Highly Precautionary Against the Danger or Deeply Influenced by the Media?

One explanation about the strong trend toward stricter GMO regulations in the EU can be given in terms of an increasingly broad and influential anti-GMO coalition. This coalition includes environmental and consumer groups, many farmers and downstream producers, and important parts of the EU's regulatory institutions.

There are cases where consumer/environmental groups or NGOs can influence not only the markets, but also the policy-making. This is a fact in cases where environmental and consumer interest groups are aware of their collective action problem and thus focus largely on issues that maximize its mobilization of membership and financial resources, especially when it comes to issues that provoke public outrage. Public outrage is the fear or anger a risk induces in a relatively large part of a country's population. It increases environmental/consumer groups' capacity to influence the regulatory process through politics and the market. According to the public choice theory, these organizations are pressure groups that offer public good: either somebody offers to participate in them or not (Salisbury 1969; Marwell and Oliver 1993; Galinsky et al. 1996).

Following this notion, many studies have been done, asserting the functioning of the mass media as the mediator between the regulation authorities and informal public opinion. Journalistic selection and the framing processes are the most profound techniques that the media apply, reinforcing their mediating nature in a way that can, to some degree, differentiate the meaning of the message that is being reported, mostly due to limitations of time and space by which they are bound. This selection of events to be reported and the way that they are portrayed may have a profound impact on the public's perceptions of those events, and may in turn even influence policy formation. But this implies that the public can construct an image of the "biotechnology reality" based only on what the media decides to convey (Gutteling et al. 2002). However, at the same time, the media have to obey both the general feeling of the receivers (the consumers of the product, press, radio, TV etc.), and the original facts that constitute the event itself. By these means, and to some credible extent, the media function as a "mirror of reality" (Kepplinger et al. 1991), since they are conditioned to "eavesdrop" into the society's wider sentiment about a specific event, not only for news-value reasons, but also for their communicating function to the authorities. Thus, if the media reflect reality, in the sense of understanding what the public feels and believes, the widely divergent regulatory approaches that the UK and USA apply in the products of biotechnology should be depicted in their respective media coverage.

16.2 Methodology

The number of newspapers, along with the years being examined which are as long as the research questions that have been set, are in accordance with the academic standards and requirements of a scientific paper derived from a thesis that has been programmed and researched for a Master's Degree. However, we believe that even a small amount of data can make scientifically precious conclusions if based on the right theory and if a thorough and well-designed theoretical and empirical model is employed.

First of all, the reason we chose the USA is because public attention over biotechnology has been synchronized across Europe, but not with the USA. This makes the country an excellent case for comparison with the EU from which inferences of great importance are expected. Furthermore, *The Guardian* and *The Washington Post* are two newspapers widely known all over the world for their credibility and trustworthiness, accessible to millions of people through the Internet without subscription fees; both assert to be highly concerned with technology and food. Such pieces led us to the immediacy of using them as our database.

The examined newspapers are also considered to be opinion-leading newspapers in the sense that they "conciliate" the aggregate picture of the media arena in a country. It is generally believed and widely accepted that in different places of the world, certain newspapers and news magazines are identifiable as opinion-leading sources of information, and that they have attained this status in relation to other media, for important decision-makers (such as politicians, civil servants, experts, and industrialists), as well as the general public (Bauer and Gaskell 2002). Thus, by analyzing the opinion-leading newspapers of the two countries, we can realistically expect to retrieve some insight into what is significant at one time and get a reasonable impression (Gutteling et al. 2002) of how each of the two societies process the meaning of GMOs. Under this theme, we can make a collage of the aggregate representations of the events by accepting a sample of observations, observations that come to the surface and become measurable. In the reality of events, each news items (in research terms, each item of coding) has a natural history and is indicative of a climate of opinions outside the news-rooms, accrediting virtually what the readers believe.

Content analysis is widely used to analyze and interpret qualitative data (Neuman 1989; Neuendorf 2002; Krippendorff 2012). The method we apply is summative content analysis with a directed approach, which is particularly useful for analytical comparative analyses when interpreting the meaning of text content. The adapted approach is directed because of the existing initial codes that we applied, and summative because it involved counting, and comparisons guided by keywords. The main strength of a directed approach to content analysis is that the existing theory can be supported and extended (Hsieh and Shannon 2005), and has been the research tool for a number of studies in multifold areas when studying psychology (i.e., Sliter et al. 2013), health insurance terminology (i.e., Politi et al. 2014), social media (i.e., Valtysson 2014), etc.

The codes we use are asserted to be, each of them and all together, a sufficient and robust justification of what the article addresses and how it is presented toward the issue being investigated. Particularly, in contrast to the key words for retrieving the articles from the Internet database, the coding list was set both before and after the text was accumulated in the software. Each article was then examined manually according to the coding set, and then asserted by the software; a random percentage of the articles were frequently re-appraised by another auditor to insure objectivity and personal impartibility. In detail, the following method was employed:

(a) first, articles were chosen in terms of our basic research questions, by providing answers to "what" and "how" is reported (frames),
(b) then, the major fields of application of biotechnology (themes) were taken into account,
(c) thereafter, the main actors were distinguished,
(d) following, the overall positive or negative message was identified, which the editors, speakers, or representatives of organizations are communicating when genetically modified procedures are described in terms of predictions for the future, or their assessments as a whole,
(e) and lastly, the findings for the two countries are all-encompassing, one beside the other, so that comparisons can be made easily.

16.3 Intensity-Content

Intensity refers to the number of all relevant articles, measured by online identification, and then all the articles in the two newspapers that contained the core key words of "biotech*", "genetic*", "engineer*", and "modificat*" were retrieved. At this stage, the unit of analysis was the "single press article".

On the other hand, content refers to what is being reported and how it is reported. After the retrieval of the articles, they were all subdivided according to the year and month of reportage and their respective newspaper. From this stage on, Atlas ti.7, software for analyzing qualitative data was used. Hereafter, the existing code structure for media analysis of biotechnology that was used in previous studies (Bauer and Gaskell 2002) was implemented, with slight modifications that correspond to modern events being reported by the media. This coding frame structure provided a framing, thematic and evaluative composition that served the purpose of indicating what was important for the media at this time in the light of questions posed by our coding frame. This coding frame process involved analysis of their; (1) size, (2) format, (3) focus, (4) thematic content, (5) framing, (6) actors, and (7) outcome. Furthermore, the coding process included one additional code, the controversial index, which was about the controversy being demonstrated by the events being presented, or being implied by the actors. Frames, themes and evaluations were further differentiated and the information was incorporated in a cross-Atlantic database.

After the completion of the coding of each article, the information was concentrated in a table of basic frequencies, which comprised the core frames for our analysis in terms of comparison of statements (see Table 16.1).

We adopted the coding frame structure that Bauer and Gaskell (Bauer et al. 1998, 1999; Bauer and Gaskell 2000, 2002) established in their analyses for two reasons: First, the mentioned codes could be used as an explicit interpretation of each article in all possible gainful (in terms of insight) dimensions; and since we limited our analysis to 2 years only, this was of great importance. Second, if we applied the same coding frame structure for media analysis that was used before for almost half a century in resolving the debate over biotechnology, we would gain an invaluable source of information in terms of comparative media interpretation of biotechnology both for the UK and US, but mostly for the two countries taken in a comparative way together, for their responsive behavior then and now.

Following this perspective, and with the advantage of hindsight, we then revisited the literature to view the two countries together in order to identify their contour both individually in the UK and US, but also for their biotechnological media profiles, over the years until the present time. At this point, we induced the media profiles of the two countries by confining the comparison between them only in those core sets of coding that we used in our analysis, which was structured and implemented previously in the literature. This allowed us to draw some inferences about how the two countries evolved in terms of their media coverage of biotechnology since the first systematic comparable interpretation of 1992–1996, the second 1997–1999, and finally 2011–2013, all taken under the same segregated and incorporated coding aspect.

Our analysis is completed with the conclusive evaluation of the three cymas that the two countries followed through the years, which we believe can be categorized in three respective media profile snaps, which in turn evidence the media, political and informal discourses that were "born" and "died" through the years, associated with the content that the mass media of each time have been communicating.

16.4 Findings

The findings show that in terms of *intensity* there is an evident difference between the two countries. Overall, 179 articles coming from the British press and 61 from the American press were found (240 in total), resulting in almost 7 articles per month in the UK and 2 per month in the US press, respectively (Fig. 16.1).

Table 16.1 Basic frequencies of characteristics of press coverage for The Guardian and The Washington Post by code, January 2011–February 2013

Values	UK		USA	
Frame (%)	Progress	50	Progress	32
	Public accountability	30	Public accountability	30
	Nature nurture	13	Nature nurture	25
	Ethical	4	Economic	12
	Economic	3	Ethical	1
Theme (%)	Agri-food	57	Agri-food	63
	Generic research	18	Generic research	21
	Genetics	15	Genetics	10
	Biomedical	8	Biomedical	3
	Cloming	2	Cloming	3
Actors (%)	Independent science	44	Independent science	25
	Interest groups	17	Media/Public voice	23
	Politics	17	Interest groups	20
	Business	12	Business	19
	Media/Public voice	10	Politics	13
Benefit/Risk (%)	Neither	40	Neither	46
	Benefit only	24	Risk and benefit	21
	Risk and benefit	18	Risk only	20
	Risk only	18	Benefit only	13

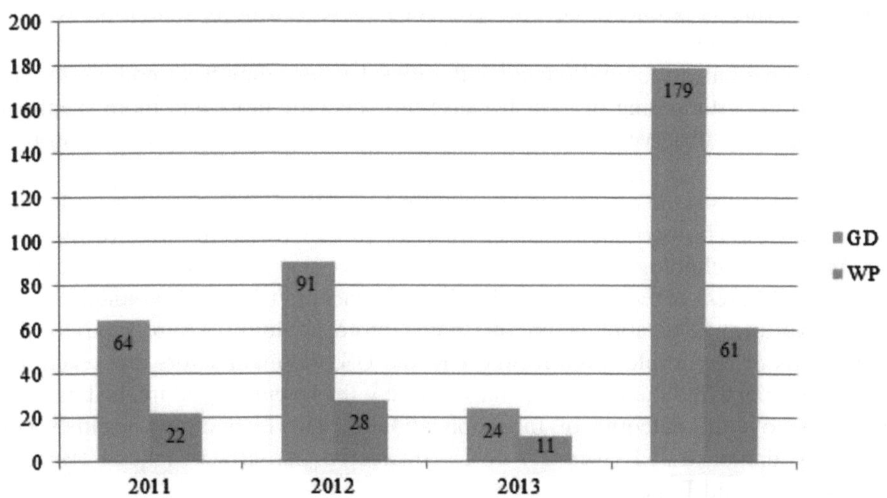

Fig. 16.1 The intensity of press coverage of GM products in *The Guardian* and *The Washington Post*: Number of articles per year and newspaper, from February 2011 until February 2013

The level of intensity, even though it cannot be an independent indicator of the direction of the debate over biotechnology, can nevertheless show how much attention, awareness and interest is dedicated to this issue at the moment, following the concept that only events and stories that keenly activate the public's attention will reach the headlines of a newspaper. Following this concept, the UK media cover, for the greatest part, agro-biotechnology framing in almost any kind of theme, extending GM expressions in relatively every aspect of reported life. This makes sense if someone retains the example of the Anglo-Saxon debate over science and its public understanding. This debate, which could be summarized in the phrase "to know it is to love it," was actually built on the notion that since ignorance breeds contempt, and knowledge breeds love (Miller 1983), the more accustomed society becomes with technology, the easier the integration step with it will be in terms of its acceptance.

However, this seems not to be the case or at least not the only case that the GM industry should have been focused on, in order to introduce their products into European society. If the notion of the perceived knowledge of a kind of technology were the basic factor of accepting that technology, European citizens, or at least the British whom we are examining, would have embraced it a decade ago, if not earlier. We thus believe that familiarity with the technology is a necessary condition in the process of introducing GMOs into society, and it is sufficient, but only if the regulation authorities act relatively quickly after society's relevance to it. This has to do with what we call the rounds of mediation (see Fig. 16.2), whereby there is a point of coverage after which consumers reach a saturation point of how much information about a certain technology they can and are willing to digest. If that point is left without establishing a concrete legislation/regulatory decision, things become more complex. This is because the public sphere of technology from this point is bound to become gradually more and more polarized, for the reason that people who do not wish to follow one or the other edge decide not to take an active part in the discourse at all, either because they have nothing to win or lose from participating in it, or because they become more and more skeptical about it as they observe the media coverage becoming more intensified through time.

The findings for the *content* do not manifest significant differences between the two countries. The two countries demonstrate similar traces in terms of the reasoning of biotechnology, since biotechnology is either framed by argument of prospect (Progress & Economic) or by that of concerns (Public Accountability or Nature/Nurture). The application of "green biotechnology" (Agro-food) is the main focus of the articles both for the UK and US, the main actors that reach the North American and the British media scenes are the scientists and interest groups, while the overall outcome of the application of this technology is either not explicitly demonstrated (Neither) or is considered partially beneficial (Benefit Only or Risk and Benefit).

In conclusion, having the two perspectives of comparison in mind, in terms of intensity, there is a difference between the two countries; however, in terms of the content of the media coverage, there seem to be two clearly defined pictures illustrating the same landscape. But if we expect the media to reflect public opinion, and since the public is diverse in the two countries, shouldn't the media also be different? Does this mean that we have come to the wrong conclusion? Most certainly not.

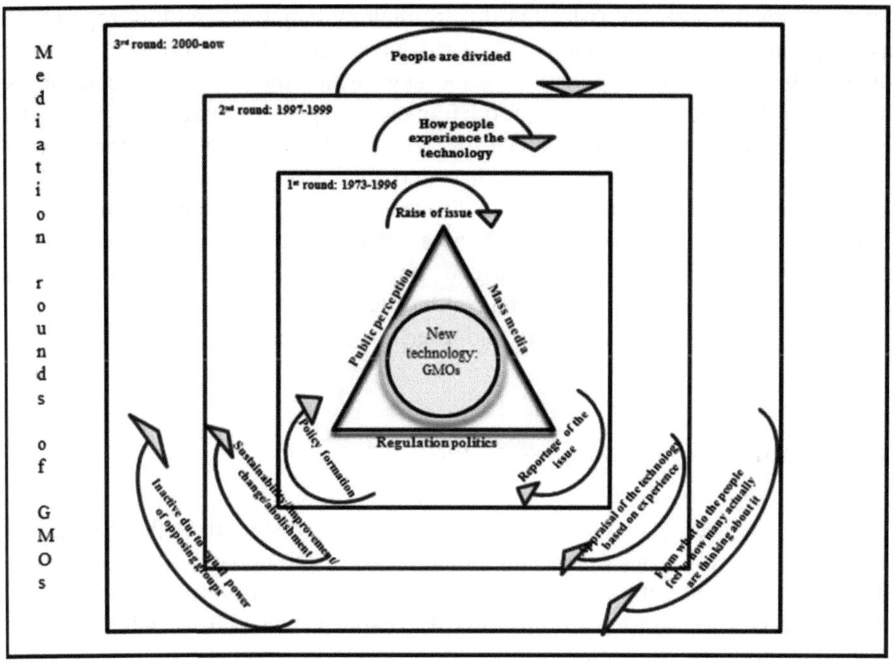

Fig. 16.2 The three rounds of the arenas of the public sphere and their mediating response through time

As mentioned previously, the procedure of introducing a new technology, such as biotechnology, into the society is a social dynamic process, affected by various historical instances and their implications in the public sphere being depicted in the economic, political, and social discourses that appear in the media. Furthermore, in order for any of our findings to make sense, we have to look at them relatively to something else; for this reason, we had to go back in the past, and actually revisit the history of the two countries in terms of their media coverage. We thus constructed two more snapshots of the basic frequencies for the UK and US by adopting the findings of Gaskell and Bauer (2001, 2002) over the media profiles of the two countries before 1992–1996, and after the so-called "watershed years", 1997–1999 (Gaskell and Bauer 2001, 2002). These two snapshots can sum up the general idea of the media discourse of that time: the media profiles of the two countries from the 1970s until 1996 were again the same in general terms. Indeed in the early days of biotechnology's introduction into our lives, the coverage was rather positive and progressive in both countries, focusing on its biomedical and genetic applications (red biotechnology). The prevailing actors were the business and science representatives, and the content was manifesting a beneficial effect on the use or the future use of this technology. During the "watershed years" of 1996–1999, the UK reportage changed dramatically, since the previous mostly progressive coverage of the use of biotechnology is dismissed by the worries over the potential risks that this technology may carry with it, represented as tempering with Mother Nature. The British then shifted in reportage, which was obvious in the field of application of the new technology, since the controversy then starts to focus on the agro-food framework (green biotechnology), and is reflected by the impact of the debate over GM food and cloning that these events had on European public opinion. At that time, the issue of biotechnology took the dimensions of a public issue, since politicians entered the debate, while the two groups of prevailing actors were not people coming from industry and science, but rather from industry and politics instead. As for the US, reportage remained the same as in the previous decades, and can, virtually, be used as an example of how the British press developed before and after the watershed years of 1996–1999.

16.5 Implications and Discussion

Following the reality of GMO representatives that we found in the media (scientists and interest groups being the protagonists of the debate), while the rest of society seems to be left out of it, another plausible explanation for this deviation in intensity could be asserted by the Mere Exposure Effect (MEE), which is especially effective when people are ambivalent about an issue. This phenomenon is derived from the field of social psychology and it describes how people show a tendency to like the one thing that they were more exposed to. Previous exposure makes you like something, and this expands to the decisions people make. Indeed, it has been proved through scientific research (Zajonc 1968) that there is something about previous exposure that makes an item, a person, or an event stand out somehow and make us support it. However, it seems not to matter in what kind of way this item, person, or event is presented; the message doesn't really seem to matter, and this is the point of the mere exposure effect. In fact, there doesn't have to be a rational explanation for the issue presented. When you're exposed to something very quickly (reading the headlines) and often enough, without really getting a chance to process the information, then later if you notice something, and you have some questions in your mind (for instance to reject GMOs or not), that previously exposed item seems to stand out as the answer. This is what mere exposure is doing: it's drawing you to something and making it seems like the answer to whatever question you have in mind. This is the insidious side of familiarity and intensity: "having preferences with no inferences." This practice can be found in manifold areas of the marketing and politics of contemporary life. It exemplifies the astute fact that there are times when previous experience influences your behavior through a familiarity-like process, but it gets transformed somewhere along the way such that you end up taking ownership of something and believing that you are making a conscious decision, when in fact you are being heavily influenced by familiarity. This may sound a little luminous, but it could make sense in the representation of GMOs by the media.

Although this theory is said not to be true for all instances in our lives, there is evidence that we tend to do that only for the most important positions. For the rest, familiarity can often play a huge role. Coming back again to our topic of concern, we often hear people from within one or the other party of the GM debate, either the scientists or the NGOs, come on TV or a newspaper and state some message, almost always a very over-simplified message, but they will state the same message over and over and they will have different people state that same message, found in almost every theme or corpus of a newspaper. The idea is that if you hear that message over and over and over, and especially if you hear it in different places spoken by different people, it starts to feel true. That familiar message starts to feel like the answer to the question. And that is the reason why it is repeated over and over again, trying to influence people with familiarity. A familiar argument feels like a correct argument.

Our evidence leads us to believe that the situation previously described is the one that we are now experiencing, as it is conceivable enough that the debate over biotechnology can be illustrated as an extremely polarized landscape of counter-players antagonizing each other over the share of publicity. This conforms to the findings over the content of the coverage.

16.5.1 A Synchronized Debate

Biotechnology is indeed a process of "challenge and counter-challenge" where progress can be made only if the interested arenas overcome their rigidity to make inroads inside each counter-arena and become more amenable to the artifacts of the others. These notions are shown in Figs. 16.3 and 16.4 and demonstrate how each public actor (regulators, public opinion, and media) have responded to it since the 1970's till now. This technology is a process through which different issues are being inadvertently or intentionally highlighted by different spokespersons at specific times—and at other times, different aspects are being ignored. In particular, setting off from nonawareness, the coverage employs some kind of evidence, mostly using the imagination and techniques of capturing the public's attention (1st round). The media are thereafter mostly well-informed and conduct the responsibility of better endorsing the public with the issue and at the same time they carry the power of demonstrating to the regulators what the wider feeling of society is about the issue (2nd round). In the third and last round, the media keeps the pace of the debate at a rate that corresponds to what the regulators address, but first and foremost to what the group of people who manages to reach the media (here scientists or NGOs) has to manifest.

Through the years different actors, scientists, activists, industry, and others kept this movement going and moving forward; but this repressed at a point where this movement became crystallized, keeping public opinion, the ordinary consumers who represent the majority of society, away from the debate, and bequeathing it only to the scientists and interest groups. This flood of opinions keeps the actively

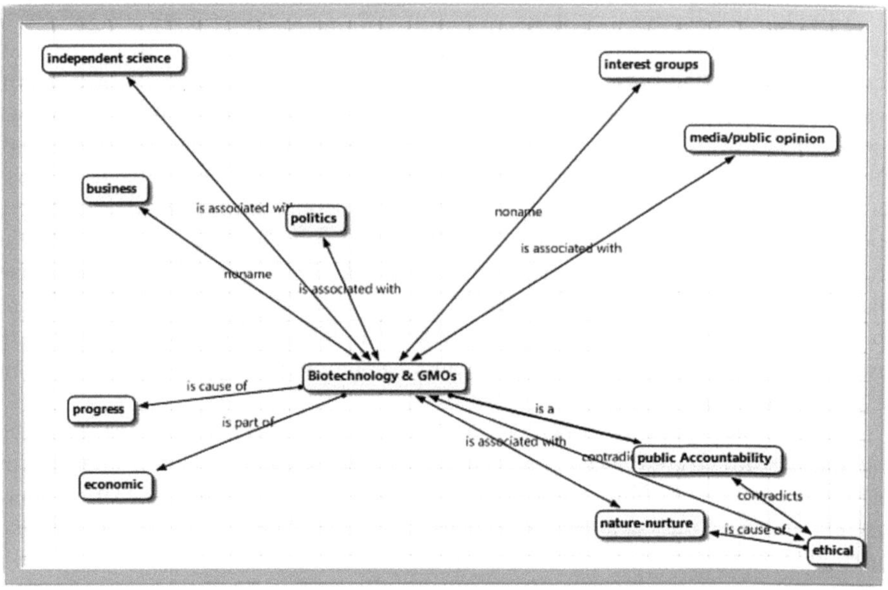

Fig. 16.3 Actors and frames represented by coding family association using Atlas.ti 7

participating groups miles away from the majority of ordinary people—not elite, or well educated, but common people, consumers who will eventually be the recipients of the product and consequently its doom or blessing—which is in fact a move to normalize policy decisions. The public perception in the British case today is somehow in the same position as it used to be when technology just emerged. Since the Asilomar conference of 1975 (rDNA), when people first heard about the potentials of genetic engineering, until now where its use and applications can be observed in almost every aspect of our lives, the voice of the public seems to be echoed only in the dramaturgy of the counterparts trying to win some space in the drama scene of biotechnology.

From the first round (1973–1996), people have turned from previously unaware-ignorants to aware-ignorants today, in the sense that they evolved from not knowing and not participating (this technology was initially more of a scientific issue, and few people except laboratory workers knew or had experienced something about it), to knowing about it and not being interested in participating (most people are familiar with the technology, but not concerned about it either by supporting or rejecting it actively). The second round of GM technology (1997–1999) changed things, but only for a while, when people started to argue and raise opinions over this technology, either by supporting or rejecting it, until the big influence groups of the media polarized the two sides of the debate, when the third round, that we are still experiencing, prevailed. It seems as though you have to either be enthusiastically for or against it, since those are the only views

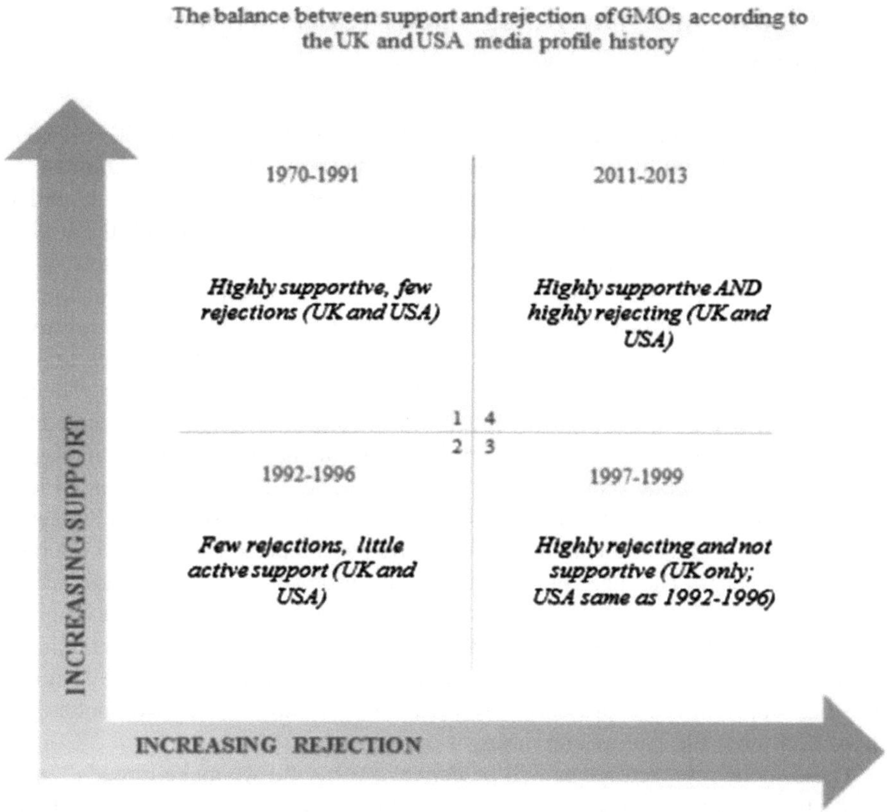

Fig. 16.4 The interplay between support and rejection of GMOs according to the UK and US media profile history

that are being manifested by the media. Hence, apart from the actors who actually keep succeeding in entering the media scene, most people, whether they have informed views or not, choose to stay away from this controversy, simply because it does not seem to affect their lives.

The overall comparison of the cross-Atlantic discourse reveals that the similarities between the two countries are far more impressive than the differences because they contest what the biotechnology debate is all about. The consistency between the one and the other press is assembled by that of a pasteboard melodrama, since it seems that the "biotechnology movement" has passed through three phases of first, being highly supportive, to highly negative, and then again to highly supportive and negative together.

16.6 Conclusion: Why Going Back Means Stepping Forward

By boldly juxtaposing the snapshots of the two countries for then (1970s till 2000) and now (2013), and by primarily seeking for questions, through the contrast, we were astounded by how the similarities were crying for the answers. It will be ultimately them that may help us understand well and deeply enough the truth of the GM debate in a cross-Atlantic context. Because the conflicting sets of the debate revealed a fundamental idea: that it is the reciprocities, and not the differences, in the media coverage between the two countries, that condition a proximate answer to what this debate has evolved into. In fact, it seems that the outset of a synchronized, hyphen debate has been outstated no less than a decade now. What our theoretical pathway allowed us to see in hindsight and recollection was the underlying story; the script and scenario of the whole "biotechnological movement" that can be seen in three main scenes. This performance was initiated by the same media profiles for the UK and US, offering space and voice for biotechnology and its applications to be reported as a representation of its potential powers and perils. The main player was in the European case the second "scene" of 1996–1999 where this extraordinary technology started to tediously reveal risks and benefits in both a hypothesized and pragmatic way. The last but not conclusive scene is the one that we are viewing now. The current scene exhibits the "actors," (see Fig. 16.3) who are presented by being fully aware of their competency to shape and mold the pathos and ethos of the audience.

These social representatives will be conveyed only "if we make inroads" in the arenas that constitute the public sphere. This process is a path where the challenge and counter-challenge of each actor is deeply dependent on the degree of influence and knowledge that each carries, being the end result of their counter-actuals, in which the public arena invests. This heuristic view of the existing public opinion as a unique and simultaneously integrated part of the arena is followed by their power to affect and to be affected by the media and the public authorities only up to the point where their sufficiency of knowledge and influence permits them to do so.

This is expressed by the struggle of the stage actors to convince the counterparts of the pro- and against-GMOs standpoints that the society seems to behold. The curtain will not fall unless one of the two protagonists wins this debate and convinces the audience; so, the last call is in the public's hand, and the play will eventually be guided by the future and most probably will not be cleared up by the situation as it is, but by the social representatives of the situation as the public anticipates it to be.

Acknowledgments It is a great pleasure to express my deepest gratitude to Dr. George Baourakis and the Mediterranean Agronomic Institute of Chania (MAICh) that presented me with a most memorable learning experience and an inside look into the world of science and research. First and foremost, I could not but thank Dr. Kostas Karantininis for his invaluable guidance, advice, support, and encouragement, both in times of excitement and disappointment. My deepest appreciation goes also to Drs. Gaskell and Bauer for all their academic work that has always been an enormous inspiration for me.

References

Bauer, M. W., Durant, J., & Gaskell, G. (Eds.). (1998). *Biotechnology in the public sphere: A European sourcebook*. London: NMSI Trading Ltd.

Bauer, M. W., & Gaskell, G. (Eds.). (2000). Qualitative researching with text, image and sound: A practical handbook for social research. Bauer, M. W., & Gaskell, G. (Eds.). (2000). *Qualitative researching with text, image and sound: A practical handbook for social research*. London: Sage.

Bauer, M. W., & Gaskell, G. (Eds.). (2002). *Biotechnology-the making of a global controversy*. Cambridge: Cambridge University Press.

Carlarne, C. (2007). Notes from a climate change pressure-cooker: sub-federal attempts at transformation meet national resistance in the USA. *Connecticut Law Review, 40*, 1351.

European Commission. Directorate-General for Economic. (2007). The joint harmonised EU programme of business and consumer surveys (No. 5-2006). European Communities.

Galinsky, E., Bond, J. T., & Friedman, D. E. (1996). The role of employers in addressing the needs of employed parents. *Journal of Social Issues, 52*(3), 111–136.

Gaskell, G., & Bauer, M. W. (Eds.). (2001). *Biotechnology, 1996–2000: The Years of Controversy*. London: NMSI Trading Ltd.

Gaskell, G., Bauer, M. W., Durant, J., & Allum, N. C. (1999). Worlds apart? The reception of genetically modified foods in Europe and the US. *Science, 285*(5426), 384–387.

Gutteling, J. M., Fjæstad, B., & Olofsson, A. (2002). Media coverage 1973–1996: trends and dynamics.

Hsieh, H. F., & Shannon, S. E. (2005). Three approaches to qualitative content analysis. *Qualitative Health Research, 15*(9), 1277–1288.

Kepplinger, H. M., Ehmig, S. C., & Ahlheim, C. (1991). Gentechnik im Widerstreit: Zum Verhaltnis von Wissenschaft und Journalismus.

Krippendorff, K. (2012). *Content analysis: An introduction to its methodology*. London: Sage.

Marwell, G., & Oliver, P. (1993). *The critical mass in collective action*. Cambridge: Cambridge University Press.

Miller, J. D. (1983). Scientific literacy: A conceptual and empirical review. *Daedalus, 112*(2), 29–48.

Neuendorf, K. A. (2002). *The content analysis guidebook*. London: Sage.

Neuman, W. R. (1989). Parallel content analysis: Old paradigms and new proposals. *Public communication and behavior, 2*, 205–289.

Politi, M. C., Kaphingst, K. A., Kreuter, M., Shacham, E., Lovell, M. C., & McBride, T. (2014). Knowledge of health insurance terminology and details among the uninsured. *Medical Care Research and Review, 71*(1), 85–98.

Reid, I., & Association, Canadian Medical. (2002). *Annual national report card on health care 2002*. Ottawa, Ont: Canadian Medical Association.

Salisbury, R. H. (1969). An exchange theory of interest groups. Midwest *Journal of Political Science, 13*, 1–32.

Sheldon, I. M. (2002). Regulation of biotechnology: Will we ever 'freely' trade GMOs? *European Review of Agricultural Economics, 29*(1), 155–176.

Sliter, M., Boyd, E., Sinclair, R., Cheung, J., & McFadden, A. (2013). Inching Toward Inclusiveness: Diversity Climate, Interpersonal Conflict and Well-Being in Women Nurses. *Sex Roles*, 1–12.

Tsioumani, E. (2004). Genetically modified organisms in the EU: Public attitudes and regulatory developments. *Review of European Community & International Environmental Law, 13*(3), 279–288.

Valtysson, B. (2014). Democracy in disguise: the use of social media in reviewing the Icelandic Constitution. *Media, Culture and Society, 36*(1), 52–68.
WHO, F. (2000). Ad-hoc intergovernmental Codex task force on animal feeding. 13 p. First Session. Dinamarca, 13–15.
Zajonc, R. B. (1968). Attitudinal effects of mere exposure. *Journal of Personality and Social Psychology*, *9*(2p2), 1.

Author Biography

Lena Galata born and raised in a mixed family of both Greek and Russian origins, is a master student of Business and Management of Economics in Mediterranean Agronomic Institute of Chania (Crete). She graduated from University of Macedonia with a specialization in International and European Economics and Diplomacy and her primal research locus is on Agricultural Economics and Agribusiness.

Chapter 17
Examining the Evolution of Agricultural Production of Three SAARC Countries: Bangladesh, India, and Pakistan

Anthony N. Rezitis and Shaikh Mostak Ahammad

Abstract This chapter examines the issue of trends, cycles, and irregular components in the per capita agricultural production of three countries—Bangladesh, India and Pakistan—which are members of the South Asian Association for Regional Cooperation (SAARC). SAARC countries constitute about 23 % of the world population, and have 15 % of the world's arable land. The selection of the aforementioned countries is based on their agricultural economic importance to the region, as they possess about 80 % of the agricultural economy. This chapter uses the unobserved components model to decompose the per capita agricultural production of each country, and investigates the relationship of each component among these countries. The time period for the study is 1961–2010, and the FAO's statistical dataset is used. The smooth trend plus stochastic cycle methodology of Koopman et al. (2009) is used to estimate the model by maximum likelihood. Primarily, the residual diagnostics will validate the model with good fit. Diagnostics of normality, auxiliary, prediction, and forecast also show that there is no deficiency in the model. Empirical results clearly demonstrated that India is positively correlated with Bangladesh in irregular components, but moderately correlated with Pakistan in growth. Finally, there is an evidence of a stronger correlation between the three countries in short cycles than in long cycles.

A. N. Rezitis · S. M. Ahammad (✉)
Department of Business Administration of Food and Agricultural Enterprises,
University of Patras, Patras, Greece
e-mail: shaikhmostak@yahoo.com

A. N. Rezitis
e-mail: arezitis@upatras.gr

C. Zopounidis et al. (eds.), *Agricultural Cooperative Management and Policy*,
Cooperative Management, DOI: 10.1007/978-3-319-06635-6_17,

17.1 Introduction

The importance of agriculture is immense. From the primitive times to the present day, it has been continuing its pace to support the living beings across the globe. It is one of the most important sources of occupation for numerous populations. Excess numbers in the labor force (inactive, seasonal, or hidden labor) are also generally absorbed by the agricultural sector. This sector provides not only food for about seven billion people around the world, but also raw materials for the jute, textile, tobacco, sugar, and leather industries, among others. In addition, it is one of the most important markets for industrial goods such as machinery, fertilizers, and agrochemicals. Consequently, it appears that people are still relying on agriculture, emphasizing a need to care for this sector.

The growth in world food production has been faster than population growth over the last four decades. As a result, hunger has reduced while per capita agricultural production has increased. This situation paves the way to improve nutrition, eradicate poverty, and achieve sustainable economic growth. Extracting biofuel from food grain is a new development of the agricultural sector, although this is not exempt from criticism. At present, the agricultural sector plays a pivotal role in international trade, transportation, the building industry, sources of household income and progressing economies, especially in developing countries. A vast number of countries' GDP is directly influenced by the agricultural sector. The countries whose GDP is not directly influenced by the agricultural sector are also indirectly relying on this sector to feed their people. Thus, for the balanced development of the economy, the agricultural sector is not to be given any less attention. The agricultural sector also faces challenges in keeping per capita agricultural production stable. These challenges can trigger increases in population, reductions in arable land, and changes in agricultural policies, with an impact on climatic change. These challenges can create unexpected consequences for the agricultural growth of the economy.

On the 8th of December 1985, seven South Asian countries formed the South Asian Association for Regional Cooperation (SAARC). The founding member countries of the SAARC are Bangladesh, Bhutan, India, Maldives, Nepal, Pakistan and Sri Lanka. At present SAARC has eight member countries since Afghanistan joined the organization in 2007. The main objective of SAARC is to boost economic, social, and cultural cooperation among the member countries. SAARC consists of about 15 % of the world's arable land and has about 23 % of the world's population, which is mostly shared by Bangladesh, India and Pakistan. The increase in population is creating obstacles to the development of the economy, and also causing a reduction in the three countries' arable land area. A recently published report ranking global climate change states that Bangladesh, India and Pakistan (among others) will be extremely affected by the effect of climatic change. Thus, when coping with the agricultural challenges of keeping its production growth stable, it is important to assess these three countries' long-term agricultural trends, short-term transitory effects and future forecasts of agricultural growth.

On the basis of these issues, this study aims to analyze the evolution of these countries' agricultural production since their economies are mostly agrarian. These three SAARC countries have been selected for analyzing the evolution of their agricultural production since they represent about 80 % of the agricultural economy of the region. The agricultural sector constitutes a significant proportion of GDP of the selected countries; for instance, according to the CIA world factbook, 17.3, 17.0 and 20.1 % of the GDP of Bangladesh, India, and Pakistan is shared by the agricultural sector, respectively. Most of the population of these three countries lives in rural areas (72, 70 and 64 % of Bangladesh, India and Pakistan, respectively). The agricultural sector is considered to be the principal source of employment, leading to an economy boom in these three countries which accounts for 45, 53, and 45.1 % in Bangladesh, India and Pakistan, respectively.

The main objective of this study is to apply the unobserved components (UC) approach to the per capita agricultural production of the aforementioned countries, in view of modeling their evolution. The second objective of this study is to investigate the presence of common trends or cycles in the per capita agricultural production of these three countries. The existence of common factors indicates that the variance matrices of the relevant disturbances are less than full rank. The presence of common trends refers to cointegration, which implies that variables are moving together, i.e., cointegrated, in the long-run; whereas the presence of common cycles refers to common features, indicating that variables might converge in the short-run. The main advantage of the UC model is that it is a flexible econometric tool that is generally used to decompose the evolution of time series data into trends, seasons, cycles, and irregular components, which is not possible to observe directly from the data set. Trends are represented in the long-run direction of the economy, which is known as a "permanent component"; cycles represent the fluctuations of the short-run economic activity, which is known as a "transitory component." An irregular component focuses on the nature of the unobserved factor. Another advantage of the UC model is that it can easily produce forecasts for the future. The prospective instances of outliers and structural breaks can also be investigated.

The stochastic characteristics of this approach provide an important econometric tool for performing richer statistical analysis, as well as improving upon traditional interpretations of the evolution behavior of the per capita agricultural production of these countries. Furthermore, in relation to literature, there is evidence of a large number of studies examining agricultural production linked to these three SAARC countries. However, there is limited evidence of examinations using the UC approach to investigate the per capita agricultural production of these countries. The findings of this chapter also strengthen the endeavor to extend the econometric application of the structural time series approach to the analysis of the evolution of per capita agricultural production in Bangladesh, India, and Pakistan.

The remainder of this study is presented as follows. The theoretical background and the literature of the study are discussed in Sect. 17.2. Section 17.3 presents details of the methodology of the unobserved components model in a long-run trend, short-run deviation (cycles) and irregular components. Section 17.4 describes the details of data used in this study. Section 17.5 reports the estimation results of this study. Section 17.6 discusses the results. Finally, Sect. 17.7 offers the conclusions of the study.

17.2 Literature Review

The analysis of the evolution of agricultural production has gained attention, especially for the countries whose economies are predominantly aided by this sector. There are a wide number of studies investigating the evolution of agricultural production in Bangladesh, India, and Pakistan. For example, Kurosaki (2007) studied how irregular agricultural growth is associated with land-use change in these countries. This study mentions that the agricultural growth trends of these countries were reversed before and after the 1947 partition: Pakistan experienced superior performance before 1947 and India showed improvements after 1947; Bangladesh, however, has shown significant growth during recent years. Kurosaki (2003) also investigated trends in agricultural productivity in India and Pakistan. He reports that institutional changes such as the establishment of an irrigation commission, land reforms, and water treaties, as well as changes in agricultural policy such as the abolition of a fertilizer subsidy and integration with the green revolution, have had significant effects on agricultural growth in India and Pakistan. Robbani et al. (2007) dealt with how recent agricultural trends and changes in the agricultural environment affected the sustainability of Bangladesh's agriculture, demonstrating that the production of the main crops subsector—rice and wheat—expanded faster than other minor crops sub-sectors.

Quddus (2009) examined crop production growth in different regions of Bangladesh. He indicated that per capita crop productions in different regions were irregular during the study period due to reasons such as population growth and use of high-yield inputs. Kannan and Sundaram (2011) specified the trends and patterns in agricultural crop production growth in India. They identified that India's crop production growth is accelerating; however, India has undergone significant changes after shifting from the cultivation of food crops to commercial crops, made possible by the high yields of crop production, capital formation enhancement, improved fertilizer, and better irrigation facilities. Rehman et al. (2011) compared the decomposition results of agricultural output growth between the pre- and poststructural adjustment program in Pakistan. The analysis of the empirical results showed that the area and yield effects had almost equal contribution to the change in output growth during the study period. Chaudhry et al. (1996) reviewed the growth performance of Pakistan's agriculture. They argued that, even though the growth rate was fluctuating from year to year, or even from decade to decade, it has been respectable.

A number of studies use the UC model to capture the permanent (trend) components and transitory (cycle) components from the time series data. Most of these studies examine GDP and price series data. For example, Harvey and Jaeger (1993) conducted a study relating to US GNP, US prices, the US monetary base and Austrian GDP. They favored the acceptability of the UC model as it can be described in a stylized fact and business cycle. Flaig (2002) decomposed the quarterly German GDP into trend, cycle and seasonal components and working-day effect. He found that the growth rate of the potential output of German GDP declined at the end of the sample period. The cycle included short-run and long-run cycles of 4 and 8 years, respectively. He also argued that seasonality was not constant, but that the working-day effect has a considerable impact on the short-term variability of output. Ferrara and Koopman (2010) worked on GDP and real house prices in the Euro area, and investigated common business and housing market cycles. Their empirical results indicated a strong relationship between business cycles in France, Italy, and Spain. On the other hand, for the house price cycle, France and Spain are strongly related. Koopman and Azevedo (2008) measured synchronization and convergence of business cycles in the Euro area, the UK, and the US. They demonstrated that, though there are some exceptions, European Union countries show synchronization in their business cycles.

Wongwachara and Minphimai (2009) investigated GDP data to extract the output gap, as well as cycles and growth for ASEAN countries. The estimated Phillips curve relation of this study concludes that relationships between the output gap and inflation in ASEAN countries are not notable. Labys and Kouassi (1996) analyzed commodity price cycles, employing two aggregate commodity price indexes—the UC index and US index—and 20 individual monthly commodity price series, collected from different international sources. Their extracted cyclical components indicate varying behaviors among the commodity price cycles. Koopman and Lucas (2005) examined business and default cycles for credit risk by using US data. They reported that strong cocyclicality between credit spreads and default rates, and between credit spreads and GDP, appeared in the 6-year cycle. However, in the 11-year cycle, only strong cocyclicality between GDP and the default rate appeared. Cuevas (2002) carried out a study testing the association of trends and cycles between Venezuela's real GDP and real oil price. He used yearly data, and found that there is a strong positive association between real GDP and real oil price in the long-run trend component. In contrast, he found a negative association in the cycle component of Venezuela's real GDP and real oil price.

Mitra and Sinclair (2010) assessed macroeconomic fluctuations in selected Asian and Latin American emerging economies. They employed the UC approach to decompose the quarterly GDP of seven Asian economies (Hong Kong, South Korea, Singapore, Taiwan, Indonesia, Malaysia, and Thailand) and five Latin American economies (Argentina, Brazil, Chile, Colombia, and Mexico) into permanent and transitory components. They argued that permanent and transitory innovations by Asian countries are roughly the same in their magnitude; however, it is widely different for Latin American countries. Additionally, correlation between permanent and transitory components within the Asian and Latin American countries shared a negative relationship. They also noted that in the cross-country relationship, Asian countries are more correlated than Latin American countries.

Taking into account the facts confirmed by the literature, and the relative properties of agricultural production and the UC model, this chapter attempts to analyze the findings into the evolution of the agricultural production of Bangladesh, India, and Pakistan in detail. Though these three countries have a similar climatic and weather history, their political and socioeconomic profile is different, and they also show variability in their agricultural production history. Thus, this chapter intends to decompose the per capita agricultural production of these countries to figure out the characteristics of the unobserved components, i.e., trends, cycles, and irregular components.

17.3 Empirical Model

This study presents the insights into the evolution of the agricultural production of three SAARC countries—Bangladesh, India, and Pakistan—by using a multivariate UC model as described in a number of studies, such as those by Harvey (1985), Carvalho and Harvey (2005a) and Carvalho et al. (2007).

The multivariate UC model, which decomposes per capita agricultural production into trends ($\boldsymbol{\mu}_t$), cycles ($\boldsymbol{\psi}_t$), and interventions ($\boldsymbol{w}_t$), is given as follows:

$$y_t = \mu_t + \psi_t + \Lambda w_1 + \varepsilon_t \quad \varepsilon_t \sim NID(0, \Sigma_\varepsilon), \quad t = 1, \ldots\ldots, T, \tag{17.1}$$

where $\boldsymbol{y}_t$ is a 3 × 1 vector with elements representing time series observations of the log per capita of agricultural production corresponding to each country studied, i.e., log per capita agricultural production of India ($lpcapi_t$), Pakistan ($lpcapp_t$), and Bangladesh ($lpcapb_t$), with $t = 1$ for 1961 and $t = T = 50$ for 2010. The smooth trend component $\boldsymbol{\mu}_t$ is a 3 × 1 vector, $\boldsymbol{\psi}_t$ represents a 3 × 1 vector of the stochastic cycle component, and $\boldsymbol{\varepsilon}_t$ denotes a 3 × 1 vector of the unobserved irregular component term, which is normally distributed with a mean of 0 and a 3 × 3 covariance matrix Σ_ε.

The smooth trend $\boldsymbol{\mu}_t$ component in $\boldsymbol{y}_t$ is defined as follows:

$$\mu_1 = \mu_{t-1} + \beta_{t-1} + \eta_1 \quad \eta_t \sim NID(0, \Sigma_\eta), \quad t = 1, \ldots\ldots, T, \tag{17.2}$$

$$\beta_t = \beta_{t-1} + \zeta_t \quad \zeta_t \sim NID(0, \Sigma_\zeta), \quad t = 1, \ldots\ldots, T, \tag{17.3}$$

where $\boldsymbol{\beta}_t$ is the slope of the trend component ($\boldsymbol{\mu}_t$), which is a 3 × 1 vector with $t = 1$ for 1961 and $t = T = 50$ for 2010. The level disturbance $\boldsymbol{\eta}_t$ and the slope disturbance $\boldsymbol{\zeta}_t$ are uncorrelated with each other. Each of the Σ_η and Σ_ζ is a 3 × 3 covariance matrix. When Σ_η is not 0 but Σ_ζ is 0, $\boldsymbol{y}_t$ is called a "random walk plus drift"; however, a deterministic linear trend is generated when Σ_η and Σ_ζ are both 0. When Σ_η is 0 but Σ_ζ is not 0, the trend is called a "smooth trend." This model is often referred to as the integrated random walk (IRW). In this chapter, the IRW smoother is applied. Carvalho and Harvey (2005b) argued that this model often allows a clearer separation into trend and cycle.

The multivariate cyclical component ($\boldsymbol{\psi}_t$) captures short-term movement, and is defined as follows:

$$\begin{bmatrix} \psi_t \\ \psi_t^* \end{bmatrix} = \left\{ \rho \begin{bmatrix} \cos \lambda_c \sin \lambda_c \\ -\sin \lambda_c \cos \lambda_c \end{bmatrix} \otimes \mathrm{I_N} \right\} \begin{bmatrix} \psi_{t-1} \\ \psi_{t-1}^* \end{bmatrix} + \begin{bmatrix} \kappa_t \\ \kappa_t^* \end{bmatrix}, \qquad t = 1, \ldots, T, \tag{17.4}$$

where $\boldsymbol{\psi_t}$ and $\boldsymbol{\psi_t^*}$ are cycle components of 3 × 1 vectors with $t = 1$ for 1961 and $t = T = 50$ for 2010, and κ_t and κ_t^* are 3 × 1 vectors of disturbances such that:

$$E(\kappa_t \kappa_t^{'}) = E(\kappa_t^* \kappa_t^{*'}) = \Sigma_\kappa \qquad \text{and} \qquad E(\kappa_t \kappa_t^{*'}) = 0 \tag{17.5}$$

where $\boldsymbol{\Sigma_\kappa}$ is a 3 × 3 covariance matrix. The cyclical frequency λ_c defines the fluctuations of cycles which satisfy $0 \leq \lambda_c \leq \pi$. ρ is the damping factor on the amplitude of the cycle which satisfies $0 \leq \rho < 1$. The important nature of the cycle is the period which is related to frequency. The period of the cycle is given as $2\pi/\lambda_c$. This chapter considers two stochastic cycles: cycle 1 may capture the short-run cyclical dynamics; cycle 2 may capture the long-run cyclical dynamics. This is called a "similar cycle model," since the frequency λ_c and damping factor ρ are the same in each time series. Note that the covariance matrix of $\boldsymbol{\psi_t}$ is given by:

$$\Sigma_\psi = \left(1 - \rho^2\right)^{-1} \Sigma_\kappa \tag{17.6}$$

In Eq. (17.1), intervention dummies have been employed to capture outlier and structural breaks by including $\boldsymbol{w_t}$, which is a 3 × 1 vector of interventions. Some elements of the parameter matrix $\boldsymbol{\Lambda}$ could be specified to 0, indicating that certain variables can be excluded from the particular equations. An outlier is a temporary event of irregular disturbance, which is modeled by taking the value of 1 at the time of the outlier, and 0 otherwise. A structural break is modeled by a step intervention variable, which takes a value of 0 before the event, but 1 at the time of the structural break and onward. A structural break is a permanent event which shifts the level of the time series up or down.

There are common cycles if $\boldsymbol{\Sigma_\psi}$ is less than full rank. In addition, if the rank of $\boldsymbol{\Sigma_\psi}$ is 1, there is a single common cycle and model (17.1) can be written as:

$$y_{it} = \mu_{it} + \theta_i \psi_t + \Lambda w_{it} + \varepsilon_{it}, \quad i = 1, 2, 3 \qquad t = 1, \ldots\ldots, T, \tag{17.7}$$

where ψ_t is a scalar cycle, and the common cycle appears in each series, i.e., i = 1,2 and 3, with a different amplitude due to the presence of θ_is (c.f. Carvalho and Harvey 2005b).

The estimation of the unobserved components model can be formulated by using a maximum likelihood (exact score) approach. When the estimation has been made, the fitted model will be checked for serial correlation, normality and heteroskedasticity by using standard time series diagnostics. In addition, graphs of residual diagnostics, auxiliary residuals, prediction tests and forecasting are also used to detect any deficiency in the model.

Table 17.1 Descriptive statistics

Variables	Means	Standard deviation	Variables (logarithms)	Means	Standard deviation
pcapi	88.05	11.33	*lpcapi*	4.47	0.13
pcapp	85.31	10.41	*lpcapp*	4.44	0.12
pcapb	93.73	10.00	*lpcapb*	4.54	0.10

Notes pcapi stands for per capita agricultural production of India, *pcapp* stands for per capita agricultural production of Pakistan and *pcapb* stands for per capita agricultural production of Bangladesh. *lpcapi, lpcapp and lpcapb* represent logarithms of per capita agricultural production of India, Pakistan and Bangladesh, respectively

17.4 Data

The data set used is yearly data on the logarithm of the per capita agricultural production of India (*lpcapi*), Pakistan (*lpcapp*), and Bangladesh (*lpcapb*) from 1961–2010 with 2004–2006 = 100; it was obtained from the FAOSTAT statistical dataset (http://faostat.fao.org), accessed in October 2012. As FAO indicates, net per capita agricultural production is derived from the gross production after subtracting quantities used for seed and feed. The descriptive statistics of the data set used in this chapter are reported in Table 17.1.

Figure 17.1 depicts the evolution of the per capita agricultural production of India (*lpcapi*), Pakistan (*lpcapp*), and Bangladesh (*lpcapb*) from 1961–2010. The per capita agricultural production of India (*lpcapi*) showed a downward trend until 1966, and then gained an upward pace with some fluctuations, before reaching a high point in 2010. The per capita agricultural production of Pakistan (*lpcapp*) kept its upward trend (with some fluctuations) up until 2010. However, during the last year, it displayed a downward trend. The per capita agricultural production of Bangladesh (*lpcapb*) experienced a downward trend (with fluctuations) until 1972. Agricultural production then increased slightly, but reached its lowest level in 1994. After that, it dramatically regained pace, and reached its highest point in 2010.

17.5 Empirical Results

Models (17.1) to (17.7) can be estimated using the maximum likelihood approach as mentioned in Harvey (1989). The smooth trend and two stochastic cycle components can be extracted by using a smoothing algorithm mentioned in Koopman (1992). The empirical results obtained using the STAMP 8.2 package of Koopman et al. (2009) indicate strong convergence.

Table 17.2 reports some diagnostics and goodness-of-fit statistics such as N (χ_2^2) (the normality test following a χ^2 distribution with two degrees of freedom), H_{15} ($F_{15,\,15}$) (the heteroskedasticity test following an F distribution with (15, 15) degrees of freedom), Q(11, 5) (the Ljung Box statistic based on the first 11

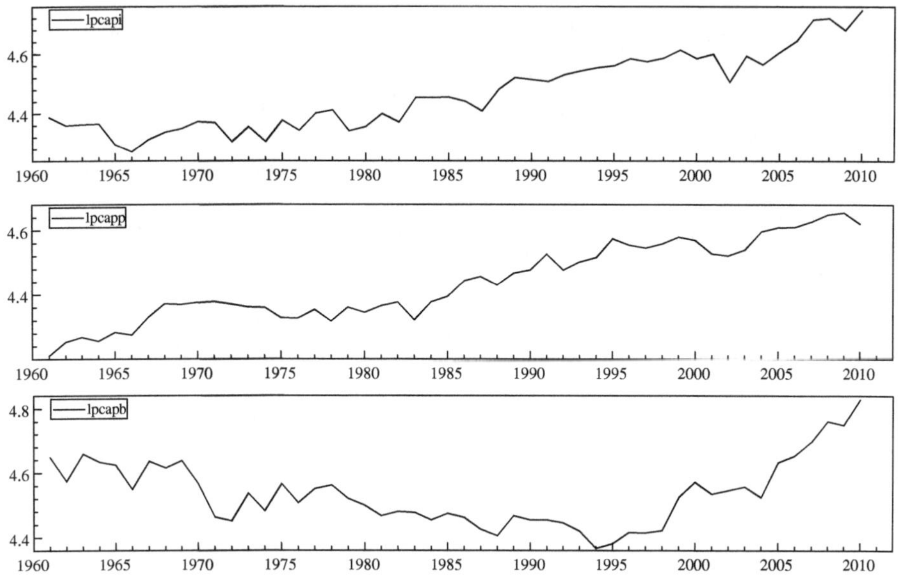

Fig. 17.1 Evolution of the FAOSTAT net agricultural production (PIN) of *lpcapi*, *lpcapp* and *lpcapb* 1961–2010

Table 17.2 Diagnostics and goodness-of-fit statistics

Statistics	*lpcapi*	*lpcapp*	*lpcapb*
$N(\chi^2_2)$	0.33970(0.8438)	0.83082(0.6601)	1.0007(0.6063)
$\mathbf{H_{15}(F_{15,15})}$	0.67644(0.7710)	0.63829(0.8028)	0.87051(0.6041)
Q(11,5)	6.2696(0.2809)	8.0679(0.1525)	7.3817(0.1938)
$\mathbf{R^2}$	0.93774	0.96904	0.8543

Note Values in parentheses are p-values

autocorrelations, which is tested against a χ^2 distribution with five degrees of freedom), and R^2 (coefficient of determination). The aforementioned statistics do not show any deficiencies in the estimated model.

Additional information about the estimated model is presented in Fig. 17.2, such as graphs of the standardized residuals, the spectral density, the residual correlogram and the density. The residuals are standardized one-step-ahead prediction errors or innovations (Koopman et al. 1999), and they are assumed to be normally and independently distributed for a correctly-specified model. The statistics presented in Table 17.2 and the residual graphs presented in Fig. 17.2 are used to check the validity of the model. In particular, the spectral density and the correlogram graph (Fig. 17.2) show that the residuals are not serially correlated.[1]

[1] The theoretical spectrum is a horizontal line for white-noise residuals.

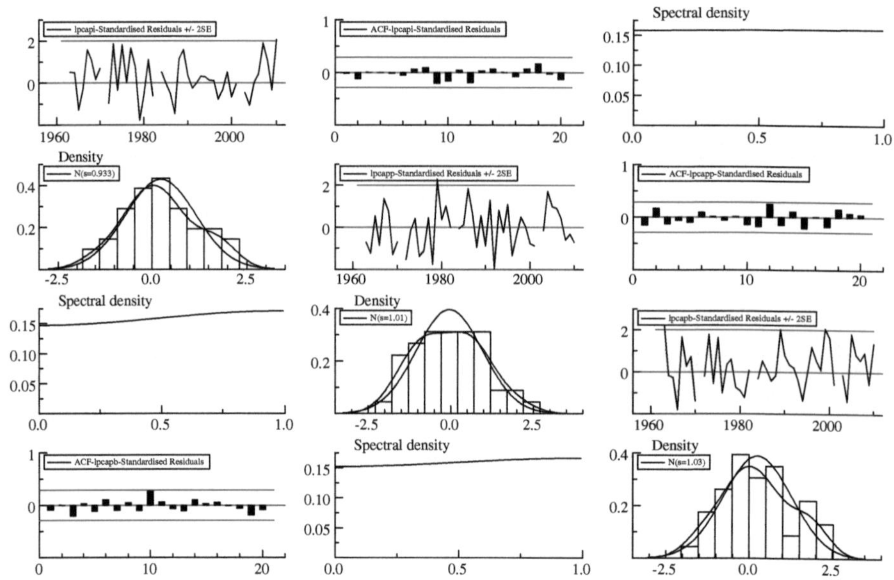

Fig. 17.2 Per capita agricultural production index residuals

Figure 17.3 presents the decomposed components of the structural time series model for trends (level and interventions) and for cycles 1 and 2. Cursory views of cycles 1 and 2 in Fig. 17.3 indicate that cyclical activities of *lpcapi*, *lpcapp,* and *lpcapb* in cycle 1 are increasing year by year, whereas cyclical activities in cycle 2 of the aforementioned variables are not so pronounced in the middle of the observation period.

As shown in Table 17.3, the q-ratio in the level of all three variables is 0, since it is deterministic. The q-ratio of cycle 1 of *lpcapb* is also 0, indicating that variations of this variable are coming from the remaining three sources (irregular, slope and cycle 2). In the case of *lpcapi*, most of the variations are coming from irregular components, followed by slope, cycle 2 and cycle 1. The q-ratios of *lpcapp* indicate that most of the variations are coming from irregular components followed by slope, cycle 2 and cycle 1, whereas large variations of *lpcapb* are attributed to irregular components, followed by slope and cycle 2. Furthermore, the q-ratios of the variables under consideration indicate that fluctuations in the irregular components are the most important sources of variations. The irregular components are due to unpredictable short-term fluctuations brought on by events such as unpredicted weather conditions, plant diseases, infestation, and unexpected changes in economic policies, among others. These unexpected events might play a pivotal role in the variation of the per capita agricultural production of these three countries as droughts, floods, cyclones, plant diseases, and infestations, like natural disasters, sometimes take place in these countries.

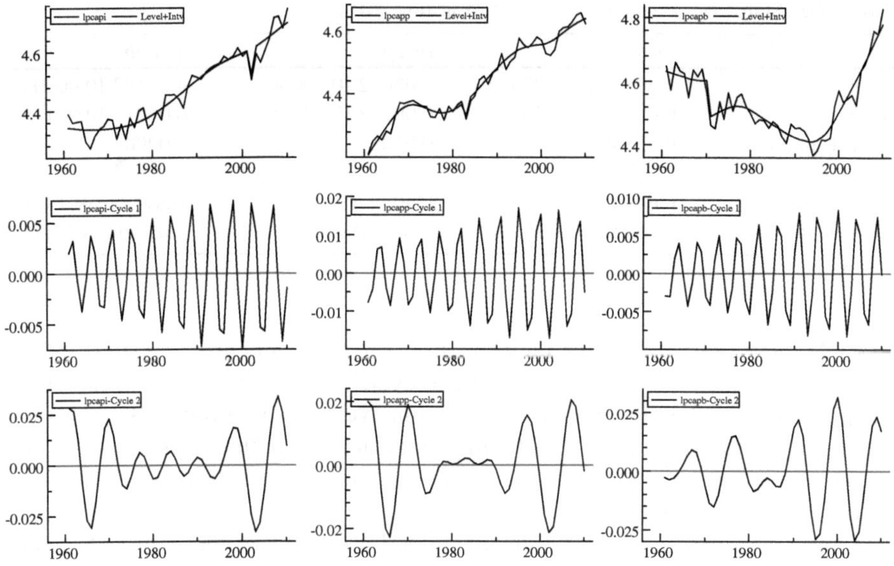

Fig. 17.3 Per capita agricultural production index decomposition

Table 17.3 Variance of disturbances: values and q-ratio

Variance of disturbances	*lpcapi*	*lpcapp*	*lpcapb*
Level	0.000000 (0.0000)	0.000000 (0.0000)	0.000000 (0.0000)
Slope	4.34041e-006 (0.006923)	2.29081e-005 (0.03654)	1.85210e-005 (0.02954)
Cycle 1	1.34113e-008 (2.139e-005)	2.55253e-007 (0.0004072)	0.000000 (0.0000)
Cycle 2	4.15169e-006 (0.006622)	2.45119e-006 (0.003910)	6.01105e-006 (0.009588)
Irregular	0.000626923 (1.000)	0.000147374 (0.2351)	0.000589656 (0.9406)

Notes Values in parentheses are the q-ratio. q-ratio is the ratio of each variance to the largest

Table 17.4 Parameters of cycle 1 and cycle 2

Parameters	Values
Number of order (n)	2
Period ($2\pi/\lambda_{c1}$) in years	4.50046
Frequency (λ_{c1})	1.39612
Damping factor ($\rho_{\psi 1}$)	0.94955
Number of order (n)	2
Period ($2\pi/\lambda_{c2}$) in years	9.61278
Frequency (λ_{c2})	0.65363
Damping factor ($\rho_{\psi 2}$)	0.85577

Table 17.5 State vector analysis in final state at time 2010

	lpcapi	*lpcapp*	*lpcapb*
Level	4.69866 [0.00000]	4.63672 [0.00000]	4.89582 [0.00000]
Slope	0.01237 [0.01257]	0.00718 [0.39608]	0.03017 [0.00438]
Cycle 1 amplitude	0.02866	0.01666	0.02004
Interventions			
	Outlier 2002	Outlier 1983	Level break 1971
Coefficient	−0.08945 [0.00035]	−0.04658 [0.00329]	−0.11687 [0.00092]

Note Values in brackets are p-values

Table 17.4 presents detailed information on the cyclical parameters of the model. The results show that the short-run cycle has a period of 4.5 years and a damping factor of 0.949, whereas the long-run cycle has a period of 9.6 years and a damping factor of 0.855. These findings indicate that both cycles exhibit a high degree of persistence, and that the series are stationary, since the damping factor of cycle 1 and cycle 2 are less than 1.

Table 17.5 reports the maximum likelihood estimates of the final state vector and intervention dummies.

The level values of *lpcapi*, *lpcapp,* and *lpcapb* are 4.698, 4.636 and 4.895, respectively, which are statistically significant, while the antilog analysis of the levels produces the values of 109.79, 103.20 and 133.72, respectively. Note that the antilog values of the levels are well above the corresponding mean values (Table 17.1). The slope yields a yearly growth rate of about 1.24, 0.72 and 3.02 % for *lpcapi*, *lpcapp,* and *lpcapb,* respectively. Note that only the growth rate of Pakistan is statistically insignificant. It is also observed that the amplitude of cycle 1 as a percentage of the trend is 2.87 % for *lpcapi*, 1.67 % for *lpcapp* and 2 % for *lpcapb*.

Bangladesh's high-growth rate might be the result of initiatives taken by the Bangladeshi government, such as the timely supply of agricultural inputs at affordable prices, an appropriate action plan for agricultural credit, agricultural mechanization, and pest management. On the other hand, policies taken by the Pakistani government, such as reducing agricultural subsidies, financial sector reforms, unfavorable agricultural price policy and bottlenecks in the agricultural inputs market, might explain Pakistan's insignificant per capita agricultural production growth. The amplitude of cycle 1 as a percentage of the trend for India indicates that India takes more time to respond against the short-run shocks of per capita agricultural production than Bangladesh and Pakistan, and also that it takes more time for India to come closer to the equilibrium level.

Table 17.6 reports the diagnostic test statistics of the auxiliary residuals, which are smoothed estimates of irregular, level and slope disturbances. These statistics show that, in general, the auxiliary residuals behave well.

The graphs of the t-values of corresponding estimated auxiliary residuals are reported in Fig. 17.4. A cursory look at the graphs shows that there is no absolute

Table 17.6 Normality tests (χ^2 tests) for auxiliary residuals: irregular, level and slope

	lpcapi			*lpcapp*			*lpcapb*		
	Irregular	Level	Slope	Irregular	Level	Slope	Irregular	Level	Slope
Skewness	0.0070408 (0.9331)	0.011647 (0.9141)	1.5105 (0.2191)	0.0024363 (0.9606)	0.0035495 (0.9525)	1.3252 (0.2497)	0.82142 (0.3648)	0.042154 (0.8373)	1.9561 (0.1619)
Kurtosis	1.9113 (0.1668)	0.41581 (0.5190)	0.84741 (0.3573)	1.1953 (0.2743)	0.72004 (0.3961)	0.42564 (0.5141)	1.3432 (0.2465)	0.44455 (0.5049)	1.1461 (0.2844)
Bowman-Shenton	1.9184 (0.3832)	0.42746 (0.8076)	2.358 (0.3076)	1.1977 (0.5494)	0.72359 (0.6964)	1.7509 (0.4167)	2.1646 (0.3388)	0.4867 (0.7840)	3.1022 (0.2120)

Note Values in brackets are p-values

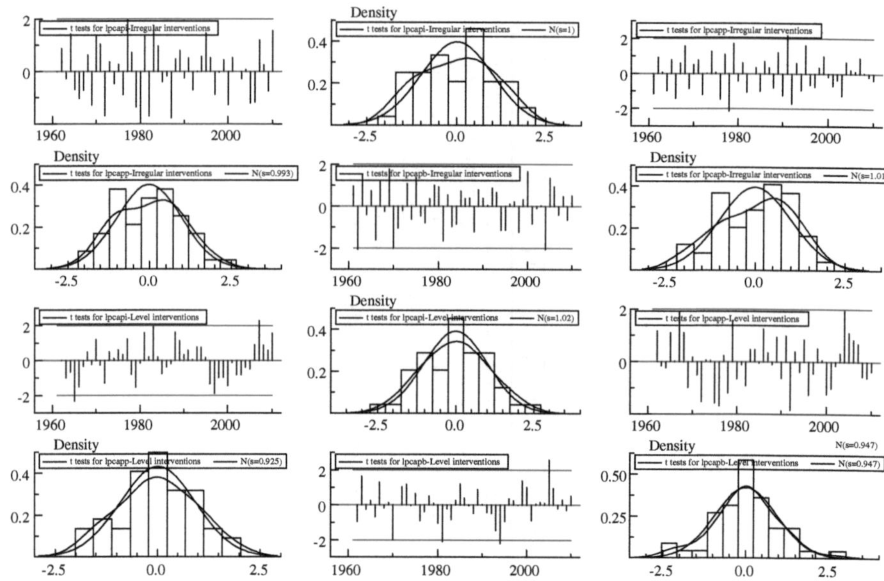

Fig. 17.4 Auxiliary residuals: irregular and level

value of more than 3, which indicates that the appropriate interventions are included in the model.

The interventions presented in Table 17.5 are negative and statistically significant, and capture decreases in per capita agricultural production. In particular, the 2002 outlier intervention which corresponds to India indicates a decrease in the country's per capita agricultural production (*lpcapi*) as a result of about 22 % less rainfall in the country that year. This drought is ranked as a hydrological drought, and it caused a decline in the production of paddy, oilseed, and wheat by 22, 23.5, and 9.5 %, respectively (Shah et al. 2009). The 1983 outlier corresponding to Pakistan indicates a drop in the agricultural production due to drought, and to the fact that during this year production was affected by infestation. The 1971 structural break corresponding to Bangladesh presents the negative effect on Bangladesh's agricultural production during the 9-month long liberation war of 1971. The war affected the structure of the country's agricultural sector, causing long-run negative effects on agricultural production, which is captured by a structural break intervention in 1971.

Figure 17.5 shows the prediction graphics of per capita agricultural production for *lpcapi*, *lpcapp,* and *lpcapb*, generated by estimating the model from 1961–2002. In this estimation, the years 2003–2010 are reserved for the out-of-sample forecast. The first predictions were made using information from 2003, and are updated each time a new observation is captured. The graphs show that the predicted values and residuals of *lpcapi*, *lpcapp,* and *lpcapb* are within the

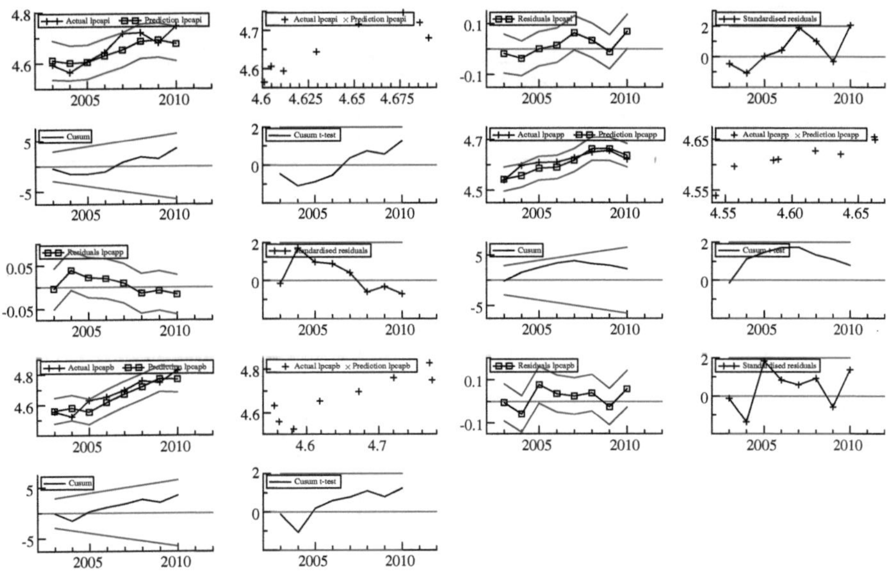

Fig. 17.5 Prediction testing for the per capita agricultural production index of *lpcapi*, *lpcapp* and *lpcapb*

Table 17.7 Post-sample prediction tests on *lpcapi*, *lpcapp* and *lpcapb*

	lpcapi	*lpcapp*	*lpcapb*
Failure χ^2_8 test	10.4894 (0.2323)	5.8714 (0.6616)	9.3550 (0.3132)
Cusum t(8) test	1.2534 (0.2454)	0.7983 (0.4477)	1.2305 (0.2535)

Note values in brackets are p-values

prediction intervals, indicating that the forecast of the trend has no more than two standard errors. The CUSUM graphs used for testing parameter stability and forecasting accuracy validate the specifications of the models of this chapter, and support their forecasting accuracy.

Moreover, the post-sample prediction tests of Table 17.7 support the aforementioned argument.

Figure 17.6 presents the forecast of the per capita agricultural production of *lpcapi*, *lpcapp,* and *lpcapb* for 5 years (2011–2015). The forecasted values are given within a band of 1 root mean square error (RMSE) on either side. The forecast graph of *lpcapi* shows that per capita agricultural production will remain static up to 2013 but will then increase. On the other hand, *lpcapp* will maintain high growth until 2013, but will be stable from then onward. The *lpcapb* forecast will be of high growth, indicating that the per capita agricultural production of Bangladesh is expected to remain above the historical production level up to the end of the forecasting period.

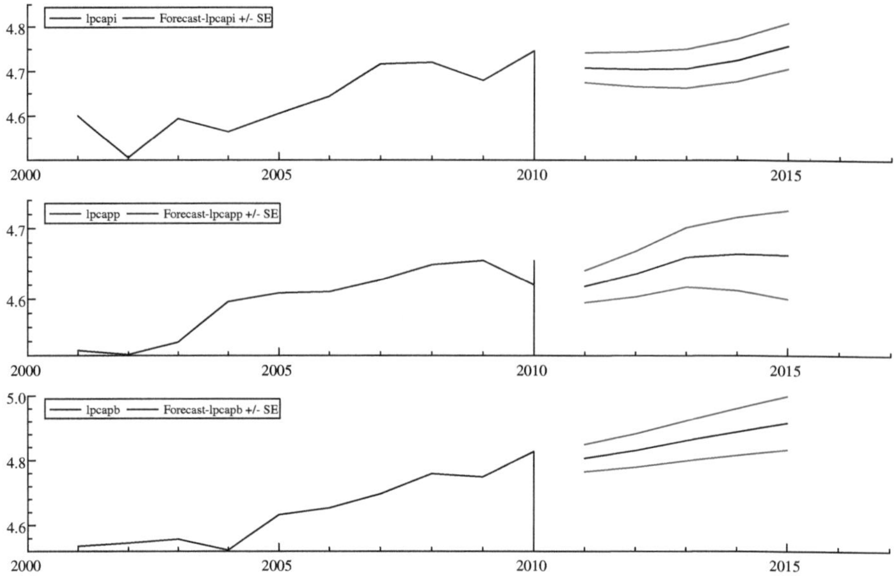

Fig. 17.6 Forecasts of the per capita agricultural production of *lpcapi*, *lpcapp* and *lpcapb*

17.6 Discussion

Disturbance covariances and corresponding correlations of irregular components, slope, cycle 1 and cycle 2 are reported in Table 17.8. Level components are not reported since this chapter uses multivariate smooth trend plus cycle models, i.e., level-fixed, slope-stochastic, and two stochastic cycles.

The irregular components (Table 17.8) do not show very strong evidence of correlation among *lpcapi*, *lpcapp*, and *lpcapb*. In particular, the correlation between Pakistan and Bangladesh (−0.057) is very weak, i.e., close to 0, indicating that an unexpected event in one country's per capita agricultural production does not affect the other country. This might lie in the fact that Pakistan is far away from Bangladesh, and thus why the unexpected factors affecting the irregular components, e.g., weather conditions, infestation, plant diseases and other unpredictable events, might be different and not affect both countries simultaneously. On the other hand, the correlation between India and Bangladesh (0.531) indicates that an unexpected event in per capita agricultural production in one country will have some effect on the other country's agricultural production in the same direction, indicating that these two countries might share some similarities in unpredictable events such as weather, plant diseases and infestation. For example, there are about 54 rivers flowing across the border of these two countries, which sometimes creates similar effects between them (e.g., floods and drought). The opposite happens in the case of the relationship between India and Pakistan: their correlation is negative (−0.371), indicating that an unexpected event in one

Table 17.8 Disturbance covariances and correlation of irregular, slope, cycle 1 and cycle 2

	lpcapi	*lpcapp*	*lpcapb*
Irregular			
lpcapi	1	−0.3711	0.5317
lpcapp	−0.0001215	1	−0.05785
lpcapb	0.0003878	−2.203e-005	1
Slope			
lpcapi	1	0.2543	−0.3135
lpcapp	2.622e-006	1	−0.5909
lpcapb	−3.563e-006	−1.595e-005	1
Cycle 1			
lpcapi	1	−0.6856	−0.9534
lpcapp	−1.728e-008	1	0.8733
lpcapb	−1.264e-008	2.176e-008	1
Cycle 2			
lpcapi	1	0.7940	0.3711
lpcapp	2.155e-006	1	−0.1125
lpcapb	1.491e-006	−2.955e-007	1

Note The lower triangular elements are the covariances and the upper triangular elements are the corresponding correlations

country negatively affects the other country. For example, a decline in India's per capita agricultural production in 2002 due to drought did not affect Pakistan; and a decline in Pakistan's per capita agricultural production in 1983 due to drought and infestation did not affect India, validating the negative correlation between India and Pakistan in their irregular component.

The correlations of slope demonstrate that there is some positive correlation (0.254) between India and Pakistan, indicating that India and Pakistan may have some similarities in their long-run per capita agricultural production trends. Kurosaki (2003) showed similar results, indicating that the treaty for the water of the Indus River between India and Pakistan—as well as agricultural institutional and policy changes such as land reforms and integration with green revolutions—might play an important role in long-run agricultural growth between India and Pakistan. On the other hand, Bangladesh displayed negative correlation with India (−0.313) and Pakistan (−0.590), indicating that the growth rate of Bangladesh has evolved in a diverse direction to the evolution of the per capita agricultural production of India and Pakistan. This indicates that the sustainable agricultural policy of Bangladesh is not that similar to the other two countries. India and Pakistan are now concentrating on satisfying export demands for agricultural products, whereas Bangladesh is concentrating on producing agricultural products to cover its domestic demand.

The estimated correlations of cycle 1 and cycle 2 are presented in the lower part of Table 17.8. The correlation of *lpcapi*, *lpcapp,* and *lpcapb* in cycle 1 ranges from −0.953 to 0.873, which strongly implies both negative and positive correlation. The correlation between Bangladesh and Pakistan (0.873) is strong; India, on the other hand, is strongly but negatively correlated with Bangladesh (−0.953) and

Pakistan (−0.685), which suggests that shocks from either country have a strong impact on the other country. This is perhaps an indication that, in the short term, Bangladesh is strongly but negatively influenced by India, since Bangladesh imports agricultural products from India, whereas this influence is not so strong between India and Pakistan, since Pakistan is less dependent on India for its imports of agricultural products. With regard to cycle 2, India is strongly correlated with Pakistan (0.794), but moderately correlated with Bangladesh (0.371). However, correlation between Pakistan and Bangladesh is weak and negative (−0.112).

An exclusive inspection of the correlation in cycles 1 and 2 reveals that these three countries are completely opposite when comparing their short and long-run cycles. For example, India is negatively correlated with Bangladesh and Pakistan in cycle 1, but positively correlated with the same countries in cycle 2: this indicates that in the short cycle Bangladesh and Pakistan might not reduce their differences with India, whereas in the long cycle they might recover their differences and positively correlate with India. Furthermore, Bangladesh is positively correlated with Pakistan in cycle 1, but negatively correlated in cycle 2, meaning that in the long cycle these two countries might diverge in their per capita agricultural production. The results further indicate that the transmission of shocks between India and Pakistan is stronger in cycle 2 (0.794) than in cycle 1 (−0.685), while the transmission of shocks between Bangladesh and Pakistan is stronger in cycle 1 (0.873) than in cycle 2 (−0.112); however, the transmission of shocks between Bangladesh and India in cycle 1 is stronger (−0.953) than in cycle 2 (0.371), indicating that in the long cycle Bangladesh is less influenced by the other two countries.

For the specification of the common cycle model, it is necessary for a restriction to be imposed in the disturbance variance matrices. The empirical results indicate that the correlation between the per capita agricultural production of Bangladesh (*lpcapb*) and India (*lpcapi*) is very strong (−0.9534) in the short-run cycle. Furthermore, the eigenvalue of Bangladesh (3.314e-023) is very small, and the actual rank of the short-run cycle disturbance variance matrix is less than full (2 instead of 3), which suggests that there may be common cycles. In this case, the common cycle model is estimated by imposing restrictions on the short-run cycle disturbance variance of Bangladesh by making it dependent on Pakistan and India, thus creating a cyclical disturbance variance matrix with a rank of 2. This specification is appropriate, since there is no deficiency in the diagnostics and goodness-of-fit statistics (Table 17.9), and the likelihood-ratio test does not reject the rank restriction on the short cycle's disturbances.[2] The findings of the two common cycles imply that the short cycle of Bangladesh is a linear function of the short cycles of India and Pakistan.

[2] Note that the log likelihood of the restricted and unrestricted models are very close (485.335 vs. 485.039), thus it does not reject the null hypothesis of the common cycle.

Table 17.9 Diagnostics and goodness-of-fit statistics

Statistics	*lpcapi*	*lpcapp*	*lpcapb*
$N(\chi^2_2)$	0.40297 (0.8175)	0.85429 (0.6524)	1.0135 (0.6025)
$H_{15}(F_{15,15})$	0.66282 (0.7825)	0.64025 (0.8012)	0.89135 (0.5867)
Q(11,5)	6.5767 (0.2541)	7.7569 (0.1702)	7.4591 (0.1887)
R^2	0.9378	0.96924	0.85343

Note Values in parentheses are p-values

Table 17.10 Disturbance covariance, correlation and factor loadings of the short cycle in the three variable models for per capita agricultural production

Cycle 1	*lpcapi*	*lpcapp*	*lpcapb*	Factor loading for *lpcapb*
lpcapi	1	−0.6896	−0.8650	−0.4396
lpcapp	−8.724e-009	1	0.9599	0.3738
lpcapb	−5.904e-009	1.379e-008	1	

Note *lpcapi* corresponds to India, *lpcapp* corresponds to Pakistan and *lpcapb* corresponds to Bangladesh. The lower triangular elements are the covariance; the upper triangular elements are the corresponding correlations; the last column shows the factor loading values

The estimated covariance, correlation and factor loadings of the short cycle are presented in Table 17.10. By comparing cycle 1's disturbance correlations in Table 17.10 to cycle 1's disturbance correlations in Table 17.8 it can be observed that the signs of the correlation remained unchanged, but the correlation between Bangladesh and India becomes weaker (−0.953 to −0.866), while the correlation between Bangladesh and Pakistan becomes stronger (0.873–0.959). It is worth noting that the correlation between Pakistan and India remained unchanged to the nearest second integer (−0.68). This is expected because only Bangladesh's short cycle depends on India and Pakistan, while the short cycle disturbances of India and Pakistan are independent. The factor loading for Bangladesh on India is −0.439, which produces a strong negative correlation of about −0.865. On the other hand, the factor loading for Bangladesh on Pakistan is 0.373, resulting in a strong positive correlation of about 0.959. These results confirm the presence of two common cycles – one between Bangladesh and India; and one between Bangladesh and Pakistan.

The estimated short-run cycle of the restricted model has a period of 4.49 years, with a frequency of 1.39 and a damping factor of 0.957—which is very close to the unrestricted model. The empirical results are also satisfactory since there is no major deviation from the unrestricted model. The decomposed components of the common cycle model for trend (level and interventions) and cycles 1 and 2 are presented in Fig. 17.7.

The graphs of cycle 1 for Bangladesh and Pakistan in Fig. 17.7 show that these two cycles have the same swings from peaks to troughs over time; however, the timings of the peaks and troughs are a little different, which validates that Bangladesh and Pakistan share a common cycle with a positive correlation of 0.959.

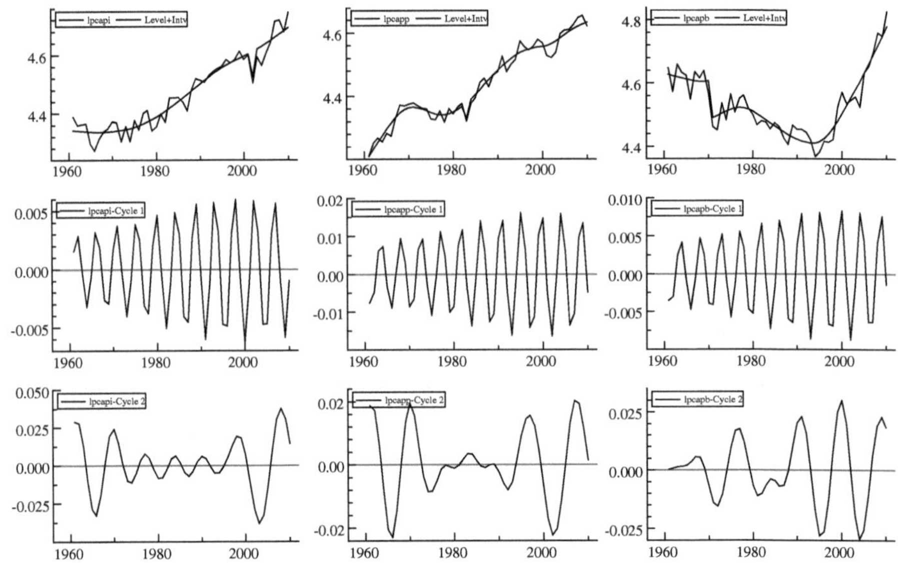

Fig. 17.7 A multivariate trend plus common cycle decomposition for *lpcapi*, *lpcapp* and *lpcapb*

On the other hand, Bangladesh and India in cycle 1 have opposite swings from peaks to troughs over time, having different peak and trough timings; this indicates that they share a negative common cycle with a correlation of −0.865.

17.7 Conclusions

The aim of this chapter was to examine the evolution of the per capita agricultural production of three SAARC countries: Bangladesh, India, and Pakistan. The SAARC is an organization representing eight South Asian nations. The aforementioned three countries from SAARC are selected since their agricultural economic importance to the region is very strong. The unobserved components model is employed to decompose the per capita agricultural production of India, Pakistan and Bangladesh into trends, cycles, interventions and irregular components. For the specification of the model, smooth trend plus stochastic cycle components are considered, i.e., fixed level, stochastic slope and two stochastic cycles and irregular components. This approach provides some significant insights into the evolution of the per capita agricultural production of these three countries. Empirical results of this chapter indicate that the per capita agricultural production of the three countries' time series are best fitted by smooth trend plus the stochastic cycle model, since there are no deficiencies in the diagnostic statistics. Furthermore, intervention dummies have been included, which accurately captured shocks.

The findings of this chapter reveal that in the irregular components, the correlation between the countries is not very strong. There is some positive correlation between Bangladesh and India, which might be the result of the presence of 54 rivers between these two countries, as well as some similarities in their weather conditions. The correlation between Pakistan and India is small but negative, indicating that any short-run unexpected events of one country will negatively affect the other country, though the effect is not strong. From the literature, it is observed that unexpected events like drought, floods, and infestation in one country do not have a significant effect on the other country. Correlation between Bangladesh and Pakistan is very weak, indicating that the irregular component between these two countries has an insignificant impact on the variation of their per capita agricultural production.

With regard to slope, Bangladesh is negatively but moderately correlated with India and strongly correlated with Pakistan, indicating that in the long-term, Bangladesh's per capita agricultural production is diverging from those of the other two countries. On the other hand, India and Pakistan have some positive correlation, indicating long-run positive movements in their per capita agricultural production trend. The growth rate of Bangladesh is high, followed by India and Pakistan, with the latter presenting insignificant growth. This is perhaps an indication that the substantial governmental agricultural policy of Bangladesh is boosting the growth of the per capita agricultural production; conversely, reducing agricultural subsidies, financial sector reforms, agricultural price policy and bottlenecks in the agricultural inputs market of Pakistan might be the causes of its insignificant agricultural growth. Knowledge about these growths could be useful when designing future agricultural development policies.

In the short cycle, India is strongly but negatively correlated with Pakistan and Bangladesh, while correlation between Bangladesh and Pakistan is also strong but positive, indicating that the transmission of transitory shocks between these countries is strong. In the case of the long cycle, India is strongly correlated with Pakistan, whereas Bangladesh is moderately correlated with India and Pakistan. These findings reveal that the transmission of shocks between India and Pakistan lasts longer, whereas the transmission of shocks between Bangladesh and India, and between Bangladesh and Pakistan, is transitory and does not last long. Furthermore, in the short cycle, the per capita agricultural production of Pakistan and Bangladesh shares a common cycle, with positive correlation indicating convergence between these two countries; the per capita agricultural production of India and Bangladesh also shares a common cycle, with negative correlation indicating divergence between these two countries. Finally, the present chapter presents yearly forecasted values of the per capita agricultural production of Bangladesh, India, and Pakistan from 2011–2015. The forecasts show that Bangladesh will maintain its upward growth until 2015, India will keep its growth stable up to 2013 before rising, whereas Pakistan will plateau at the end of the forecasting period.

Acknowledgements Financial support by the state scholarship foundation (IKY), Greece is gratefully acknowledged.

References

Carvalho, V. M., & Harvey, A. C. (2005a). Convergence in the trends and cycles of euro-zone income. *Journal of Applied Econometrics, 20*, 275–289.

Carvalho, V. M., & Harvey, A. C. (2005b). Growth, cycles and convergence in US regional time series. *International Journal of Forecasting, 21*, 667–686.

Carvalho, V. M., Harvey, A. C., & Trimbur, T. (2007). A note on common cycles, common trends, and convergence. *Journal of Business & Economic Statistics, 25*(1), 12–20.

Chaudhry, M. G., Chaudhry, G. M., & Qasim, M. A. (1996). Growth of output and productivity in pakistan's agriculture: trends, sources, and policy implications. The Pakistan Development Review *35*(4)Part II, 527–536 (1996).

Cuevas, M. A. (2002). Potential GDP growth in Venezuela: A structural time series approach. Policy Research Working Papers, 2826, The World Bank.

Ferrara, L., & Koopman, S. J. (2010). Common business and housing market cycles in the euro area from a multivariate decomposition. Banque de France, Working Paper N275.

Flaig, G. (2002). Unobserved components models for quarterly German GDP. Center for Economic Studies & IFO Institute for Economic Research, Munich, Germany, Working Paper no. 681(5).

Harvey, A. C. (1985). Trends and cycles in macroeconomic time series. *Journal of Business & Economic Statistics, 3*(3), 216–227.

Harvey, A. C. (1989). *Forecasting, Structural Time Series Models and the Kalman Filter.* Cambridge: Cambridge University Press.

Harvey, A. C., & Jaeger, A. (1993). Detrending, stylized facts and the business cycle. *Journal of Applied Econometrics 8*(3), 231–247.

Koopman, S. J. (1992). Disturbance smoother for state space models. *Biometrika, 80*(1), 117–126.

Koopman, S. J., & Azevedo, J. V. (2008). Measuring synchronisation and convergence of business cycles for the Euro area, U.K. and U.S. *Oxford Bulletin of Economics and Statistics, 70*(1), 23–51.

Koopman, S. J., & Lucas, A. A. (2005). Business and default cycles for credit risk. *Journal of Applied Econometrics, 20*, 311–323.

Koopman, S. J., Shephard, N. & Doornik, J. A. (1999). Statistical algorithms for models in state space using SsfPack 2.2. *Econometrics Journal, 2*, 113–166. Retrieved from http://www.ssfpack.com.

Koopman, S. J., Harvey, A. C., Doornik, J. A. & Shephard, N. (2009). Structural time series analyser and modeller and predictor STAMP 8.2. London: Timberlake Consultants Ltd.

Kannan, E., & Sundaram, S. (2011). Analysis of trends in India's agricultural growth. The Institute for Social and Economic Change, Bangalore, India, Working paper 276.

Kurosaki, T. (2003). Long-term agricultural growth and agricultural policy in India and Pakistan. Reform of the agricultural sector, 5th GDN annual conference, Delhi, India. Retrieved from January 28 2004.

Kurosaki, T. (2007). Land-use changes and agricultural growth in India, Pakistan, and Bangladesh, 1901–2004. Conference PAPER, Indian Statistical Institute, New Delhi, India.

Labys, W. C., & Kouassi, E. (1996). Structural time series modeling of commodity price cycles. Research Paper 9602.

Mitra, S., & Sinclair, T. M. (2010). Macroeconomic fluctuations in emerging economies: An unobserved components approach. Macroeconomics Research Group discussion paper series, no 3911. University of Queensland, Australia.

Quddus, M. A. (2009). Crop production growth in different agro-ecological zones of Bangladesh. *Journals of Bangladesh Agricultural University, 7*(2), 351–360.

Rehman, F., Saeed, I., & Salam, A. (2011). Estimating growth rates and decomposition analysis of agriculture production in Pakistan: pre and post sap analysis. *Sarhad Journal of Agriculture, 27*(1), 125–131.

Robbani, M., Siddiquee, S. U., Zaman, S., & Nakamura, H. (2007). Agriculture in Bangladesh—Recent Trend and Agroenvironment toward Sustainability. *Journal of the Faculty of Agriculture, Shinshu University 43*(1–2),17–25.

Shah, T., Kishore, A. & Hemant, P. (2009). Will the impact of the 2009 drought be different from 2002?. *Economic & Political Weekly*, *XLIV*, 37. Retrieved from September 12, 2009.

Wongwachara, W., & Minphimai, A. (2009). Unobserved component models of the Phillips relation in the ASEAN economy. *Journal of Economics and Management, 5*(2), 241–256.

Chapter 18
Assessing the Evolution of Technical Efficiency of Agriculture in EU Countries: Is There a Role for the Agenda 2000?

G. Vlontzos and S. Niavis

Abstract One of the major goals of the Agenda 2000 for EU agriculture was to increase its market orientation and improve the competitiveness level of the primary sector of member states. In order to evaluate if this goal has been reached, both Stochastic Frontier Analysis (SFA) and Data Envelopment Analysis (DEA) models have been applied. This research focuses on the 2003–2011 period. Inputs include agricultural land, labour and fixed capital consumption, and output is the total agricultural output of each country. Both models prove that there is an increasing trend on mean efficiency levels of the primary sector of member states of the EU. The application of the Mann-Whitney U-test prove that eastern European countries, which recently became EU members, are significantly less efficient, compared with the older member states. This finding applies for both efficiency measurement methodologies, strengthening in this way the validity of the results.

18.1 Introduction

The Agenda 2000 and the 2003 Common Agricultural Policy (CAP) reform established a totally new operational framework for the EU agricultural holdings. The cornerstones of this new policy scheme are the decoupling of subsidies from farm production and a holistic approach to the rural development issue. One of the main goals of CAP since its establishment, as part of the Treaty of Rome, is to *"increase productivity by promoting technical progress and ensuring the optimum*

G. Vlontzos (✉)
Department of Agriculture Crop Production and Rural Environment, University of Thessaly, Volos, Greece
e-mail: gvlontzos@agr.uth.gr

S. Niavis
Department of Planning and Regional Development, University of Thessaly, Volos, Greece

C. Zopounidis et al. (eds.), *Agricultural Cooperative Management and Policy*, Cooperative Management, DOI: 10.1007/978-3-319-06635-6_18,

use of factors of production". Efficiency increase of primary sectors is synonymous with the upgrading of economic performance of agricultural holdings in participant countries. It was common belief that the single payment scheme, which is the outcome of the latest CAP reform, should increase the agricultural holdings' flexibility, improving in this way their ability to exploit the investment and market opportunities, without jeopardising the loss of EU subsidy funds which until 2004 were tailored to specific agricultural products. After several decades of implementation of subsidy schemes tailored to production, it was decided that it was time to apply the economic theory which suggests that decoupling subsidies will lead to a reduction of efficiency loses (Serra et al. 2006).

The issue of the impact of subsidies decoupling on the primary sector performance materialized in the US after the implementation of the 1996 Federal Agricultural Improvement and Reform Act and the 2002 Farm Security and Rural Investment Act. The general conclusion is that there still is distorted farm behaviour, even when subsidies are decoupled. Perhaps the most characteristic findings of that are from Hennessy (1998), where decoupled support policies affect the decisions of risk-averse farmers when there is uncertainty. Chau and Gorter (2005) proved that decoupled payments provide the opportunity to some agricultural holdings to remain in business, without being able to gain profits, by using part of these payments to cover production costs. A different approach regarding managing practices of farms is mentioned by Femenia et al. (2010). According to this, the single payment scheme encourages a prosperity effect which diversifies farmers' attitude towards risk, which leads to production responses. Finally, it is proven that participation in the 1996 FAIR Act increased production among participant farmers (Key et al. 2005).

Regarding the EU CAP, there is some empirical work on assessing the impact of coupled payments on agriculture efficiency. All of these studies verify that the previous model of subsidy administration had a negative impact on technical efficiency (TE) or productivity. Rezitis et al. (2003) verify that subsidies, off-farm family income and hired labour factors are factors that have a negative impact on TE on Greek farms participating in EU farm credit programs. Iraizoz et al. (2005) showed that CAP works against efficiency, regarding beef production in Spain. Similar results were presented by Karagiannis and Sarris (2005) regarding tobacco growers in Greece. The degree of TE was found to be lower than the degree of scale efficiency. The fact that the majority of farms in the research sample exhibited suboptimal scale implies that their output should be expanded. In England and Wales factors influencing positively the technical change at the farm-level were farm or herd size, farm debt ratios, farmer age, levels of specialization, and ownership status, without EU subsidies having a positive impact on this (Hadley 2006). Skuras et al. (2006) prove that capital subsidies affect total factor productivity through technical change and not through scale efficiency. Zhu et al. (2008) suggest that the latest CAP reforms had a significant impact on the TE and TE change in Germany, Sweden and Netherlands. Additionally, decoupled subsidies create delusion regarding gaining income, reducing in this way motivation for changes on production activities and practices.

The present work is an attempt to assess the TE of the primary sectors of EU member states, by implementing both parametric and non-parametric methodologies. This research focuses on the 2003–2011 period, covering both the pre- and post-CAP reform periods. In order to examine the TE evolution of the period under analysis, dynamic models of Stochastic Frontier Analysis (SFA) and Data Envelopment Analysis (DEA) are being used. The utilized agricultural land, labour and fixed capital consumption are used as inputs and the total agricultural output of each country as output.

18.2 Methodological Review

SFA and DEA can be regarded as the most widely used approaches for performance measurement in the primary sector, as they encompass the concept of benchmarking (Headey et al. 2010). Both methodologies rely on the theory of production frontiers which was firstly introduced by Farrell (1957) but are based on different assumptions concerning their application.

DEA focuses on the comparison of the TE of a number of Decision Making Units (DMUs). It involves the solution of a series of linear programming problems, in which both the inputs and outputs of the production process are employed to calculate the relative efficiency of each DMU. The methodology was first suggested by Farrell (1957) and it was then extended by Charnes et al. (1978) and Banker et al. (1984). The two basic DEA models refer to the DEA-CCR model, which assumes constant returns to scale (CRS), and the DEA-BCC model, which assumes variable returns to scale (VRS). SFA methodology is based on specific production functions, and hence, it incorporates the basic principles of production economic theory. SFA is an econometric technique which attributes the deviation from the frontier to the loss of efficiency in the production process and to the randomness or statistical noise. Thus, inefficiency of each DMU is determined by its deviation of the production frontier (Coelli et al. 2005).

The two categories of methodologies display specific advantages and weaknesses. The basic strength of DEA is its simplicity, as it constitutes a non-parametric analysis, which is independent from assumptions on production functional form and error distribution. However, the basic DEA models cannot incorporate statistical errors in efficiency estimations. Additionally, SFA estimations encompass statistical properties, as SFA adapts econometric techniques for efficiency estimation. The main disadvantage of SFA methodology arises from the fact that, as an econometric approach, it comes up with strong a priori assumptions about the production technology of the DMU set under evaluation, since the true production technology is unknown. Taking into account the strengths and limitations of the two methodologies, the choice between them relies on the specific characteristics of the problem at hand. When the sample size allows the implementation of parametric methods then it is advisable that both of the methodologies are employed in order to assess the efficiency of DMUs. The comparative application

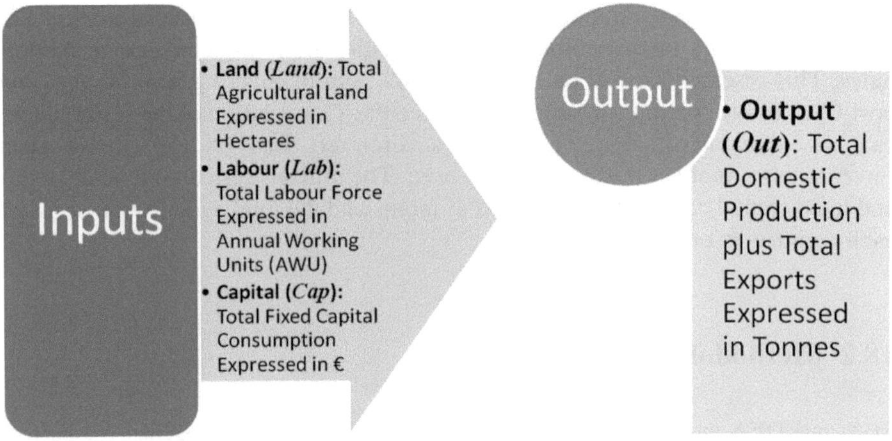

Fig. 18.1 EU's agricultural production process

of the two methodologies leads to more accurate results and allows the identification of possible differences in the efficiency estimates which mainly arise from the different assumptions that are used in order for the production frontier to be estimated by DEA and SFA methodologies (Banker et al. 1986; Sharma et al. 1997; Wang et al. 2005).

There are previous attempts to apply similar methodologies to assess efficiency in agriculture. The assessment of TE in MENA countries by both DEA and SFA concluded with similar efficiency scores, with the best performance in both models being achieved by Qatar (Zamanian et al. 2013). Comparison of efficiency scores, regarding technical and scale efficiency of Italian citrus farming, proved to be similar for the first case, while SFA scores appear to be larger, compared to the DEA ones for the second (Madau 2012). Finally, application of the same methodological approach has been applied for cattle farming too. The main findings are that both methodologies verify that an integrated vegetable-based system is more efficient, compared to a milk-based system in terms of milk revenue as well as total revenue (Serasinghe et al. 2003).

Taking into account the aforementioned, the present chapter will rely both on the non-parametric method DEA and on the parametric method SFA in order to capture the efficiency trends of EU countries' agricultural sectors. SFA will be based on the simple Cobb-Douglas production form and the DEA on the CRS output model. Both models use panel data which covers the period 2003–2011. The selected period provides adequate information about the efficiency levels before and after the implementation of the new CAP, which has as a cornerstone the decoupling of subsidies from the production of specific agricultural products. The conceptual model employed by the present chapter in order to assess the efficiency of EU's agricultural sectors and the variables used in the evaluation process is presented in Fig. 18.1.

Regarding DEA, estimation will be conducted with the use of Window DEA Analysis. DEA Window Analysis constitutes a dynamic DEA model which can capture the efficiency trends of EU countries' agricultural sectors. For the execution of a DEA Windows Analysis problem the basic DEA models should be calculated. For an extended analysis of the basic DEA models' formulation, readers are referred to Cooper et al. (2000) and Zhu (2003). Furthermore, the model considers a country under evaluation as it was a different country over time. Then, a number of windows are constructed and the DEA models are adjusted to the data of each window.

In order to construct a DEA Window Analysis problem several assumptions should be made about the number and the size of the windows. Moreover, the procedure requires the fragmentation of the period analysis T into sub-periods or windows. Each window (w) has the same length (p) with the others. Supposing that the problem focuses on the evaluation of $\boldsymbol{N}$ countries, then the first window contains $\mathrm{N_w}$ countries under evaluation for the sub-period $(1,\ldots,p)$. The second window includes N_w countries under evaluation for the sub-period $(2,\ldots,p+1)$. The procedure is repeated until the efficiency of the countries that are included in the last window of the sub-period $(T-p+1,\ldots,T)$ is estimated. The number of the countries that are finally evaluated in each window is given by $N_w = N * p$ and the total number of countries under evaluation is equal to $N_{tw} = N * p * w$ (Cooper et al. 2000).

The adoption of DEA Windows Analysis has significant advantages because its results provide more detailed information about the performance of each country than the information gathered from the basic DEA models. Furthermore, the consideration of each country as it was another country during the window process leads to repeated measures of efficiency. The higher the price of (w) and (p), the more the repeated measures are conducted. The rows and columns of the windows can then be used in order to test the stability of each country's efficiency and the identification of efficiency time trends, respectively. However, it should be noted that the major drawback of the method is that there is an inconsistency in the number of times that each period is tested. Thus, the efficiency of countries in the middle periods of the panel are tested more intensely than in the starting and ending periods (Cooper et al. 2000; Polyzos et al. 2012).

Based on the output-oriented DEA envelopment model (Zhu 2003), the formulation of the empirical DEA Window model for the first window of the EU27-country data and the values of the variables that will be used to construct the windows are as follows:

$$\max_{\varphi,\lambda} \varphi^*$$

$$\text{s.t.}$$

$$\sum_{j=1}^{N_1} \mathrm{Land}_j \lambda_j \leq \mathrm{Land}_o$$

$$\sum_{j=1}^{N_1} \mathrm{Lab}_j \lambda_j \leq \mathrm{Lab}_o$$

$$\begin{aligned}
&\sum_{j=1}^{N_1} \text{Cap}\lambda_j \leq \text{Cap}_o \\
&\sum_{j=1}^{N_1} \text{Out}_j\lambda_j \geq \varphi\text{Out}_o \\
&\sum_{j=1}^{N_1} \lambda_j = 1,\ \lambda_j \geq 0\,\forall j,\ N_1 = (1, 2, \ldots, 135) \\
&T = 9,\ N = 27,\ p = 5,\ w = 5
\end{aligned} \tag{1}$$

Additionally, the parametric estimation of EU countries' efficiency will be based on the time-varying SFA model that has been proposed by Battesse and Coelli (1992). The model is based on the assumption that efficiency levels for countries vary across time as the stochastic component of the error term (U) is a parametric function of time. The specification of this parametric function is the following:

$$U_t = \eta_t U = \{\exp[-\eta(t - T)]\}U$$

where

η an unknown parameter to be estimated

t the time periods of estimations (t = 1, 2, …, T)

The inefficiency term satisfies the non-negativity condition ($U_t \geq 0$), and must either increase at a decreasing rate if $\eta < 0$, decreasing in an increasing way if $\eta > 0$, or remaining constant if $\eta = 0$. The assumption of varying efficiency levels renders the capture of efficiency trends possible and furthermore the results are comparable to the respective results of the Window DEA model. Finally, in order to separate inefficiency change from technology progress we incorporate time dummies into the deterministic part of the equation (Battesse and Coelli 1992). Taking into account the initial year under research (2003), eight dummies representing the respective years are entered into the model. The formulation of the SFA model is presented as follows:

$$\begin{aligned}
\ln \text{Out}_{it} = {} & \beta_0 + \beta_1 \ln \text{Land}_{it} + \beta_2 \ln \text{Lab}_{it} + \beta_3 \ln \text{Cap}_{it} + \beta_4 D_{2004} \\
& + \beta_5 D_{2005} + \beta_6 D_{2006} + \beta_7 D_{2007} + \beta_8 D_{2008} + \beta_9 D_{2009} \\
& + \beta_{10} D_{2010} + \beta_{11} D_{2011} + (v_{it} - u_{it}) \\
& (i = 1, 2, 3, \ldots\ldots, 27), (t = 1, 2, 3, \ldots, 9)
\end{aligned} \tag{2}$$

Table 18.1 Descriptive statistics of the inputs and output

		Output	Inputs		
		Out	Land	Lab	Cap
2003	Mean	16240.4	6784.9	490.5	1689.3
	Coefficient of variation	133 %	119 %	138 %	158 %
2004	Mean	16970.8	6790.7	467.9	1749.9
	Coefficient of variation	131 %	118 %	135 %	157 %
2005	Mean	16260.6	6709.5	468.0	1818.9
	Coefficient of variation	130 %	117 %	141 %	157 %
2006	Mean	16202.8	6642.3	458.3	1884.4
	Coefficient of variation	128 %	117 %	142 %	156 %
2007	Mean	17599.7	6578.5	436.1	1992.5
	Coefficient of variation	128 %	117 %	140 %	154 %
2008	Mean	18610.2	6595.6	425.0	2087.4
	Coefficient of variation	127 %	119 %	142 %	155 %
2009	Mean	16520.5	6829.9	411.3	2086.3
	Coefficient of variation	129 %	126 %	144 %	157 %
2010	Mean	17538.7	6567.5	384.6	2134.4
	Coefficient of variation	128 %	119 %	137 %	154 %
2011	Mean	19276.0	6560.9	376.2	2191.0
	Coefficient of variation	127 %	120 %	137 %	153 %

Source Eurostat (2013) and unpublished data from the Greek statistic Bureau

18.3 Empirical Application

The descriptive statistics of the inputs and outputs employed by the present chapter are presented in Table 18.1. It is noticeable that the data being used for this research cannot be characterised as homogenous, because the CV value is exceeding 100 % in each year. One positive issue, providing hints for the final outcome, is the constant increase of the Agricultural Output after the year 2007, as well as the gradual decrease of the variability of observations. Regarding the inputs being utilised, the Land factor is being reduced, the Labour factor is being reduced even faster, and the Fixed Capital factor is significantly increased.

The results of the SFA model application are presented in Table 18.2. The value of the Likelihood Ratio Test statistic with three restrictions is exceeding the critical value from the χ^2 distribution. Thus, we can reject the null hypothesis that a model without variables would perform better than the one selected for this chapter, at the <0.01 significance level. Furthermore, the statistically significant estimation of the γ coefficient leads to the conclusions that inefficiency exists in the present data. Additionally, the good fit of the model is testified from the statistically significant estimations of the σ^2 and μ parameters.

The estimation of the regressors of the chosen variables is statistically significant, showing the expected positive sign. Thus, all of the inputs positively contribute to the output and the selected variables fulfil the isotonicity condition.

Table 18.2 Results of the SFA model parameters' estimation

Coefficient	Estimate	Standard error	$P > \|z\|$
β_0	4.481	0.356	0.001
$\beta_1(\ln \text{Land})$	0.274	0.067	0.001
$\beta_2(\ln \text{Lab})$	0.220	0.066	0.001
$\beta_3(\ln \text{Cap})$	0.327	0.081	0.001
$\beta_4(D_{2004})$	0.048	0.020	0.017
$\beta_5(D_{2005})$	−0.007	0.022	0.754
$\beta_6(D_{2006})$	−0.021	0.024	0.391
$\beta_7(D_{2007})$	0.046	0.027	0.085
$\beta_8(D_{2008})$	0.092	0.030	0.002
$\beta_9(D_{2009})$	−0.054	0.034	0.117
$\beta_{10}(D_{2010})$	−0.010	0.038	0.780
$\beta_{11}(D_{2011})$	0.075	0.041	0.070
σ^2	0.288	0.103	0.005
γ	0.982	0.012	0.001
μ	1.018	0.185	0.001
η	0.014	0.005	0.007
Log likelihood	212.9		
Likelihood ratio test	398.05	$P(\chi^2)$	0.001

Taking into account the estimated production elasticities of inputs, it is evident that the Cap production factor is the most important one, compared with the other two being used in this research. This is a quite positive outcome, because it is signalling the transformation of European agricultural sectors towards more entrepreneurial forms of operation. On the contrary though, the Lab production factor appears to achieve the lowest score. This finding verifies the need for further improvement of the skills of agricultural employees in order for their efficiency to be increased. Moreover, the estimated coefficients for the time dummy variables vary significantly, both in terms of sign and statistical significance. Thus, we are not in a position to spot any pattern of technical change through the period 2003–2011. Additionally, the estimation of the time varying parameter η is found to be positive and statistically significant at the 0.01 level. This result indicates that the TE of the agricultural sector in Europe gradually increased in the period 2003–2011.

Focusing on the analysis of TE, the results after the implementation of the SFA analysis will be comparatively presented with the DEA analysis results. The implementation of both parametric and non-parametric methodologies increases the credibility of the results, because they are being extracted by totally different methodological approaches.

Figure 18.2 presents the average estimated efficiency of the EU agricultural sector. Additionally, the average TE of the two sub-periods before and after the implementation of Agenda 2000 and the average efficiency change are highlighted. The estimated mean TE highlights that there is a small but significant efficiency change after the implementation of the new CAP. The estimation of this change is

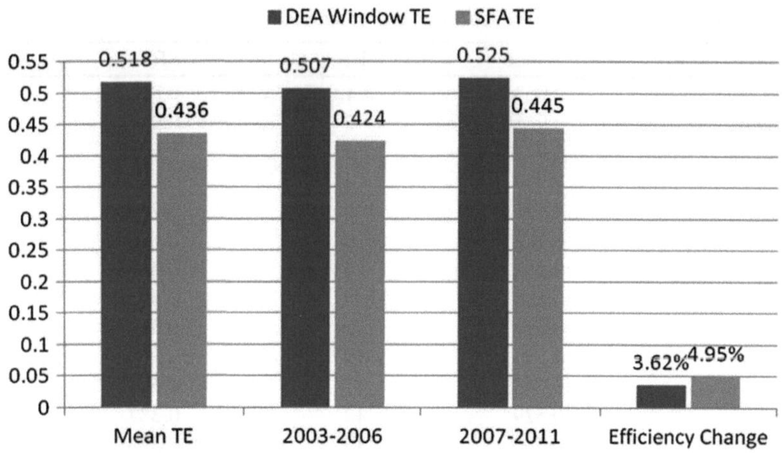

Fig. 18.2 EU agricultural sector efficiency estimation and efficiency change

Table 18.3 EU agricultural sector efficiency estimation and efficiency change

Model	Mean efficiency		
	2003–2007	2008–2011	Change (%)
DEA	0.514	0.525	2.14
SFA	0.427	0.448	4.80

similar for both DEA Window and SFA. If the two sub-periods are changed, by including the year 2007 into the first sub-period, there are slightly different mean efficiency scores for both models, as expected, as shown in (Table 18.3). The most important finding though is that the mean efficiency is being improved in both cases, providing hints that the implementation of the new CAP subsidy administration strengthens this efficiency upgrading procedure.

The consistency of the TE estimations produced by the two models is verified by the high and positive value of the estimated Spearmann correlation coefficient (0.770) which is statistically significant at the 0.01 level and the estimated Pearson correlation coefficient (0.837) which is also statistically significant at the 0.01 level. Furthermore, although the average estimated TE seems to be lower when SFA results are taken into account, the results of the non-parametric Mann-Whitney test do not reveal significant differences between the two models' outcomes (Mann and Whitney 1947). More specifically, the value of the z statistic (1.237) produced by the implementation of the Mann-Whitney test doesn't lead us to the rejection of the null hypothesis that the average of the two models' TE values significantly differs, as the estimation is not statistically significant ($P > |z| = 0.216$).

Table 18.4 presents the efficiency scores of every EU member state and its ranking. The vast majority of countries achieve similar efficiency scores. The most efficient countries are the northern European ones. On the contrary eastern

Table 18.4 Ranking and efficiency scores of EU countries

	Window	Rank 1	SFA	Rank 2
Belgium	0.963	1	0.831	2
Netherlands	0.955	2	0.944	1
Cyprus	0.947	3	0.658	6
Malta	0.940	4	0.563	9
Denmark	0.925	5	0.700	3
United Kingdom	0.782	6	0.630	7
Ireland	0.642	7	0.464	10
Germany	0.588	8	0.695	4
Spain	0.577	9	0.591	8
Luxembourg	0.553	10	0.255	22
France	0.542	11	0.673	5
Sweden	0.529	12	0.387	14
Czech Republic	0.500	13	0.323	17
Slovakia	0.437	14	0.260	21
Estonia	0.406	15	0.199	26
Bulgaria	0.384	16	0.277	19
Portugal	0.356	17	0.347	15
Greece	0.352	18	0.396	12
Austria	0.345	19	0.325	16
Finland	0.340	20	0.301	18
Italy	0.338	21	0.448	11
Lithuania	0.313	22	0.215	25
Hungary	0.295	23	0.272	20
Poland	0.281	24	0.393	13
Slovenia	0.271	25	0.223	24
Latvia	0.248	26	0.160	27
Romania	0.179	27	0.252	23

Table 18.5 Mann-Whitney U-test results

Ranks[a]			Test statistics[a]	
Ho hypothesis	N	Rank sum	Z value	$P > \|z\|$
Mean efficiency of new member states	12	124 (106)	−2147 (−3025)	0.032 (0,002)
Mean efficiency of old member states	15	254 (272)		

[a] Numbers in parentheses are for SFA results

European countries, which accessed the EU relatively recently, achieve low efficiency scores. This result provides hints that the new members are more inefficient than the member states that formed the EU before 2004.

In order to check this difference a relevant hypothesis is tested by implementing the Mann Whitney U-test. This hypothesis is that there is a significant difference in efficiency between old and new EU member states; thus, the period of EU accession is positively correlated with efficiency. The Mann Whitney U-test is being applied for both DEA Window and SFA methodologies.

Table 18.5 presents the results of these assessments. The hypothesis testing by the Mann-Whitney U-test showed that EU member states with longer implementation of the CAP are more efficient, compared with the newcomer countries.

18.4 Conclusion

This chapter is an attempt to evaluate the TE and TE changes of the agricultural sector of EU member states for the 2003–2011 period. Two methodologies have been implemented: the DEA Window analysis and the SFA. For this, Agricultural Land, Labour and Fixed Capital were used as inputs and the agricultural output of each country as output. The results of the two models suggest that there is an overall small but significant efficiency change. There are also noticeable differences regarding the efficiency levels among EU member states, with northern European countries achieving the highest scores and Eastern European countries being the most inefficient ones.

Quite important is the ranking similarities of EU countries, regarding their efficiency. This importance lies in the fact that they have been provided by totally different methodological approaches, one parametric and one non-parametric. In both cases there is efficiency improvement after the implementation of the new CAP and the new subsidy management framework based on the decoupled scheme. Due to the fact that up till now the available data referring to the new subsidy administration is rather limited, the efficiency improvement cannot be directly linked to the new subsidy scheme. After the completion of this period this issue should be a topic of research, because efficiency improvement is a goal of the ongoing CAP. Nevertheless, the present results provide hints for a positive relationship between CAP reform and efficiency.

The application of the Mann-Whitney U-test proves that eastern European countries, which recently became EU members, are significantly less efficient compared with older member states. This finding applies for both efficiency measurement methodologies, strengthening in this way the validity of the results. Thus, the policy guidelines should take into account the efficiency gap among the older and newer member states.

Finally, taking into consideration the input and output statistics, it is evident that the agricultural output is continuously increasing after the year 2007 and there is a substantial turn of labour-intensive agriculture to a capital-intensive one. This tendency provides hints for a considerable increase of entrepreneurship in EU agriculture in the near future.

References

Banker, R. D., Charnes, A., & Cooper, W. (1984). Some models for estimating technical and scale inefficiencies in Data Envelopment Analysis. *Management Science, 30*(9), 1078–1092.

Banker, R. D., Conrad, R. F., & Strauss, R. P. (1986). A comparative application of data envelopment analysis and translog methods: an illustrative study of hospital production. *Management Science, 32*(1), 30–44.

Battese, G. E., & Coelli, T. J. (1992). Frontier production functions, technical efficiency and panel data: With application to paddy farmers in India. *Journal of Productivity Analysis, 3*, 153–169.

Charnes, A., Cooper, W. W., & Rhodes, E. (1978). Measuring the efficiency of decision making units. *European Journal of Operational Research, 2*(6), 429–444.

Coelli, T. J., Rao, D. S. P., O Donnel, C. G., & Battese, G. E. (2005). *An introduction to efficiency and productivity analysis* (2nd ed.). New York: Springer.

Chau, N., & de Gorter, H. (2005). Disentangling the consequences of direct payment schemes in agriculture on fixed costs, exit, and output. *American Journal of Agricultural Economics, 87*(5), 1174–1181.

Cooper, W. W., Seiford, L. M., & Tone, K. (2000). *Data envelopment analysis: A comprehensive text with models, applications, references and dea-solver software*. Boston: Kluwer.

Eurostat (2013) Retrieved from http://epp.eurostat.ec.europa.eu/portal/page/portal/statistics/search_database.

Farrell, M. J. (1957). The measurement of productive efficiency. *Journal of the Royal Statistical Society A, 120*(3), 253–281.

Femenia, F., Gohin, A., & Carpentier, A. (2010). The decoupling of farm programs: Revisiting the wealth effect. *American Journal of Agricultural Economics, 92*(3), 836–848.

Hadley, D. (2006). Patterns in technical efficiency and technical change at the farm-level in england and wales, 1982–2002. *Journal of Agricultural Economics, 57*(1), 81–100.

Headey, D., Alauddin, M., & Prasada Rao, D. (2010). Explaining agricultural productivity growth: an international perspective. *Agricultural Economics, 41*, 1–14.

Hennessy, D. (1998). Production effects of income support under uncertainty. *American Journal of Agricultural Economics, 80*(February), 46–57.

Iraizoz, B., Bardaji, I., & Rapun, M. (2005). The Spanish beef sector in the 1990s: Impact of the BSE crisis on efficiency and profitability. *Applied Economics, 37*(4), 473–484.

Karagiannis, G., & Sarris, A. (2005). Measuring and explaining scale efficiency with the parametric approach: the case of Greek tobacco growers. *Agricultural Economics, 33*(s3), 441–451.

Key, N., Lubowski, R., & Roberts, M. (2005). Farm-level production effects from participation in government commodity programs: Did the 1996 federal agricultural improvement and reform act make a difference? *American Journal of Agricultural Economics, 87*(5), 1211–1219.

Madau, F. (2012). *Technical and scale efficiency in the Italian citrus farming: A comparison between stochastic frontier analysis (SFA) and data envelopment analysis (DEA) models*. Munich Personal RePEc Archive MPRA Paper No. 41403.

Mann, H. B., & Whitney, D. R. (1947). On a test of whether one of two random variables is stochastically larger than the other. *Annals of Mathematical Statistics, 18*(50), 50–60.

Polyzos, S., Niavis, S., & Pnevmatikos, T. (2012). Longitudinal evaluation of Greek regional policies using window data envelopment analysis. *MIBES Transactions 6*(1), 53–65.

Rezitis, A. N., Tsiboukas, K., & Tsoukalas, S. (2003). Investigation of factors influencing the technical efficiency of agricultural producers participating in farm credit programs: The case of Greece. *Journal of Agricultural and Applied Economics, 35*(03), 529–542.

Serasinghe, R., Mahipala, M., & Gunaratne, L. (2003). Comparison of stochastic frontier analysis (SFA) and data envelopment analysis (DEA) to evaluate technical efficiency: Illustrated by efficiency analysis of cattle farming systems in up- country wet-zone of Sri Lanka. *Tropical Agricultural Research, 15*, 217–225.

Serra, T., Zilberman, D., Goodwin, B. K., & Featherstone, A. (2006). Effects of decoupling on the mean and variability of output. *European Review of Agricultural Economics, 33*(3), 269–288.
Sharma, K. R., Leung, P., & Zaleski, H. M. (1997). Productive efficiency of the swine industry in Hawaii: Stochastic frontier vs. data envelopment analysis. *Journal of Productivity Analysis, 8*(4), 447–459.
Skuras, D., Tsekouras, K., Dimara, E., & Tzelepis, D. (2006). The effects of regional capital subsidies on productivity growth: A case study of the Greek food and beverage manufacturing industry. *Journal of Regional Science, 46*(2), 355–381.
Wang, T.-F., Cullinane, K., & Song, D.-W. (2005). *Container Port Production and Economic Efficiency*. New York: Palgrave-McMillan.
Zamanian, Gh., Shahabinejad, V., & Yaghoubi, M. (2013). Application of DEA and SFA on the measurement of agriculturaltechnical efficiency in MENA countries. *International Journal of Applied Operational Research, 3*(2), 43–51.
Zhu, J. (2003). *Quantitative models for performance evaluation and benchmarking: data envelopment analysis with spreadsheets*. Boston: Kluwer.
Zhu, X., & Lansink, A. L. (2008). *Technical efficiency of the crop farms under the various CAP reforms: Empirical studies for Germany, The Netherlands and Sweden*. Paper prepared for presentation at the 107th EAAE seminar modelling of agricultural and rural development policies, Sevilla, Spain.

Author Biographies

George Vlontzos holds since 2009 the position of Lecturer of agricultural economics at the Department of Agriculture Crop Production and Rural Development of the University of Thessaly in Greece. He graduated as an agronomist from the Aristotle University of Thessaloniki. From the University of Wales he had an MBA Agribusiness degree and from the Department of Planning and Regional Development of the University of Thessaly he had his PhD. His research interests focus on efficiency assessment of agricultural production, evaluation of agricultural policies, and food consumption behaviours.

Spyros Niavis is an economist. He earned his PhD from the Department of Planning and Regional Development (University of Thessaly). His research interests cover the fields of efficiency analysis, regional development, and economic and regional policy.

Chapter 19
Agriculture Commodity Prices Forecasting Using a Fuzzy Inference System

George S. Atsalakis

Abstract The objective of this chapter is to present a forecasting model of agricultural commodity prices using a Fuzzy Inference System. Recent studies have addressed the problem of commodity prices forecasting using different methods including artificial neural network and conventional model-based approaches. In this chapter, we proposed the use of a hybrid intelligent system called the Adaptive Neuro Fuzzy Inference System (ANFIS) to forecast agri-commodity prices. In ANFIS, both the learning capabilities of a neural network and reasoning capabilities of fuzzy logic are combined in order to provide enhanced forecasting capabilities compared to using a single methodology alone. Point accuracy of four agri-commodity prices (wheat, sugar, coffee, and cocoa) is appraised by computing root-mean-squared forecast errors and other well-known error measures. In terms of forecasting performance, it is clear from the empirical evidence that the ANFIS model outperforms over a feedforward neural network and two other conventional models (AR and ARMA).

19.1 Introduction

The rational expectations competitive storage theory is the basic economic theory on commodity prices. It originated in the work of Gustafson (1958) on the optimal demand for commodity stocks and Muth (1961) rational expectations hypothesis. The theory was subsequently developed by Samuelsson (1971), Danthine (1977), Schechtman and Escudero (1977), Kohn (1978), Newbery and Stiglitz (1981, 1982), Scheinkman and Schechtman (1983), Salant (1983), Wright and Williams (1982, 1984), Williams and Wright (1991), and Hart and Kreps (1986). The inability of

G. S. Atsalakis (✉)
Department of Production and Management Engineering, Technical University of Crete, Crete, Greece
e-mail: g.atsalakis@yahoo.gr

C. Zopounidis et al. (eds.), *Agricultural Cooperative Management and Policy*, Cooperative Management, DOI: 10.1007/978-3-319-06635-6_19,

competitive speculators to hold negative inventories is the main feature of the model. The asymmetry in storage behavior feeds through the commodity prices render the price process nonlinear. Thus, the work of Deaton and Laroque (1992, 1996) can be studied for an excellent exposition on a basic variant of rational expectations competitive storage theory, as well as a structural empirical implementation.

Soft computing techniques like neuro-fuzzy models and neural networks can be used to develop nonlinear agri-commodity prices forecasting models.

In recent years, the fuzzy set theory has been successfully applied in many different areas of engineering including automatic control, system identification, pattern recognition, design of structures, structural modeling, and many more (Adam 2003). The property that makes the fuzzy set theory particularly interesting is its ability to handle the imprecision inherently present in a system. Fuzzy reasoning has become a powerful tool for solving problems when human expert knowledge is available.

Even more attractive is the idea of utilizing the fuzzy set theory in data-driven extraction of easy to understand rule-based models (Adam 2003). In a more general context, this concept is based on the fact that certain fuzzy systems possess the universal approximation property (Wang 1992). For the most complex systems where a few numerical data exist and where only ambiguous and imprecise information may be available, fuzzy reasoning provides a way to understand system behavior by allowing us to interpolate approximately between observed input and output situations. The imprecision in fuzzy models is generally quite high (Ross 1997). Numerous examples of fuzzy and neuro-fuzzy systems, capable of data-driven function approximation can be found in literature (Jang et al. 1997).

Commodity time series are very complex for identification and forecasting purposes because of their volatile behavior. If it is considered that the agri-commodity time series prices only have an interior relation, the future prices can be forecasted by applying the following formula:

$$y_{t+1} = f(y_{t-k}, \ldots, y_t) \quad (19.1)$$

where y_{t+1} is the rate to be forecasted and $y_{t\text{-}k}$ is the influence factor. Traditional models that have been used to forecast time series commodity prices are all based on probability theory and statistical analysis with a certain of distributions assumed in advance. In most cases these assumptions are unreasonable and non-realistic. Also, the linear structure of these models does not guarantee the accuracy of forecasting.

Recent studies have addressed the problem of time series forecasting using different methods including artificial neural network and model-based approaches due to the significant properties of handling nonlinear data with self-learning capabilities (Hornik 1991; Jain and Naq 1997; Skapura 1996). Neural Networks (NN) are essentially black box models which, when used alone as the basis of a model, should not be applied at sites other than those for which they are trained. The Neural Networks have been accused by researchers of not being able to determine the degree to which an input influences the output of the model (Shapiro 2002;

Pao 1989). Fuzzy logic is an effective rule-based modeling in soft computing, that not only tolerates imprecise information, but also makes a framework of approximate reasoning. The disadvantage of fuzzy logic is the lack of self-learning capability. However, when combined with neural network in the form of neuro-fuzzy models, the logic in the model may be transferable to another site, as the neural network aspect is limited to being a rule training tool in conjunction with heuristic knowledge. On this note, this chapter proposes the use of a hybrid intelligent system called the Adaptive Neuro Fuzzy Inference System (ANFIS) for forecasting the agri-commodity prices. ANFIS combines both the learning capabilities of a neural network and the reasoning capabilities of fuzzy logic in order to offer enhanced forecasting capabilities, compared to using a single methodology alone. ANFIS has been used by many researchers to forecast various time series (Jang et al. 1997; Atsalakis and Valavanis 2009; Atsalakis 2007; Atsalakis et al. 2007; Atsalakis and Minoudaki 2007; Atsalakis and Ucenic 2006; Atsalakis 2005; Ucenic and Atsalakis 2006; Atsalakis and Valavanis 2009, 2010). Studies to forecast commodity process are: Azadeh et al. (2012), El Hédi Arouri et al. (2012) He Kaijian et al. (2012), Hu et al. (2012), Xu and Ouenniche (2012), Yu (2012), and Zhou et al. (2012).

Nevertheless, the structure of the chapter arranged in the following order: Sect. 19.2 describes the nonlinear models while Sect. 19.3 discusses the agricultural commodities data set. Models evaluation was carried out in Sect. 19.4 where an out-of-sample forecasting procedure was conducted. Finally Sect. 19.5 concludes.

19.2 Model Presentation

19.2.1 Feedforward Neural Network

Neural Networks are an information processing technique based on the biological nervous systems, such as the brain, process information. The fundamental concept of neural networks is the structure of the information processing system. Neural networks can differ on: the way their neurons are connected, the specific kinds of computations in their neurons, the way they transmit patterns of activity throughout the network; and the way they learn including their learning rate. Nevertheless, their primary advantage is that they can solve problems that are too complex for conventional technologies as well as problems that do not have an algorithmic solution or for which an algorithmic solution is too complex to be defined.

Given a training set of data, the neural network can learn the data with a learning algorithm through the use of back propagation. Through back propagation, the neural network forms a mapping between inputs and desired outputs from the training set by altering weighted connections within the network.

A feedforward neural network is similar to the types of neural networks. Just like many other neural network types, the feedforward neural network begins with an input layer. This input layer must be connected to a hidden layer. The hidden layer on the other hand can then be connected to another hidden layer or directly to the output layer. There can be as many number of hidden layers, but so long as at least one hidden layer is provided. In common use, most neural networks will have only one hidden layer.

The representation capability of a network can be defined as the range of mappings it can implement when the weights are varied. Thus, single-layer networks are capable of representing only linearly separable functions or linearly separable decision domains. However, two hidden layered networks can represent an arbitrary decision boundary to arbitrary accuracy with threshold activation functions and could approximate any smooth mapping to any accuracy with sigmoid activation functions. One hidden layered network can approximate arbitrarily well any functional continuous mapping from one finite-dimensional space to another, provided that the number of hidden units is sufficiently large. To be more precise, feedforward networks with a single hidden layer trained by least-squares are statistically consistent estimators of arbitrary square-integral regression functions if assumptions about samples, target noises, number of hidden units, and other factors are all met. Feedforward networks with a single hidden layer using threshold or sigmoid activation functions are universally consistent estimators of binary classifications under similar assumptions (McNelis 2005).

19.2.2 ANFIS Architecture

With the ANFIS approach, implementation of the model design differs in form from the more traditional NN, in that it is not fully connected, as well not all the weights or nodal parameters are modifiable. Essentially, the fuzzy rule base is encoded in a parallel fashion so that all the rules are activated simultaneously, so as to allow network training algorithms to be applied. Similar to Jang's original work, here a back propagation algorithm is used to optimize the fuzzy sets of the premises in the ANFIS architecture, while a least squares procedure is applied to the linear coefficients in the consequent terms.

Let X be a space of objects and x be a generic element of X. A classical set $A \subseteq X$ is defined as a collection of elements or objects $x \in X$ such that each x can either belong or not belong to the set A. By defining a characteristic function for each element x in X, we can represent a classical set A by a set of ordered pairs $(x, 0)$ or $(x, 1)$ which indicates $x \in A$ or $x \notin A$, respectively. On the other hand, a fuzzy set expresses the degree to which an element belongs to a set. Hence the characteristic function of a fuzzy set is allowed to have values between 0 and 1, which denotes the degree of membership of an element in a given set. In addition, a fuzzy set A in X is defined as a set of ordered pairs:

$$A = \{(x, \mu_A(x)) \mid x \in X\} \tag{19.2}$$

Here $\mu_A(x)$ is called the membership function (MF) for the fuzzy set A.

The MF maps each element of X to a membership grade (or a value) between 0 and 1. Usually, X is referred to as the universe of discourse or simply the universe. Notwithstanding, the most widely used MF is the generalized bell MF (or the bell MF), which is specified by three parameters $\{a_i, b_i, c_i\}$ and is defined as follows (Loukas 2001; Jang and Chuen-Tsai 1995):

$$\mu_{A_i}(x) = \frac{1}{1 + \left[\left(\frac{x - c_i}{a_i}\right)^2\right]^{b_i}} \tag{19.3}$$

Parameter b is usually positive. A desired bell MF can be obtained using a proper selection of the parameter set $\{a_i, b_i, c_i\}$. During the learning phase of ANFIS, these parameters are continuously changing in order to minimize the error function between the target output values and the calculated ones (Lee 1990a, b).

Moreover, the proposed neuro-fuzzy model of ANFIS is a multilayer neural network-based fuzzy system. Its topology is depicted in Fig. 19.1, although the system has a total of five layers. In this connected structure, the input and output nodes represent the training values and the predicted values, respectively. While in the hidden layers, there are nodes functioning as membership functions (MFs) and rules. Thus, the benefit of the architecture is that it eliminates the disadvantage of a normal feedforward multilayer network, where it is difficult for an observer to understand or modify the network.

For the purpose of simplicity, it is assumed that the examined fuzzy inference system has two inputs x and y as well as one output. For a first-order Sugeno fuzzy model (Jang et al. 1997), a common rule set with two fuzzy if-then rules is defined as

$$\text{Rule1: If } x \text{ is } A_1 \text{ and } y \text{ is } B_1 \text{ then } f_1 = p_1 \cdot x + q_1 \cdot y + r_1 \tag{19.4}$$

$$\text{Rule2: If } x \text{ is } A_2 \text{ and } y \text{ is } B_2 \text{ then } f_2 = p_2 \cdot x + q_2 \cdot y + r_2 \tag{19.5}$$

As can be seen from Fig. 19.1, different layers of ANFIS have different nodes. Hence, each node in a layer is either fixed or adaptive (Jang 1993). Similarly, different layers with their associated nodes are described below:

Layer 1: Every node i in this layer is a square node with a node function.

$$\begin{aligned} O_{1,i} &= \mu_{A_i}(x) \quad \text{for } i = 1,\ 2, \text{ or} \\ O_{1,i} &= \mu_{B_{i-2}}(y) \quad \text{for } i = 3,\ 4, \end{aligned} \tag{19.6}$$

where x is the input to node i and A_i is the linguistic label (small, large, etc.) associated with this node. In other words, $O_{1,i}$ is the membership function of a fuzzy set A_i and it specifies the degree to which the given input x satisfies the

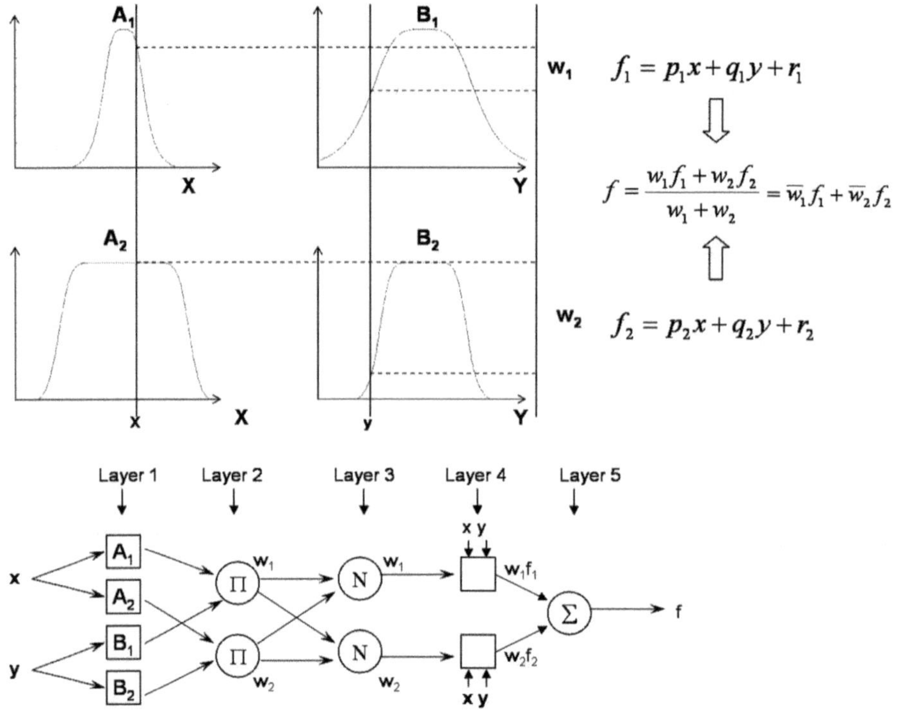

Fig. 19.1 An illustration of the reasoning mechanism for a Sugeno-type model and the corresponding ANFIS architecture (Jang et al. 1997)

quantifier A_i. Usually $\mu_{A_i}(x)$ is set as a bell-shaped curve with a maximum equal to 1 and a minimum equal to 0, such as the generalized bell function:

$$\mu_{A_i}(x) = \frac{1}{1 + \left[\left(\frac{x-c_i}{a_i}\right)^2\right]^{b_i}} \tag{19.7}$$

where a_i, b_i, c_i is the parameter set. As the values of these parameters change, the bell-shaped functions vary accordingly, thereby exhibiting various forms of membership function on linguistic label A_i. Parameters in this layer are referred to as premise parameters.

Layer 2: Every node in this layer is a circle node labeled $\prod$, which multiplies the incoming signal and sends the product out.

$$O_{2,i} = w_i = \mu_{Ai}(x) * \mu_{Bi}(y), \quad i = 1, 2. \tag{19.8}$$

Layer 3: Every node in this layer is a circle-fixed node labeled N. The *i*th node calculates the ratio of the *i* th rule's firing strength to the sum of all the rules' firing strengths:

$$O_{3,i} = \bar{w}_i = \frac{w_i}{w_1 + w_2}, \quad i = 1, 2. \tag{19.9}$$

For the sake of convenience, the outputs of this layer will be referred to as normalized firing strengths.

Layer 4: Every node i in this layer is an adaptive square node with a node function

$$O_{4,i} = \bar{w}_i \cdot f_i = \bar{w}_i(p_1 \cdot x + q_i \cdot y + r_i) \tag{19.10}$$

where: $\bar{w}_i$ is a normalized firing strength from layer 3 and $\{p_i, q_i, r_i\}$ is the parameter set in this layer. Parameters in this layer are referred to as consequent parameters.

Layer 5: The single node in this layer is a circle fixed node labeled $\sum$ that computes the overall output as the summation of all incoming signals:

$$\text{overall output} = O_{5,i} = \sum_i \bar{w}_i \cdot f_i = \frac{\sum_i w_i \cdot f_i}{\sum_i w_i} \tag{19.11}$$

This architecture develops an adaptive network that is functionally equivalent to a two inputs fist-order Sugeno fuzzy model with four rules, where each input has two membership functions. The main advantage of this model is its transparency and efficiency.

19.2.3 Learning Algorithm of ANFIS

The learning algorithm for ANFIS is a hybrid algorithm, which is a combination of gradient descent and the least-squares method. More specifically, in the forward pass of the hybrid learning algorithm, node outputs go forward until layer 4 and the consequent parameters are identified by the least-squares method (Jang 1993). In the backward pass, the error signals propagate backwards and the premise parameters are updated by gradient descent. Hence, Table 19.1 summarizes the activities in each pass.

The consequent parameters are optimized under the condition that the premise parameters are fixed. The main benefit of the hybrid approach is that it converges much faster since it reduces the search space dimensions of the original pure back propagation method used in neural networks. In addition, the overall output can be expressed as a linear combination of the consequent parameters. While the error

Table 19.1 Errors for one-step-ahead forecasting results

	Forward pass	Backward pass
Premise parameters	Fixed	Gradient descent
Consequent parameters	Least-squares estimator	Fixed
Signals	Node outputs	Error signals

measured to train the above-mentioned ANFIS is defined as follows (Jang et al. 1997):

$$E = \sum_{k=1}^{n} (y_k - \hat{y}_k)^2 \tag{19.12}$$

where y_k and $\hat{y}_k$ are the kth desired and estimated output, respectively, n is the total number of pairs (inputs–outputs) of data in the training set.

19.2.4 Auto Regression Model

The autoregressive (AR) models are used in time series analysis to describe stationary time series. These models represent time series that are generated by passing the white noise through a recursive linear filter. Nevertheless, the output of such a filter at the moment t weighs the sum of m previous values of the filter output. The integer parameter m is called the order of the AR-model. The AR-model of a random process $y(t)$ in discrete time t is defined by the following expression:

$$y(t) = \sum_{i=1}^{m} a(i) \cdot y(t-i) + \varepsilon(t) \tag{19.13}$$

where α_1, α_2,…,α_m are the coefficients of the recursive filter, m is the order of the model and $\varepsilon(t)$ are output uncorrelated errors.

The moving average (MA) models represent time series that are generated by passing the white noise through a nonrecursive linear filter. The MA-model of a random process $y(t)$ in discrete time t is defined by the following expression:

$$y(t) = \sum_{i=1}^{n} b(i) \cdot x(t-i) + \varepsilon(t) \tag{19.14}$$

where b_i, $i = 0,1,\ldots,n$ are the coefficients of the linear non-recursive filter, n is the order of the MA-model, $x(t)$ is the element of the (input) white noise and $\varepsilon(t)$ is output uncorrelated errors.

19.2.5 Auto Regression Moving Average Model

The auto regression and moving average (ARMA) models are used in time series analysis to describe stationary time series. These models represent time series that are generated by passing white noise through a recursive and non-recursive linear filter, consecutively. In other words, the ARMA model is a combination of an autoregressive (AR) model and a moving average (MA) model.

The order of the ARMA model in discrete time t is described by two integers (m, n), that are the orders of the AR- and MA- parts, respectively. The general expression for an ARMA-process $y(t)$ is as follows:

$$y(t) = \sum_{i=1}^{m} a(i) \cdot y(t-i) + \sum_{i=0}^{n} b(i) \cdot x(t-i) + \varepsilon(t) \tag{19.15}$$

where m is the order of the AR-part of the ARMA model, α_1, $\alpha_2,\ldots,\alpha_m$ are the coefficients of the AR-part of the model (of the recursive linear filter), n is the order of the MA-part of the ARMA model, while b_0, $b_1,\ldots,b_n$ are the coefficients of the MA-part of the model (of the nonrecursive linear filter), $x(t)$ are elements of the (input) white noise whereas $\varepsilon(t)$ is output uncorrelated errors.

19.3 Data and Model Parameters

The experimental data consisted of a monthly time series of four agricultural commodities (wheat, sugar, coffee, and cocoa) demand from January 1969 until October 2010 (502 samples). Figure 19.2 depicts the first 400 samples of sugar that have been used for training the model, whereas the other 101 have been used to test the performance of the resulting model. The structure of ANFIS consists of one input and one output indicating that the forecasting system is used to forecast the next month's computer demand based on the value of one month ago. After many trial and errors attempts, it was decided that the input variable must have two generalized bell MFs. Figure 19.3 depicts the initial forms of the MFs before the training and the final MFs after the training (fine-tuning).

The system uses two rules and thus has been trained for 500 epochs to converge with the optimal fuzzy inference. Figure 19.4 depicts the function of the rules. Table 19.2 describes the type and the values of the ANFIS parameters.

Figure 19.5 illustrates at the upper graph the root-mean-square reduction against the number of epochs. While, the lower graph, represents the changes of the step size during the training of neuro-fuzzy system.

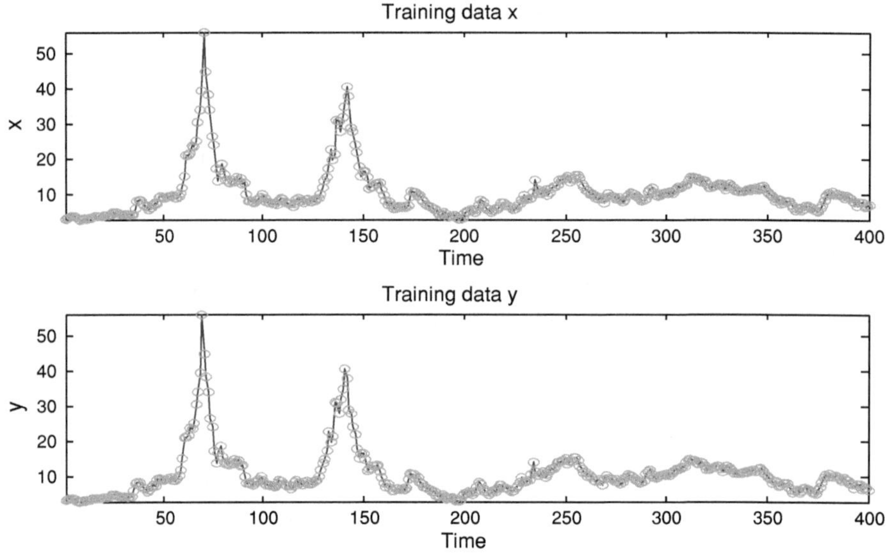

Fig. 19.2 Graphical representation of the sugar training data

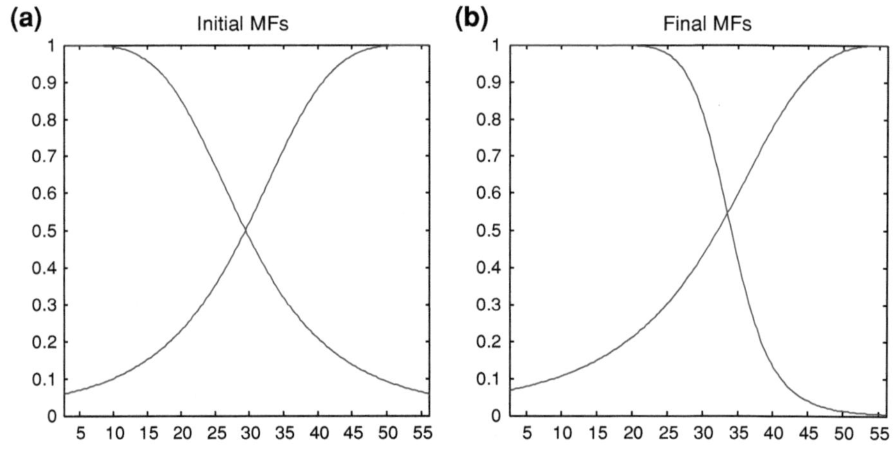

Fig. 19.3 Illustration of MFs before and after the training (for sugar)

19.4 Model Performance Evaluation

To evaluate the performance of the models, an error analysis, using some well known statistical errors: i.e. Mean Square Error (MSE), Root Mean Square Error (RMSE), Mean Absolute Error (MAE), and Mean Absolute Percentage Error (MAPE) was carried out (Makridakis et al. 1983).

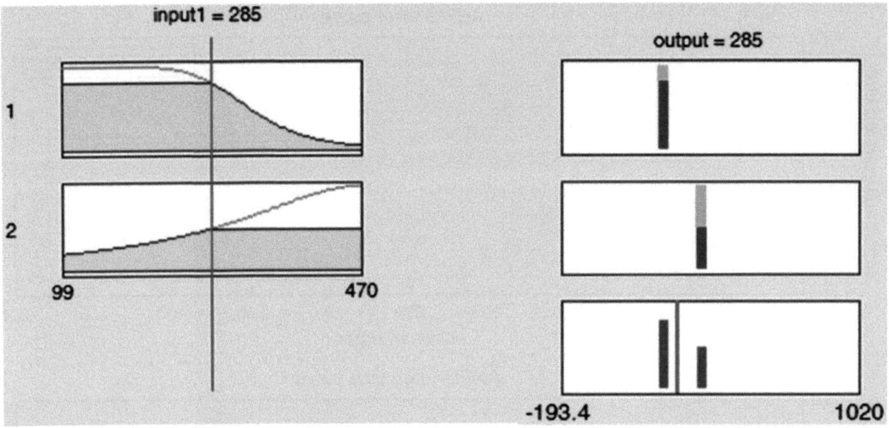

Fig. 19.4 A view of the rules: A sample depicting the prediction of strength by the layer-1 of the model for a given input for sugar (experimental value)

Table 19.2 ANFIS parameter types and their values used for training

ANFIS parameter type	Value
MF type	Bell function
Number of MFs	2
Output MF	Linear
Number of Nodes	12
Number of linear parameters	4
Number of nonlinear parameters	6
Total number of parameters	10
Number of training data pairs	400
Number of evaluating data pairs	101
Number of fuzzy rules	2

The data were obtained for four worldwide known commodities: wheat, sugar, coffee, and cocoa. The 101 observed monthly data sets, starting from June of 2002 until October of 2010, that have not been used during the training phase, were used throughout the evaluation phase. The same data were also used to test a feedforward model, an Autoregressive (AR), and Autoregressive Moving Average (ARMA) forecasting model.

Tables 19.3, 19.4, 19.5 and 19.6 that illustrate the four models were evaluated based on their performance in testing sets. It was also noted that the models were trained using nontransformed data. Thus, the models have shown significant variations in the errors of the performance evaluation for the four commodities. Hence, it appears that the ANFIS model is more accurate, where all the values of errors for all the commodities are smaller. It also indicates that the lower value of the RMSE is 1.29 (for sugar) in the ANFIS model and the highest value of the RMSE is 31.57 (in the AR model for wheat). While, the values of the MAPE (4.82 % for coffee) in the ANFIS forecasting model are much lower than those

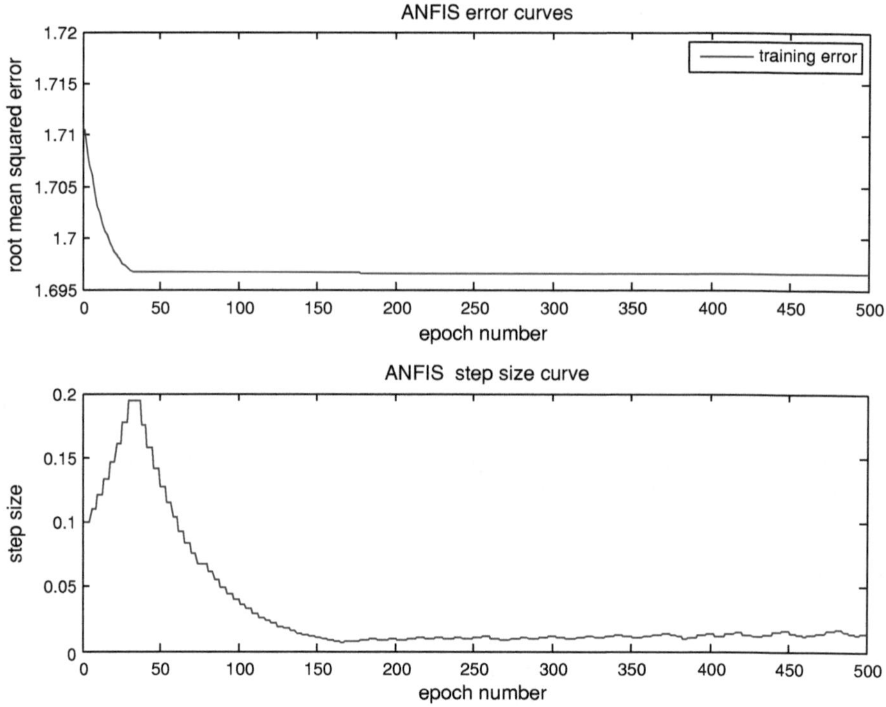

Fig. 19.5 Fuzzy surface as depicted by the model representing approximate relationship among the inputs

Table 19.3 Errors for wheat one-month-ahead forecasting results

Errors wheat	ANFIS	NN	AR	ARMA
MSE	398.69	754.59	996.96	968.10
RMSE	19.96	27.50	31.57	31.11
MAE	12.78	9.5	21.03	20.86
MAPE	5.38	3.8	8.87	8.83

Table 19.4 Errors for sugar one-month-ahead forecasting results

Errors sugar	ANFIS	NN	AR	ARMA
MSE	1.65	1.92	4.74	4.38
RMSE	1.29	1.39	2.18	2.09
MAE	0.82	0.88	1.43	1.34
MAPE	6.16	6.89	10.60	10.15

derived in the other models. In addition, the values of the MSE (1.65 for sugar) and MAE (0.82 for sugar) in the ANFIS model are lower than those found in the result of the other models.

Table 19.5 Errors for coffee one-month-ahead forecasting results

Errors coffee	ANFIS	NN	AR	ARMA
MSE	18.46	20.82	47.24	46.45
RMSE	4.30	4.56	6.87	6.82
MAE	3.13	3.29	4.91	4.93
MAPE	4.82	5.06	7.47	7.47

Table 19.6 Errors for cocoa beans one-month-ahead forecasting results

Errors cocoa	ANFIS	NN	AR	ARMA
MSE	40.55	50.22	95.42	97.07
RMSE	6.37	7.08	9.77	9.85
MAE	4.85	5.29	7.64	7.69
MAPE	5.11	5.45	7.91	8.01

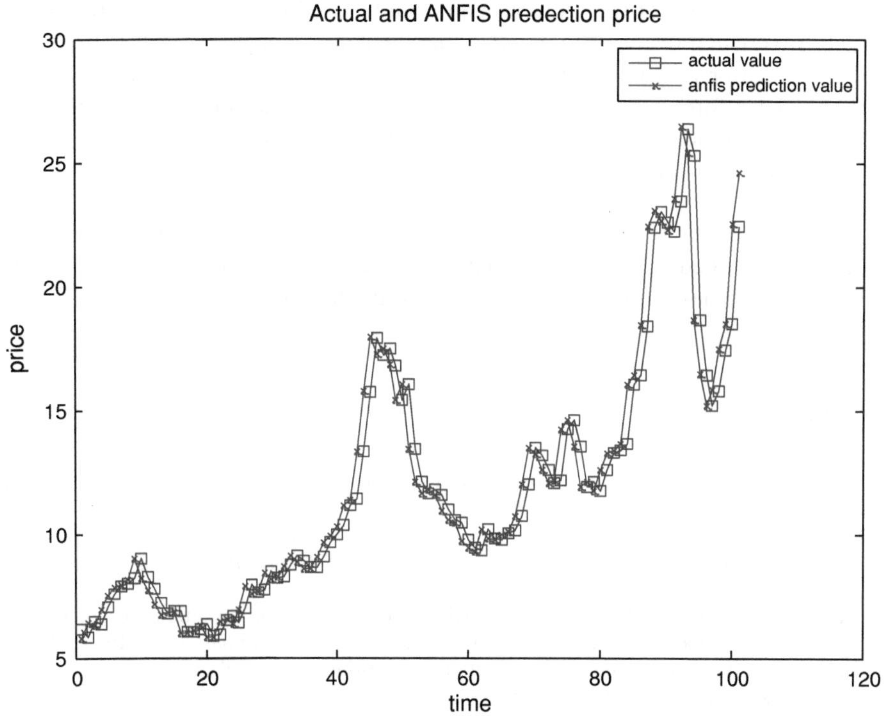

Fig. 19.6 Out of sample actual and ANFIS forecasted sugar prices

The results and observations from the ANFIS model were compared with the observed values in order to evaluate its performance. Figure 19.6 depicts the scattered diagram of the estimate and observes values derived when testing the ANFIS (for sugar). Therefore, the forecasting performance of ANFIS was satisfactory as can be seen in the research and thus was acceptable in practice as shown in Fig. 19.6. Nevertheless, the square signs in the blue line depict the actual sugar monthly prices and the asterisks in the red line depict the forecasted monthly sugar prices.

19.5 Conclusion

This chapter was based on the comparative analysis of neuro-fuzzy network, a feedforward neural network and two conventional forecasting methods. Four different models were investigated, namely, ANFIS, NN, AR and ARMA. Thus, this research aimed at proving that a neuro-fuzzy approach can be used to forecast the monthly prices of four agricultural commodities. In addition, the weak aspects of other forecasting methodologies for time series could be overcome with the proposed Adaptive Network with Fuzzy Inference System (ANFIS), while the data available in the form of input/output pairs can be used in the ANFIS with relative ease. Experimental results also indicate that the neuro-fuzzy approach outperforms the other feedforward neural network and the two conventional models (AR and ARMA). Therefore, it cannot entirely be claimed that the problems arising from forecasting of agricultural commodities can be completely solved, but it is obvious that the findings of this chapter have important managerial and practical implications.

Further evaluations improvements are still possible if the directional accuracy could be examined. Hence, directional accuracy is assessed by calculating the proportion of times the model was correct in its directional prediction.

References

Adam, E. G. (2003). Data-driven linguistic modeling using relational fuzzy rules. *IEEE Transactions on Fuzzy Systems, 11*(1), 121–134.

Atsalakis, G., & Valavanis, K. (2010). Forecasting stock trends using a combined technical analysis and a neuro-fuzzy based approach. *Journal of Financial Decision Making, 6*(1), 79–94.

Atsalakis, G., & Valavanis, K. (2009). Forecasting stock Market short-term trends using a neuro-fuzzy based methodology. *Journal of Expert Systems with Applications, 36*(7), 10696–10707.

Atsalakis G. (2007). Wind energy production forecasting by neural networks and genetic algorithms. Athens: European Computer conference.

Atsalakis, G., Skiadas, C., Braimis, I. (2007). Probability of trend prediction of exchange rate by neuro-fuzzy techniques. *Recent advances in stochastic modeling and data analysis* (pp. 414–422). London: World Scientific Publishing.

Atsalakis, G., Minudaki, C. (2007). *Prediction of Daily irrigation water demand using Adaptive Neuro-fuzzy inference system (ANFIS), International Conference on energy, environment, ecosystems and sustainable development* (pp. 368–373). Ag. Nikolaos, Greece: Greece Water Economics, Statistics and Finance.

Atsalakis, G., & Ucenic, C. (2006). Forecasting the production level for wind energy using a neuro-fuzzy model. *Journal of WSEAS Transactions on Environment and Development, 2*(6), 823–828.

Atsalakis, G. (2005). Exchange rate forecasting by Neuro-Fuzzy Techniques. *Journal of Financial Decision Making, I* (2), 15–26.

Azadeh, A., et al. (2012). A flexible neural network-fuzzy mathematical programming algorithm for improvement of oil price estimation and forecasting. *Computers & Industrial Engineering, 62*, 421–430.

Danthine, J.-P. (1977). Martingale, market efficiency and commodity prices. *European Economic Review, 10*, 1–17.

Deaton, A., & Laroque, G. (1996). Competitive storage and commodity price dynamics. *Journal of Political Economy, 104*, 896–923.

El Hédi, Arouri M., et al. (2012). Forecasting the conditional volatility of oil spot and futures prices with structural breaks and long memory models. *Energy Economics, 34*, 283–293.

Gustafson, R. L. (1958). *Carry over levels for grains*. Washington: US Department of Agriculture, Technical Buttletin. 1178.

Hart, O. D., & Kreps, D. M. (1986). Price destabilizing speculation. *Journal of Economy, 94*, 927–952.

He, K. et al. (2012). Crude oil price analysis and forecasting using wavelet decomposed ensemble model. *Energy, 46*, 564–574.

Hornik, K. (1991). Approximation capabilities of multi-layer feed-forward networks. *Neural Networks, 4*, 251–257.

Hu J. et al. (2012): Applying neural networks to prices prediction of crude oil futures. *Hindawi Publishing Corporation Mathematical Problems in Engineering, 2012*, Article ID 959040.

Jain, B. A., & Nag, B. N. (1997). Performance evaluation of neural network decision models. *Manage Information Systems, 14*, 201–216.

Jang, J. S. R. (1993). ANFIS: Adaptive network-based fuzzy inference system. *IEEE Trans Systems, Man Cybern, 23*(3), 665–685.

Jang, J. S. R., & Chuen-Tsai, S. (1995). Neuro-fuzzy modeling and control Proc. *IEEE, 83*, 378–406.

Jang, J-S.R., Sun, C-T.E., Mizutani, E. (1997). *Neuro-fuzzy and soft computing: a computational approach to learning and machine intelligence*, New Jersy: Prentice Hall.

Kohn, M. (1978). Competitive speculation. *Econometrica, 46*, 1061–1076.

Lee, C. C. (1990a). Fuzzy logic in control systems: fuzzy logic controller. I. *IEEE Transaction on System, Man Cybern, 20*, 404–418.

Lee, C. C. (1990b). Fuzzy logic in control systems: fuzzy logic controller. II. *IEEE Transaction on System, Man Cybern, 20*, 419–435.

Loukas, Y. L. (2001). Adaptive neuro-fuzzy inference system: an instant and architecture-free predictor for improved QSAR studies. *Journal of Medicinal Chemistry, 44*, 2772–2783.

Makridakis, S., Weelwright, S. C., & McGee, V. E. (1983). *Forecasting: methods and applications* (2nd ed.). New York: Wiley.

McNelis, D. P. (2005). *Neural networks in finance: gaining predictive edge in the market* (p. 21). USA: Elsevier Academic Press.

Muth, J. F. (1961). Rational expectations and the theory of price movements. *Econometrica, 29*, 315–335.

Newbery, D. M. G., & Stiglitz, J. E. (1982). Optimal commodity stockpiling rules. *Oxford Economic Papers, 34*, 403–427.

Skapura, D.(1996). *Building neural networks* (1), New York: Addison Wesley.

Shapiro, A. F. (2002). The merging of neural networks, fuzzy logic, and genetic algorithms. *Insurance: Mathematics and Economics, 31*, 115–131.

Pao, Y. H. (1989). *Adaptive pattern recognition and neural networks*. New York: Addison Wesley.

Ross, T. J. (1997). *Fuzzy logic with engineering Applications* (International Edition). New York: McGraw-Hill.

Salant, S. W. (1983). The vulnerability of price stabilization schemes to speculative attack. *Journal of Political Economy, 91*, 1–38.

Samuelson, P. A. (1971). Stochastic speculative price. *Proceedings of the National Academy of Science, 68*, 335–337.

Schechtman, J., & Escudero, V. (1977). Some results on An income fluctuation problem'. *Journal of Economic Theory, 16*, 151–166.

Scheinkman, J. A., & Schechtman, J. (1983). A simple competitive model with production and storage. *Review of Economic Studies, 50*, 427–441.

Ucenic, C., Atsalakis, G. (2006). Forecasting the electricity demand using a neuro-fuzzy approach versus traditional methods, *Journal of WSEAS Transactions on Business and Economics*, *3(1)*, pp.9–17.

Wang L. X. (1992). Fuzzy systems are universal approximators, *Proceedings of the 1st IEEE Conference Fuzzy Systems* (Vol. 8, Issue 12, pp. 1163–1170), San Diego, CA.

Williams, J. C., & Wright, B. D. (1991). *Storage and commodity markets*. Cambridge: Cambridge University Press.

Wright, B. D., & Williams, J. C. (1982). The economic role of commodity storage. *Economic Journal, 92*, 596–614.

Xu, B. X., & Ouenniche, J. (2012). A data envelopment analysis-based framework for the relative performance evaluation of competing crude oil prices' volatility forecasting models. *Energy Economics, 34*, 576–583.

Yu, W. (2012). Forecasting volatility of fuel oil futures in China: GARCH-type, SV or realized volatility models? *Physica A, 391*, 5546–5556.

Zhou, S., et al. (2012). A dynamic meta-learning rate-based model for gold market forecasting. *Expert Systems with Applications, 39*, 6168–6173.

Part VII
Greening Criteria for Agricultural and Rural Policy Management

Chapter 20
Assessment of CAP Reform 2014–2020 in the Emilia-Romagna Region

R. Gigante, F. Arfini and M. Donati

Abstract The aim of this contribution is to evaluate the impacts of the European Commission proposals on rural areas of the Emilia-Romagna Region in Italy. The model considers the three main characteristics of the CAP 2014–2020 reform, and in particular measures the impact of greening criteria on land allocation in different farm systems and economic effects on rural areas. It will show which rural systems and types of farm will be favored or penalized by the reform. The model will also provide results on dynamics in land use. The assessment is made using an "integrated" regional model based on Positive Mathematical Programming (PMP).

20.1 Introduction

The debate on the Common Agriculture Policy (CAP) for the years 2014–2020 started in 2011, and at the time of writing is nearing conclusion. The reform was approved by the final European plenary session on 13th March 2013 by 25 of the 27 Ministers of Agriculture. On April 11th, it was discussed by the Trilogue in informal tripartite meetings of representatives of the European Parliament, the Council and the Commission. The goal was to reach a common political agreement by the end of June. But on the other hand, the European budget, including agriculture funding, is at present still undefined. In this uncertain context, what remains of the European Commission's official proposal of 2011 (European Commission 2011a) is that future CAP will continue its main focus on environmental measures directing European farms toward a model of sustainable agriculture responsible for the management of environmental resources and attentive to public financial resources and well-being of consumers (Matthews 2013). The aim of this contribution is to

R. Gigante (✉) · F. Arfini · M. Donati
Department of Economics, University of Parma, Parma, Italy
e-mail: gigante@inea.it

C. Zopounidis et al. (eds.), *Agricultural Cooperative Management and Policy*, Cooperative Management, DOI: 10.1007/978-3-319-06635-6_20,

evaluate the impacts of the European Commission proposals (European Commission 2011b) on rural areas in the Emilia-Romagna Region in terms of province, altitude, and farm type.

The study will consider the impact of greening criteria, capping and the regionalization scheme on land allocation, and the relative economic effects on different regional farm systems. It will focus on the regional effect of the CAP reform, considering the nonhomogeneity of the Emilia-Romagna Region, and show the main consequences in terms of land use and income distribution among different farm types and rural systems across one of the most important Italian farming regions.

The evaluation will be made by a regional model based on the use of Positive Mathematical Programming (PMP). Data are extracted from an integrated regional database that matches FADN Data with IACS information for the year 2010. The chapter is organized as follows; Sect. 20.1 describes the evolution of CAP in environmental measures, changes in distribution of aid, and the system of transferring resources from Pillar 1 to Pillar 2; Sect. 20.2 focuses on the methodology and describes the AGRISP model and its implications; Sect. 20.3 describes the new measures taken into account by the model, detailing amounts and restrictions imposed by the reform; Sect. 20.4 shows results from the model and their implications for changes in land use and economic indicators. Results are shown at regional and province level and also by altimetry. The last section discusses the new agriculture policy and the possible effects on rural areas in Emilia-Romagna.

20.2 The Policy Setting

This section discusses the main innovations in the CAP 2014 reform, and how these have evolved from recent policy reforms. The new CAP in fact fulfills many medium- and long-term expectations both within and outside the EU. The Health Check of 2008 enforced a series of provisions of the 2003 reform, and the new reform will continue to enhance intervention on environmental protection (Anania 2008). The main concerns are the new greening measures, agricultural development and modulation methods, with new proposals oriented toward rural development and setting CAP as a more equal policy between Member States (MS) with a new single payment scheme. Reform paths can be viewed from either an economic point of view or as a requirement to renew environmental strategies at a higher level. Economically, the necessity to reduce CAP expenditure and the eradication of the old system of direct support will cut the EU budget, bringing down both surplus production and international tensions (De Filippis and Frascarelli 2012). Environmentally, the measures reflect institutional concerns and widespread awareness of limits to availability of natural resources, and are a prerequisite for acceptance of expenditure by European taxpayers (European Commission (a) and (b)).

20.2.1 Evolution of Environmental Focus in the CAP

From the start, the CAP has been characterized by reforms that have tried to align the varying objectives of the Community in agriculture in European society. In the beginning, at the time of the Treaty of Rome (1957), CAP focused on agricultural policy with the priority of food security. In subsequent years, the emphasis moved to a new concept, linking agriculture to environmental issues. The first concrete attempt at this was in the MacSharry reform (1992) which gave EU farm policy a new role in an attempt to re-establish a proper relationship between agriculture and the environment and also to revise CAP expenditure limits. The MacSharry reform introduced the concept of "green" to the CAP, and its "accompanying measures" (Reg. EC no. 2078/92) contained predominantly agri-environment schemes, such as the adoption of environmentally friendly farming methods (e.g., organic farming), reducing the use of chemical products and processes and the extensification of farming. This led to the 1999 reform, Agenda 2000, which aimed to strengthen public intervention in agriculture for sustainability, especially environmental sustainability, and then to the Fischler Reform of 2003 (Reg. EC no. 1782/2003). The Fischler Reform introduced the principle of decoupling aid and the cross-compliance mechanism: the farmer is obliged to comply with minimum environmental standards in order to obtain agricultural support (Sorrentino et.al. 2011).

In 2008, a further review of the CAP (Health Check or Fisher-Boel Reform) completed the decoupling process and amended specific additional intervention, reinforcing its environmental objectives. Among other measures, the controversial Article 68 of EC Regulation No. 73/2009 provides annual additional payments for forms of agricultural management of the environment, such as extensive animal farming. The last step in this evolution on environmental issues will be implemented in programming period 2014–2020 with the new greening measures. Although at the time of writing, the European Commission has yet to take final decisions on the new CAP, this contribution attempts to make a preliminary evaluation of the effects on the agriculture sector in the Emilia-Romagna region.

20.2.2 A More Equal CAP

Since the early 1990s, CAP policy has been gradually reformed toward market orientation in the two reform packages of 2003 and 2007. These replaced a large share of the price support by direct payments per hectare of land and per head of livestock. These direct payments were only paid to certain crops and certain types of livestock. In 2003, the Fischler reform substantially changed European policies for supporting farmers with decoupling of direct payments. The new "*Single Payment Scheme*" introduced payment per hectare of agricultural land, independently of the individual farmer's output. It is paid regardless of whether the farmer produces or not, as long as the land is kept in good agricultural and environmental condition. However, there are exceptions to the general principle of decoupling, since individual member states are currently allowed to keep limited coupled payments for some products (partial decoupling). The reform was intended to make European agriculture more competitive and market-oriented as required by WTO, and at the same time to provide support to farmers with less distortion of production and trade. Decoupled payments allow farmers to respond better to signals from the market, to supply the food sector, and to create a basis for providing public goods. The scheme was amended slightly in 2007 and has been in force until today. CAP 2014–2020 is expected to provide for a fully direct payment system for all MS and by the beginning of 2019, all MS will move to a uniform payment per hectare scheme, applied at the national or regional level. In Italy as in some other MS, and consequently in Emilia-Romagna, this reform will be a big challenge for farmers accustomed to the concept of acquired rights. The change is to be accompanied by other measures; cross-compliance is maintained and greater modulation is introduced to the new coupled scheme. It could also lead to redistribution between agrarian regions and farms, and between production sectors, which could affect the competitiveness of different farm activities and sectors. There will very likely be variations in the competitiveness of farms and sectors.

An example of the effect of aid redistribution at the local level is shown in Fig. 20.1. Using administrative regional boundaries, it shows the present situation and the effects of the new regionalization, with the future single payment scheme in the different provinces of Emilia-Romagna (Gigante 2013). The shrinkage of financial resources and decoupled payments reduce the average aid level in most provinces. Because of the type of production (cereal, tomato, milk), many provinces lose resources in favor of areas producing fruits and wine, which were not eligible for payments in the past.

The effect of aid redistribution can also be differentiated by altimetry (Fig. 20.2). The effect is much stronger on plain areas, where resources are cut for the benefit of mountain areas. At altimetry levels too, the type of production is important: plain areas used to receive aid for almost all hectares, so the redistribution operates a linear reduction. Mountain and some hill areas gain advantage thanks to smaller cultivated surface area and because most farms specialize in milk production. In fact, despite the milk quota system, mountain areas have until now been less favored.

20.2.3 New Resources to Pillar 2

The new capping mechanism appears to be very different from the scheme introduced by the Health Check in 2007 and in force today. The current scheme applies the mechanism of modulation only to beneficiaries receiving more than €300,000 in direct payments, which are subject to an additional cut of 14 %. But under the new scheme, the upper limit of direct payments to farmers will be

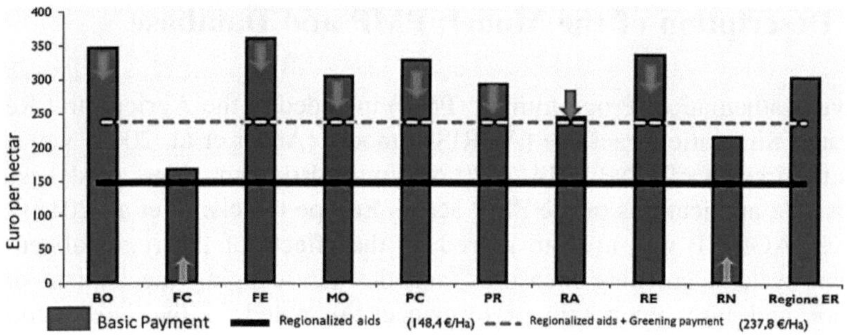

Fig. 20.1 Redistribution of aid in Emilia-Romagna provinces

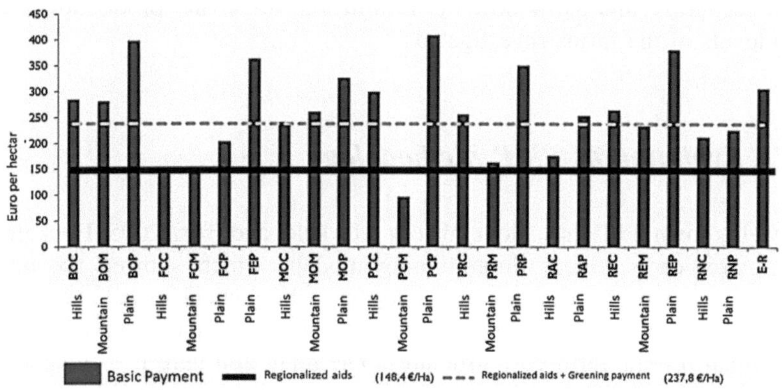

Fig. 20.2 Redistribution of aid in Emilia-Romagna provinces by altimetry

€300,000. However, in order to preserve and stimulate the application of environmental measures and practices, the cost of greening will not be considered in the budget. In order to maintain employment in the sector, the capping mechanism will be mitigated for farms employing waged labor. The EU Commission draft states that the direct payments scheme will take into account the employment level on farms, and the amount of wages actually paid and declared by farmers for the previous year, including social security contributions and employment taxes, will be added to the total amount of direct payments due.

As the result of capping, more resources will be available for transfer to Pillar 2 (Rural Development). Moreover, given that each MS has a different agriculture system, each MS is given the option of transferring up to 10 % of national financial resources assigned for direct payments (Pillar 1) to Rural Development (Pillar 2). In addition, MS which receive less than 90 % of the EU average direct payments can now transfer up to 5 % of the funds assigned for Rural Development to the Direct Payment System.

20.3 Description of the Model: PMP and Database

Positive Mathematical Programming (PMP) included in the Agricultural Regional Integrated Simulation Package (AGRISP) model (Arfini et al. 2005) was used to assess the impact of CAP 2014–2020 on Emilia-Romagna. This model is one of the possible applications of the PMP across Europe (Heckelei et al. 2012). In this research, AGRISP was used to reproduce the effects of the regionalized single payment system, greening measures, and the new capping mechanism on farm behavior and farm economic performance. As noted in the introduction, the simulations are based on the draft by the European Commission, so rather than certain consequences they indicate possible potential consequences. The simulations demonstrate a capability differential of farms in reacting to new policy and market scenarios, and show how the reform will affect the production and economic levels of the farms investigated.

20.3.1 Evolution of PMP Methodology

As is well known, PMP has a long history (Heckelei and Britz 2005; Heckelei et al. 2012). After early Linear Programming models, which showed the optimum combination for production according to the technological matrix, the next step was positive models where the optimum is considered at the observed production level, which reveals farm cost structures. The main aim here is to give as true a picture as possible of the current situation, then simulate the behavior of farm producers as agricultural policy intervention is shifted (Arfini and Donati 2013). The versatility of PMP means it can be fitted to valuation models with different levels of detail, so it can be applied to a single farm (business model) or for simulations of the dynamics of a territory (regional model) or a production sector (sectorial model). It is also possible to structure a mixed model using an integrated database including regional and sector aspects. Despite their differences, these models have a common matrix deriving from their microeconomic formulation and embodied in the use of information collected at the enterprise level, regardless of the "scale" (corporate, regional, or sectorial) of the simulation. In business models the scope is limited to the single firm, but the models used here provide regional or sectorial results for aggregate geographical areas or entire productive sectors. Sectorial models are usually used by decision-makers in assessing agricultural policy for a single and specific production. But although their primary objective is to analyze supply changes of products subject to intervention, the models can also have a regional significance to the extent they quantify the impact of the policy at the regional, national, or international level (Arfini and Donati 2011). The element that distinguishes the sectorial from the regional model is the aggregation criteria of the farms; for sectorial models, the aggregation criteria is the farm type.

The application of mathematical programming to agricultural policy entails defining from the outset a reference to a regional area (in the case of a regional model) or to a specific productive sector (sectorial model). Regional and sectorial models are not necessarily alternatives; in general, a single productive segment is analyzed individually with reference to a territorial area, in contemporaneous regional and sectorial studies. However, the variables of interest and the PMP model need to be defined a priori.

So in a complex scenario like local rural development the model needs to cover different farm types and truthfully represent land use and productivity levels of the farmers in the area. It is useful to minimize the amount of data, although there needs to be enough detail to describe both individual farmer behavior and technologies and production decisions at the farm level. AGRISP (Arfini et al. 2005), which shows in detail the use of land, farm specialization, and different farm classes by size, is an appropriate compromise. AGRISP in fact represents a fusion of two databases: the AGEA database on land use of each farm, and the FADN database of the profitability of each production process activated. So rather than a "model," AGRISP is a tool of analysis that overcomes limitations of other similar tools used to simulate agricultural policies at the regional level. The combined use of AGRISP data and PMP methodology means that all output models can be calculated precisely.

In Europe as in Italy, at present, the main source of statistical data on structural characteristics of farm production and economics is the Agricultural Accountancy Data Network (FADN). Data are obtained by survey and the database is structured as a statistical sample of all farms. It can be considered "ideal" for coefficients on farm production techniques and farm economic characteristics. FADN has, however, three major limitations: (i) a lack of technical information on the amount of input used for each process; (ii) the representativeness of data is based on the standard gross margin, while the presence of a specific process reflects the land use; (iii) the level of representativeness of farms decreases significantly from regional level to provincial level.

In particular, the first aspect (the lack of quantitative data on inputs) is an essential element to allow simulation models (especially those based on mathematical programming) in order to define the technology used by that of single farms. Consequently, the lack of such data excludes the use of FADN for the purposes of analysis of agricultural policy through the use of mathematical programming. The limits on the representativeness of FADN data also make it difficult to represent production systems in areas smaller than NUTS3 and perform statistical inferences to the statistical universe. To overcome this limitation, AGRISP integrates FADN with the Italian administrative database AGEA.[1] The integration of FADN and AGEA thus makes it possible to measure the exact dimension of agricultural production systems, gross marketable output, subsidies distributed, the

[1] AGEA is the Italian official body entitled to pay farmers eligible for CAP payments. Farmers have to provide AGEA with all the information related to agricultural activities including land use.

volume of variable costs attributable to each process, and the gross income for each type of activity.

The two databases combined in a single database FADN-AGEA give a complete dataset of land use and technical and economic parameters for production processes of all the farms included in the analysis. The aggregation of information is performed at the level of macro-farm (farms in the AGEA database grouped by size) and farm specialization by each agricultural area (a homogeneous altitude area belonging to the same province). More precisely, in each province three altitude levels, seven size classes (0–10 ha, 10–20 ha, 20–30 ha, 30–50 ha, 50–100 ha, 100–300 ha, >300 ha) and three economic sectors (Fruit and vegetables, Animal production, and Others) were considered, where each class represents the minimum farm type reference. Naturally each macro-farm considers all agricultural activity present in the territory as they are registered in the AGEA database. The integration of the two databases was effected by specific software able to perform statistical analysis and yield information on farmer choices on production and economic indicators.

20.3.2 AGRISP Model Overview

The AGRISP model is able to estimate for each province and at the overall regional level the effects of CAP measures on farmer production plans and farmer income. It gives insight into production decisions for the current observed situation (baseline) and into future decisions after the CAP reform, and thus models farmer strategy. The AGRISP model consists of three main phases: (i) Extraction of data on farms in the sample; (ii) PMP estimation of cost functions at macro-farms level, calibration to observe reality and simulations; and (iii) Analysis of results. AGRISP can be defined as a regional tool, because in a single resolution it can simulate the effects of agricultural policies on different homogeneous areas (agricultural regions) constituting the administrative regions (provinces). It can also be defined as "integrated" because it includes modules that manage the flow of information for the functional analysis of agricultural policy. The process of organizing information is probably the most innovative element. AGRISP provides information on the production choices made by individual farmers, capturing their strategies in the prereform situation and in projected results organized by agrarian regions and at the regional level. The use of a single database covering land use (from AGEA) and the profitability of single processes (from FADN) combined with PMP methodology allows analysis of the impact of agricultural policies both at the micro- and macro-level. It is useful for both setting rural policies and estimating changes in supply at the regional level. In simulating the effects of agricultural policy at the regional level, AGRISP aggregates cost functions into a single regional model, and constructs a set of constraints able to simulate the policies for the whole region (Fig. 20.3).

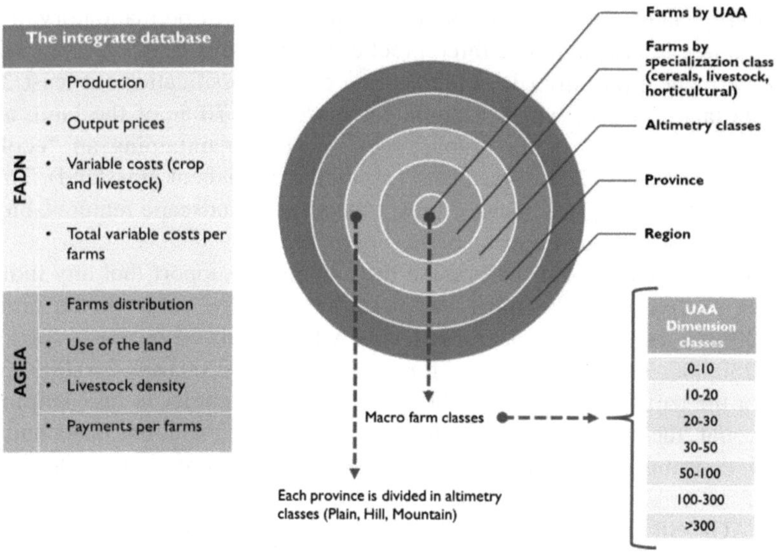

Fig. 20.3 Data structure in the AGRISP model

20.4 The Policy Scenario

In this evaluation of the CAP reform, three main aspects have been considered based on the regulation proposal of the European Commission n. 625-COM2011. These are basic payments, greening measures, and capping.

Direct Payments will follow the new Basic Payment Scheme. Until today, the EU-15 was covered by a Single Payments Scheme allowing for historical references, or a payment per hectare, or a "hybrid" combination of the two, and most of the EU-12 was covered by the Single Area Payments Scheme (SAPS). From 2013, a single new "Basic Payment Scheme" applies. The aim is to significantly reduce discrepancies between the levels of payments between farmers, between regions (internally) and between MS through full implementation of current legislation. All MS will be obliged to move toward a uniform payment per hectare at the national or regional level by the start of 2019. In line with the Commission proposals in the Multi-Annual Financial Framework, the national envelopes for direct payments will be adjusted so that those who receive less than 90 % of the EU-27 average payment per hectare will receive more. The gap between current payments and 90 % of the EU-27 average is reduced by one-third. The Commission is committed to discussing a longer term objective of achieving "complete convergence" through equal distribution of direct support across the European Union in the next Financial Perspectives after 2020.

In addition to the Basic Payment, each farm will receive a payment per hectare for following farm practices beneficial for the climate and the environment. MS will use 30 % of the national envelope in order to pay for this. The draft proposal of the European Commission states that this payment will be mandatory, and will not be subject to capping. The three practices eligible for payment are: (i) maintaining permanent pasture, (ii) implementing crop diversification (at least 3 crops on arable land, none of which account for more than 70 % of the land, and the third crop at least 5 % of the arable area); and (iii) maintaining an "ecological focus area" of at least 7 % of farmland (excluding permanent grassland). This area may include field margins, hedges, trees, fallow land, landscape features, biotopes, buffer strips, and wooded areas.

The capping mechanism will define the amount of support that any individual farm can receive from the Basic Payment Scheme. The sum will be limited to €300,000 per year. Current payment levels will be reduced by 70 % for the part from €250,000–300,000; by 40 % for the part from €200,000–250,000, and by 20 % for the part from €150,000–200,000. However, in order to take employment into account, the farm can deduct the costs of salaries, including taxes and social security contributions, declared the previous year, before these reductions are applied.

The AGRISP model is now applied to Emilia-Romagna to evaluate the effects of the new EU support measures in the agricultural sector. Results are detailed at the regional and provincial level, and by altimetry. The following scenarios are identified:

1. *Baseline* The base scenario on which the comparison is carried out is the situation recorded in 2010, obtained by updating the 2007 calibrated solution with the market price variation for 2007–2010.
2. *Greening* This scenario simulates full application of the CAP reform. All constraints and the new policy aids are activated (new Basic Payment Scheme with distribution to all Utilized Agricultural Areas (UAA), mandatory greening scheme, and new capping mechanism).

For the value of aid per hectare for the Emilia-Romagna region, the amount calculated by the National Institute of Agricultural Economics is a basic regionalized payment of 148.4€/ha, while for the greening component they calculate 89.4€/ha. These figures are used in our simulations, which also consider the reduction of total aid under the mechanism of gradual reduction according to the scale described above.

20.5 The Impact of Policy

20.5.1 Variation in Land Use

Results of land use by single process simulations show changes between the current baseline scenario and after reform in the greening scenario. Figure 20.4 reports the results in hectares, and Table 20.1 reports the results in percentages.

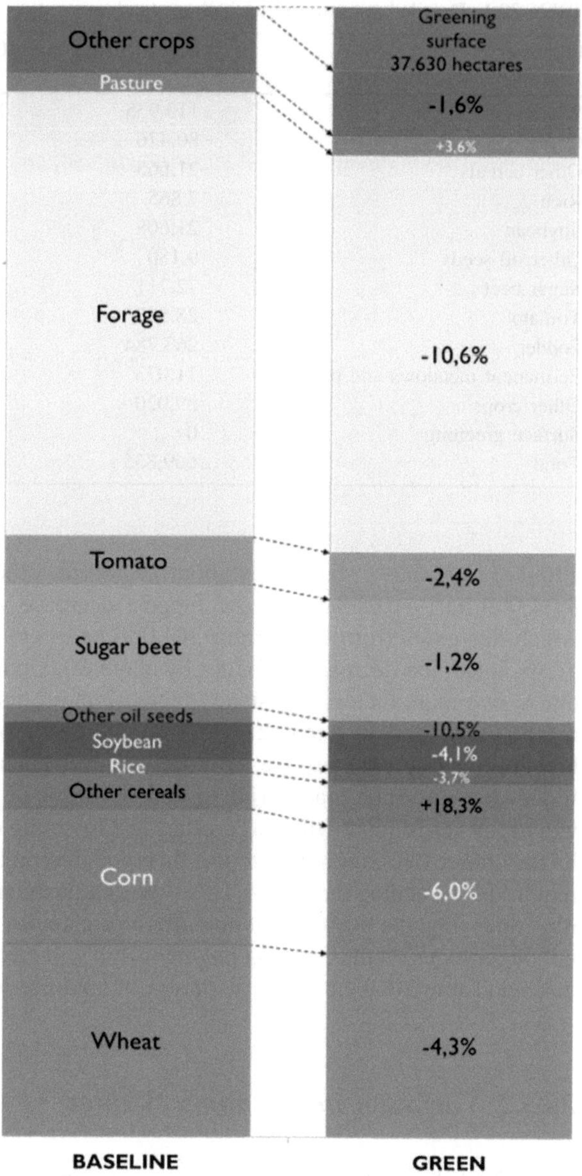

Fig. 20.4 Variation in agricultural land use

The results show that at the regional level, greening will lead to a big reduction of almost 38,000 ha in cultivated areas. There will be different impacts on various crops as farmers adjust production choices to market prices.

Basically there will be a decrease in almost all crops, except for certain cereals (barley, oats, etc.) which show a big increase (+18 %). But there will be big decreases in surface areas of wheat (−4.3 %) from the current 114,000 ha to about

Table 20.1 Detailed variation in agriculture land use

Processes	Baseline (ha)	Green (ha)	Base/green (Variation in %)
Wheat	113,935	109,008	*−4.3*
Corn	80,476	75,665	*−6.0*
Other cereals	21,665	25,639	*+18.3*
Rice	7,865	7,576	*−3.7*
Soybean	21,608	20,732	*−4.1*
Other oil seeds	9,180	8,214	*−10.5*
Sugar beet	72,711	71.863	*−1.2*
Tomato	28,518	27,837	*−2.4*
Fodder	263,784	235,802	*−10.6*
Permanent meadows and pastures	11,073	11,473	*+3.6*
Other crops	39,020	38,397	*−1.6*
Surface greening	0	37,630	–
Total	669,835	669,835	–

110,000 ha, maize (−6.0 %) by about 75,500 ha, other seed crops (e.g., sunflower) by 1,000 ha (−10.5 %), and the biggest decrease will be in fodder (−10.6 %), which decreases from the current 264,000 ha to about 236,000 ha.

As noted, the decrease in fodder by about 30,000 ha is the largest. It is probably due to two main factors; the steady increase in profits on cereal crops especially in recent years, and low market profits on fodder crops excluding crops used for breeding. These changes are extremely significant for Emilia-Romagna because fodder tends to take place prior to abandonment of the land. Greening measures would further bring down farm profits. The lowering of fodder crop is justified also by the choice of entrepreneur to use this type of crop for environmental purposes as required by greening measures. This type of adjustment entailed by the CAP reform illustrates how the market will now drive production choices for farmers, who will no longer focus on maximizing payments but will have to maximize farming profits instead. Table 20.2 shows the variation in land use by altimetry bands.

20.5.2 Variation in Economic Values

Table 20.3 shows the changes in economic components per hectare at regional and altitude levels. Gross Salable Production (GSP) falls by 7.4 % from the current 3,276€/ha to 3,033€/ha. This is mainly because part of the land is taken out of production to meet greening requirements. The values of GSP by altimetry bands present a heterogeneous distribution: contraction on plains is about −5.2 %, while hills and mountains decrease by −10.8 and −20.5 %, respectively. The reduction in variable costs (−7.7 %) is in line with the changes in GSP as farms make the mandatory adjustments required by greening and halt production of certain crops.

Table 20.2 Variation of agriculture land use by altimetry

	Plain			Hill			Mountain		
Processes	Greening (ha)		VAR (%) (ha)	Greening (ha)		VAR (%) (ha)	Greening (ha)		VAR (%) (ha)
Wheat	91,274	82167	−9.9	21,867	22,473	2.8	795	4,367	449.5
Corn	72,678	67.456	−7.2	7,744	8,072	4.2	54	137	154.1
Other cereals	11,556	10,745	−7.0	9,278	10,558	13.8	831	4,336	422.0
Rice	7,865	7,576	−3.7	–	–	–	–	–	–
Soybean	21,354	20,495	−4.0	254	237	−6.9	–	–	–
Other oilseeds	8,882	7,946	−10.5	298	268	−10.1	–	–	–
Sugar beet	70,629	69,759	−1.2	2,082	2,104	1.1	–	–	–
Tomato	23,558	22,837	−3.1	4,928	4,965	0.8	32	34	6.2
Fodder	122,733	115737	−5.7	93,720	83,484	−10.9	47,331	36,580	−22.7
Pastures	1,916	1,966	2.6	5,291	5,485	3.7	3,866	4,022	4.0
Other crops	36,110	35,485	−1.7	2,882	2,852	−1.0	28	60	114.9
Surface greening	–	26,385	–	–	7,844	–	–	3,401	–

Table 20.3 Variation of economic indicators at regional and altimetry levels

			Baseline (€/Ha)	Green (€/Ha)	VAR (%)
Region level		GSP	3,276	3,033	−7.4
	–	Total variable costs	2,356	2,176	−7.7
	=	Gross margin (1° level)	920	857	−6.9
	+	Total AID	307	237,8	−22.5
	=	Gross margin (2° level)	1,227	1,095	−10.8
Plain areas		GSP	3,416	3,238	−5.2
	–	Total variable costs	2,494	2,354	−5.6
	=	Gross margin (1° level)	922	884	−4.1
	+	Total AID	339	237	−29.9
	=	Gross margin (2° level)	1,261	1,122	−11.1
Hills areas		GSP	2,922	2,606	−10.8
	–	Total variable costs	2,051	1,801	−12.2
	=	Gross margin (1° level)	871	804	−7.6
	+	Total AID	246	237,8	−3.4
	=	Gross margin (2° level)	1,117	1,042	−6.7
Mountain areas		GSP	3,029	2,409	−20.5
	–	Total variable costs	1,989	1,646	−17.2
	=	Gross margin (1° level)	1,041	763	−26.7
	+	Total AID	186	237	27.6
	=	Gross margin (2° level)	1,227	1,001	−18.4

These dynamics appear mainly in plain and hill areas, while in mountain areas the structural rigidity of production does not permit a reduction in variable costs proportional to the fall in GSP. The gross margin at the 1st level, calculated by subtracting variable costs from GSP, which is an indicator of business efficiency, is

Fig. 20.5 Variation in GSP

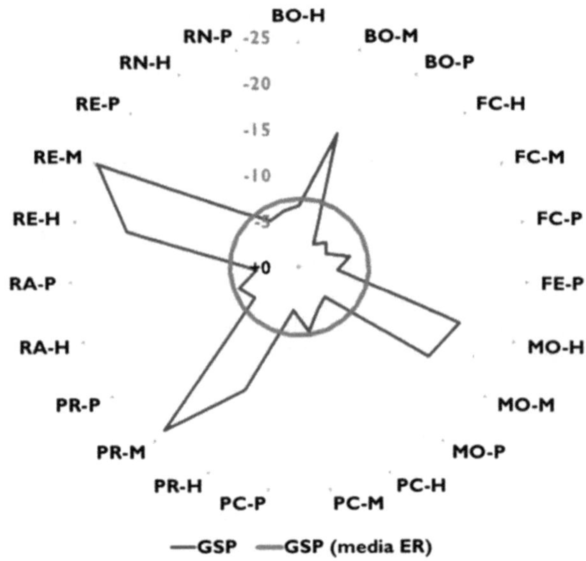

thus affected in different ways. Overall in the region it falls by 6.9 %; but while in plain areas it falls −4.1 %, and in hill areas it falls −7.6 %, in mountain areas it falls by −26.7 %.

As noted above, total payments, now consisting of regionalized basic payment + greening payments, are reduced across the region by 22.5 %, from €307/ha to €237.8/ha on average. Taking this decrease into account, the new distribution of aid over plain and hill areas leads to a contraction which poorly affects what we might call the 2nd level gross margin or 1st level gross margin + aid. In mountain areas, the new distribution of payments is advantageous given that currently average aid per hectare stands are only 190€/ha. Overall, however, across the region, the reduction of 2nd level gross margin falls by nearly 11 %. The reorganization imposed by regionalization and greening measures will clearly have a negative impact on the overall agricultural sector of the Emilia-Romagna region.

Data disaggregated by province shows differing situations across the region. At the province level, without considering altimetry: GSP/ha decreases by −11 % in the provinces of Bologna (BO), Modena (MO), Parma (PR), and by −14.3 % in Reggio-Emilia (RE), but there is a smaller contraction of between 5 and 7 % in the remaining provinces. The bigger reduction in the first group of provinces is closely linked to the decline in livestock at each altimetry.

Figures 20.5, 20.6, 20.7, and 20.8 report a more detailed analysis of reform effects, and show values specified for different altitude levels and provinces. The GSP (Fig. 20.5) is affected by a big reduction in the mountainous areas of Parma, Reggio-Emilia, Modena and Bologna, with decreases in values that range from −15 to −25 %, while in hill areas (except for Bologna) show decreases ranging

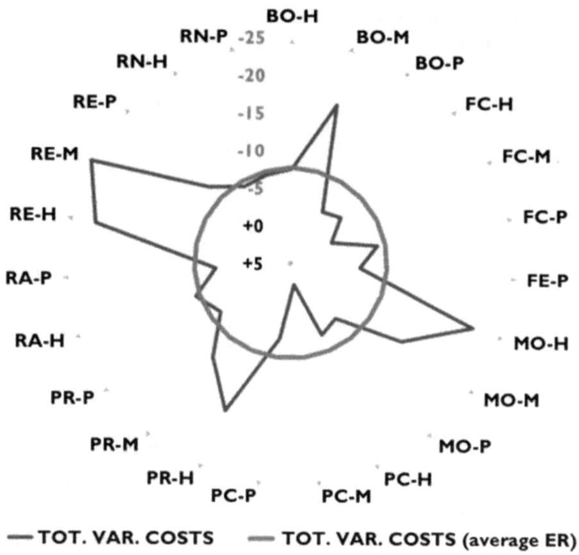

Fig. 20.6 Variation in variable costs

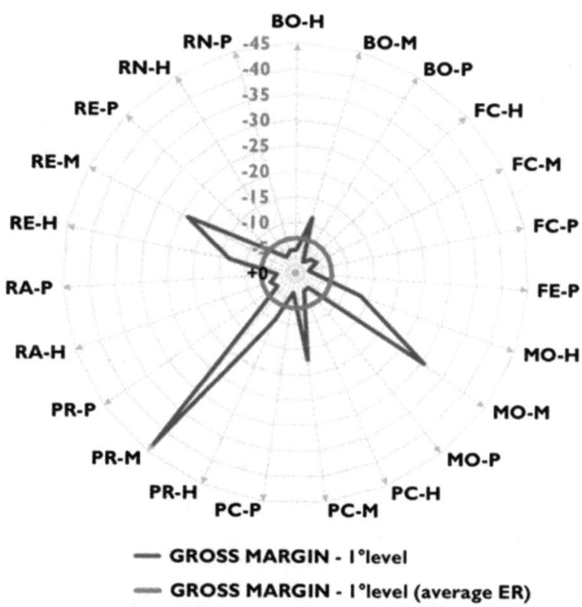

Fig. 20.7 Variation in 1st level gross margin

from −15 to −20 % approximately. The second graph (Fig. 20.6) reports dissimilarity in total variable costs. In general these are aligned with the decreases in GSP as processes are adjusted. But in some areas, such as mountain areas of Parma and Modena, where farms show more structural rigidity, the realignment is less

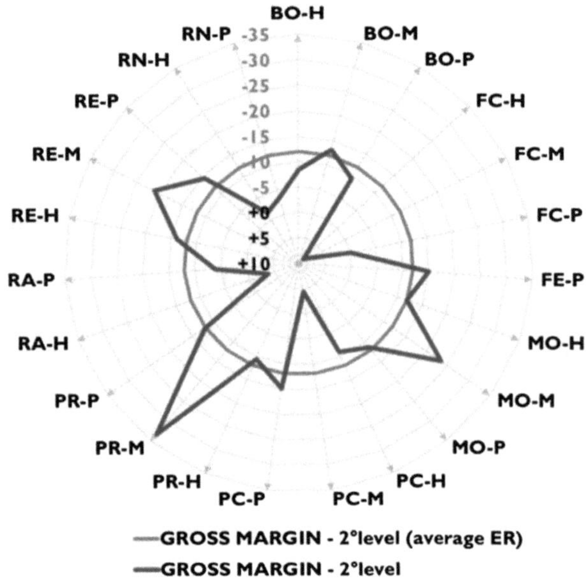

Fig. 20.8 Variation in 2nd level gross margin

proportional. For this reason, 1st level gross margin (Fig. 20.8) downsizes in mountainous areas: Parma −45 %, Reggio-Emilia −25 %, Modena −30 %.

Finally, Fig. 20.8 shows the changes in the 2nd level gross margin. This margin includes the effect of aid redistribution where mountain areas receive the most advantage. But mountain areas and some hilly areas confirm the biggest falls: Parma (−33.4 %), Reggio-Emilia (−21.1 %) and Modena (−23.4 %), and all show values well below the regional average of 11.9 %.

20.6 Conclusion and Policy Recommendations

The model shows that greening measures combined with the regionalized distribution of basic payments will lead to substantial reductions in terms of GSP and farm income in Emilia-Romagna, assuming constant prices. The biggest consequences at farm level will be covering fixed investment costs. Greening generates a double effect: a contraction of harvested surfaces (e.g., forage, wheat, and maize) and a big shift in land use and resources toward higher price crops (e.g., cereals and tomatoes). The extension of regionalized aid to almost all UAAs with a single payment level in all regions, accompanied by a reduction in the amount of aid, will lower average aid per hectare for the plain areas, where at present farmers are accustomed to higher levels of support, in favor of mountain areas. But despite the increase in direct payments for mountain areas, these are the most badly hit by the reform. Ongoing discussion between the DG-AGRI Committee and the Trilogue is currently focusing on adjusting some of the greening criteria, both in terms of practical application and access to specific aid. As noted previously, this analysis was made on the basis of the European Commission draft proposal, so numerical measurements may be taken as provisional and attention should be focused more on the structural and territorial weaknesses.

But the potential impacts of the CAP reform require discussion and debate. As currently formulated, the proposal offers fewer guarantees to specific local and territorial farmers in that it delegates to an individual MS the decision to apply for aid as "less-favored areas," or to maintain coupled aid for "productions with a local relevance." This entails direct intervention by MS policy makers in order to activate specific and voluntary schemes. Given that "the market" will be the new driver for production choices by farmers, the strategic choices of European agriculture will have to take account of the need to protect local farms and entire production sectors in disadvantaged areas, such as diary or livestock farming in hill and mountain areas. A new mindset will be necessary; the lack of competitiveness of farms in regional mountain areas is due in most cases to territorial characteristics and economic environment rather the competitiveness of the farm itself.

References

Anania, G. (2008). Il futuro dei pagamenti diretti nell'health Check della Pac: regionalizzaizone condizionalità e disaccoppiamento, In F. De Filippis (Ed.), L'Health Check della Pac: una valutazione delle prime proposte della Commissione, Quaderni Gruppo 2013, Edizioni Tellus, pp. 29–40.

Arfini F., & Donati M. (2011). *Impact of the Health Check on structural change and farm efficiency: A comparative assessment of three European agricultural regions—in Disaggregated Impacts of CAP Reforms* Proceedings of an OECD Workshop OECD Publishing.

Arfini, F., & Donati, M. (2013). Organic production and the capacity to respond to market signals and policies: An empirical analysis of a sample of FADN farms. *Agroecology and Sustainable Food Systems, 37*(2), 149–171.

Arfini, F., Donati, M., & Zuppiroli, M. (2005). Un modele quantitatif pour l'evaluation des effets de la riforme de la PAC en Italie. *Economie Rurale, 285*, 70–87.

De Filippis, F., & Frascarelli, A. (2012). Il nuovo regime dei Pagamenti diretti in agricoltura In F. De Filippis (Ed.) (A cura di), La nuova Pac 2014–2020-Un'analisi delle proposte della Commissione, Quaderni gruppo 2013, Edizioni Tellus.

European Commission. (2011a). *Proposal for a Regulation of the European Parliament and of the Council establishing rules for direct payments to farmers under support schemes within the framework of the common agricultural policy*. Brussels, 19.10.2011 COM(2011) 625 final/2 2011/0280 (COD).

European Commission. (2011b). *Impact Assessment. Common Agricultural Policy towards 2020.* Annex 2: Greening the CAP, Commission Staff Working Paper, Brussels, February.

Gigante R. (2013). *Assessment of Cap reform post 2013 in rural areas of Emilia-Romagna region*. Ph.D. thesis.

Heckelei, T., Britz, W., & Zhang, Y. (2012). Positive mathematical programming approaches—recent developments in literature and applied modelling. *Bio-based and Applied Economics, 1*, 109–124.

Heckelei, T. & Britz, W. (2005), Models based on positive mathematical programming: State of the art and further extensions In F. Arfini (Eds.), Modelling agricultural policies: state of the art and new challenges. *Proceedings of the 89th European seminar of the European Association of Agricultural Economics,* (pp. 48–73). Monte Università Parma.

Matthews, A. (2013). Greening agricultural payments in the EU's Common Agricultural Policy. *Bio Based and Applied Economics, 2*(1), 1–27.

Sorrentino, A., Henke, R., & Severini, S. (2011). *The common agricultural police after the fischler reform, national implementation, impact assessment and the agenda for future reforms.* Burlington: Ashgate.

Chapter 21
Measuring Biodiversity of Cropping Structure with the Use of FADN Data

Adam Was and Paweł Kobus

Abstract Greening of the Common Agricultural Policy as proposed by the European Commission for the 2014–2020, CAP reform raised interest in measuring crop diversity. Based on a sample of the 12,258 farms recorded in the Polish 2009 FADN (FADN—Farm Accountancy Data Network—an instrument for evaluating the income of agricultural holdings and the impacts of the Common Agricultural Policy. For details see http://ec.europa.eu/agriculture/rica/ (FADN 2013)), the authors verify the suitability of the most popular biodiversity indices for measuring the level of diversification of cropping structure for assessing fulfillment of CAP greening criteria. None of the most known biodiversity indicators provided a possibility of a proper delimiting of "green" farms based on FADN data. Modification of the Simpson index was proposed to allow proper distinguishing of "not-green" farms within FADN records. Using biodiversity indices for measuring crop diversification on large areas is related to spatial aggregation. Different indices have been calculated for Polish NUTS 2 (Nomenclature of Units for Territorial Statistics—geocode standard for referencing the subdivisions of countries for statistical purposes. The standard is developed and regulated by the European Union, and thus only covers the member states of the EU in detail. Poland consists of 16 NUTS 2 regions called voivodeships.) regions, based on FADN single-farm records as well as official statistical data on cropping structure for NUTS 2 regions. Results show that there is a very small correlation between regional indices calculated based on aggregated crop structure data and the share of "not-green" farm area in the regions. This suggests that biodiversity indices calculated using regional data are strongly biased and should not be used for verification of fulfilling the EC crop diversification criteria.

A. Was (✉) · P. Kobus
Faculty of Economic Sciences, Warsaw University of Life Sciences,
Nowoursynowska 166, 02-787 Warsaw, Poland
e-mail: adam_was@sggw.pl

P. Kobus
e-mail: pawel_kobus@sggw.pl

C. Zopounidis et al. (eds.), *Agricultural Cooperative Management and Policy*,
Cooperative Management, DOI: 10.1007/978-3-319-06635-6_21,

21.1 Introduction

The basic principles of the Common Agricultural Policy were established in the late 1950s and over the years were subject to constant changes. The previously formed CAP model as well as the budget perspective currently in force will remain binding until the end of the budgeting period in 2013. At present, preparations are underway to finalize the concept of a reformed agricultural policy for the new budgeting period for the years 2014–2020.

The Commission proposal of October 2011 for a reform of the CAP after 2013 (European Commission 2011) focuses on sustainability and the environmental performance of agriculture to an extent that has never existed before.

Three mandatory "greening" components proposed by the Commission which have to be fulfilled at the farm level are the following: retaining areas under permanent grassland (PG) as declared in 2014[1]; crop diversification[2]; and ecological focus areas (EFA).[3] Basic requirements are as follows:

- minimum of three crops in rotation, with a maximum proportion of one of them at the level of 70 % and a minimum proportion in the crop structure at the level of 5 %;
- maintaining the existing areas of permanent grassland, with the right to reduce the area by not more than 5 % compared to the base year;
- allocation of 7 % of arable land to the ecological focus area, including ecological land such as land left fallow, terraces, landscape features, buffer strips, and afforested areas.

In this chapter measurement of crop diversification is addressed. There are different approaches used to assess the effects of this requirement of the EC proposal. There are studies in which fulfillment of crop diversification criteria is verified at the single-farm level (Majewski et al. 2013), while in other cases biodiversity indices aggregated to the farm-type level representing a large number of farms are used (Britz et al. 2012).

The main topic of this study is the verification of compliance with the EC crop diversification criteria. Introduction of obligation to grow at least three crops at the same time does not directly influence biodiversity of the ecosystem. The concept of measuring crop diversity applied in this chapter is based on one of the biodiversity definitions stating that it is "the number of different native species and

---

[1] It means maintaining the areas of permanent grassland (PG) with the right to reduce the area by not more than 5 % compared to the base year.

[2] It means a minimum of three crops in rotation, with maximum participation of one of them at the level of 70 % and a minimal share in the crop structure at the level of 5 %.

[3] An area equivalent to at least 7 % of a farmer's eligible hectares (permanent grassland is excluded from the calculation) should be used for ecological purposes. Habitats and features that would be eligible to fulfill the EFA requirement may include: fallow land, terraces, landscape features, buffer strips, and areas afforested under CAP Pillar 2.

individuals in a habitat or geographical area" (Jones and Stokes 1987). Following that idea, the most popular biodiversity indices are used to measure crop diversity. Additionally, authors also examine the possibility of using aggregated regional data for this purpose.

21.2 Data and Methods

The basic source of data used in the analysis was the 2009 Polish Farm Accountancy Data Network (FADN). For assessing biodiversity based on regional cropping structures for NUTS 2, the EUROSTAT data (EUROSTAT 2013) on the area of particular crops harvested in Poland in the year 2009 have been used. Some missing values have been supplemented with data from the Polish Central Statistical Office (GUS 2011).

Indices concerning the number of species in the sample cannot be included in the analysis as there are no data on the number of individual plants growing on fields where specific crops are cultivated. This limits the scope of biodiversity indices which might be used for this analysis, i.e., the indices which use shares of species as arguments.

The following indices have been used in the study:

Shannon–Weiner (H). This is an index applied to biological systems by Shannon in 1948 (Shannon 1948; Hiep and Engels 1974). It is the most commonly used among the diversity indices. Results are generally between 1.5 and 3.5, very rarely exceeding 4.5. The Shannon index is widely used for comparing diversity between various habitats (Turkmen and Kazanci 2010).

$$H = -\sum \frac{n_i}{N} \ln \frac{n_i}{N}$$

where

n_i area of i-crop

N total area of arable land

Simpson Indices

There are three variants of the Simpson index.

Simpson's Index (SI). This diversity index derived by Simpson (Simpson 1949; Hiep and Engels 1974) measures the probability that two individuals randomly selected from a sample will belong to the same species (or some category other than species). The value of the index ranges between 0 and 1. The higher the value, the lower the diversity.

$$\mathrm{SI} = \sum \left(\frac{n_i}{N}\right)^2$$

Simpson's Index of Diversity 1 − SI (*Simpson a*) (Hiep and Engels 1974; Khan 2006)

The value of this index also ranges between 0 and 1, but now, the greater the value, the greater the sample diversity. This seems to be more appropriate and similar to other indices. In this case, the index represents the probability that two individuals randomly selected from a sample will belong to different species.

$$1 - \mathrm{SI} = 1 - \sum \left(\frac{n_i}{N}\right)^2$$

Simpson's Reciprocal Index 1/SI (*Simpson b)* (Hiep and Engels 1974; Khan 2006)

The value of this index starts with 1 as the lowest possible figure. This figure would represent a community containing only one species. The higher the value, the greater the diversity. The maximum value is the number of species (or other category being used) in the sample. For example, if there are five species in the sample, then the maximum value is 5.

$$1/\mathrm{SI} = 1/\sum \left(\frac{n_i}{N}\right)^2$$

As all Simpson indices are a monotone transformation of a prior version, there is therefore no reason to use all of them. In this study the second version (Simpson b) is used in further calculations.

McIntosh (MI) (McIntosh 1967). The values are between 0 and 1. The value of the indicator equal to 1 means that the organisms in a community are evenly distributed:

$$\mathrm{MI} = 1 - \frac{\sqrt{\sum n_i^2}}{N}$$

Prior to the calculation, farms with less than 0.5 ha of arable land had been removed from the FADN sample of 12,258 farm records. Finally, 11,820 farm records were left for the analysis.

For this study the crops in FADN records have been categorized into the 16 following groups, each treated in the analyses as a single species: wheat, rye, barley, oats, triticale, corn (grain), other cereals, pulses, sugar beets, rapeseed, industrial crops, maize (silage), fodder crops, fruits and vegetables, and other crops.

Grouping crops in such categories was necessary to achieve comparability with regional statistics on cropping structure which includes 16 crops only.

The chosen subset of 11,820 farms has been divided into two clusters of farms: "green" and "not-green" in accordance with the EC crop diversification criteria. The possibility of delimiting these two groups using the considered biodiversity indices was then tested. Finally, a modification of the Simpson index is proposed.

For all considered indices, basic descriptive statistics are calculated. McFadden R^2 value (1973) is used for verification of the predictive quality of indices for delimited farms based on the greening rules.

In the next stage, all indices are calculated and analyzed at the regional level. Biodiversity indices calculated for each of the FADN farms are aggregated to the regional (NUTS 2) level using FADN weights.[4]

In contrast to aggregating single-farm records, considered indices are calculated, also based on aggregated crop structure for all NUTS 2 regions in Poland. In this case, each of the NUTS 2 regions is treated as a single farm.

Then, based on FADN records, the share of area in "not-green" farms for each NUTS 2 has been estimated using the following formula:

$$NGj = \frac{\sum_{kj} ng_{kj} SYS02_{kj}}{TAj}$$

where

j	NUTS 2 region
k	farm within j NUTS 2 region
ng	not-green area
$SYS02$	number of real farms represented by FADN farm
TA	total area represented by FADN sample in j region

Finally, the correlations between aggregated biodiversity indices and the share of "not-green" farms at the NUTS 2 level area are calculated.

21.3 Results

The whole dataset is divided into two groups according to crop diversity criteria proposed by the European Commission. Among the selected FADN subsample, 1,934 farms were selected as "not-green" while 9,887 fulfilled diversification criteria. For every considered index, a kernel density estimator has been plotted both for "green" and "not-green" farms. The results are shown in Figs. 21.1, 21.2, and 21.3.

As shown in Figs. 21.1, 21.2 and 21.3, none of the indices is suitable for delimitation of green farms. Values of all indices are strongly overlapping. This shows that there is a need for modification of the current indices used for this purpose. Concentrating on delimitation of green and not-green farms, a small modification of the Simpson index could be proposed by adding to the formula elements concerning the minimal number of crops and number of crops exceeding a maximum threshold.

[4] Each farm in the FADN sample represents a number of farms in the population.

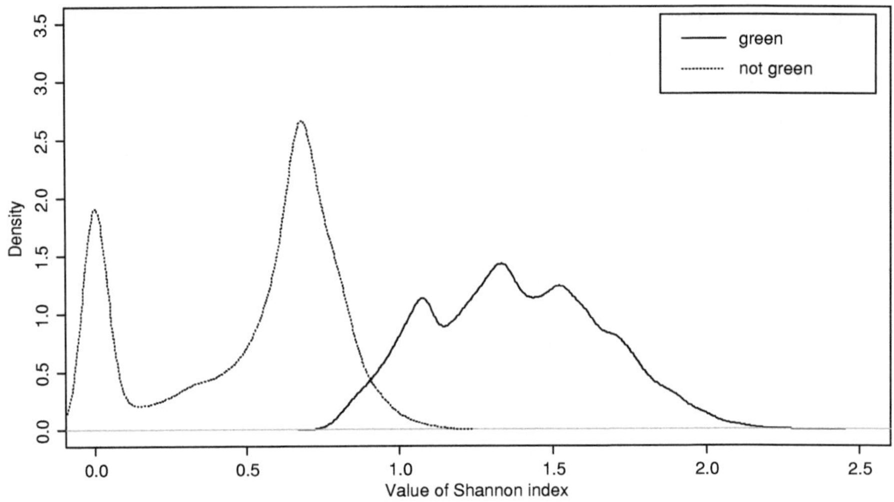

Fig. 21.1 Kernel density estimate for distributions of Shannon index for the sample of Polish FADN farms, based on crop diversification criteria. *Source* own calculations based on FADN data

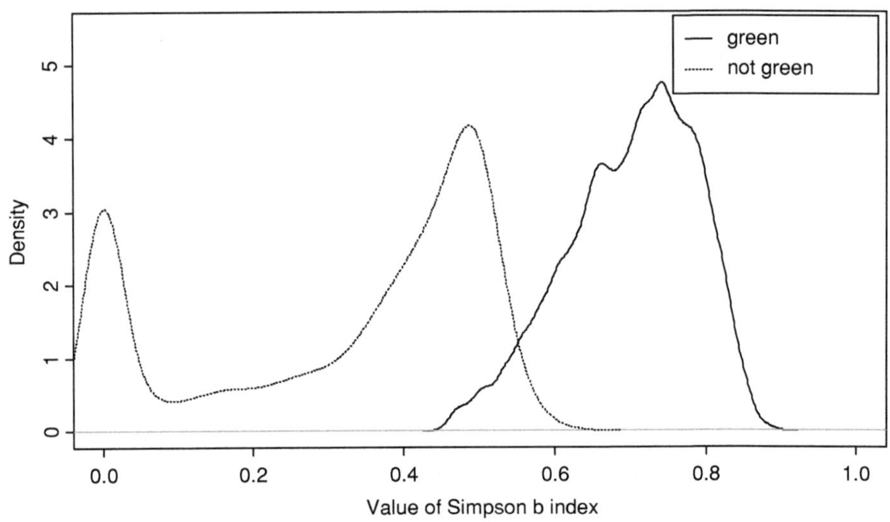

Fig. 21.2 Kernel density estimate for distributions of Simpson b (1-SI) index for the sample of Polish FADN farms based on crop diversification criteria. *Source* own calculations based on FADN data

One of the possibilities is to use the following formula:

$$SI\_mod = \left(1 - \sum \left(\frac{n_i}{N}\right)^2\right)(\ln n.\text{min} - n.\text{max})$$

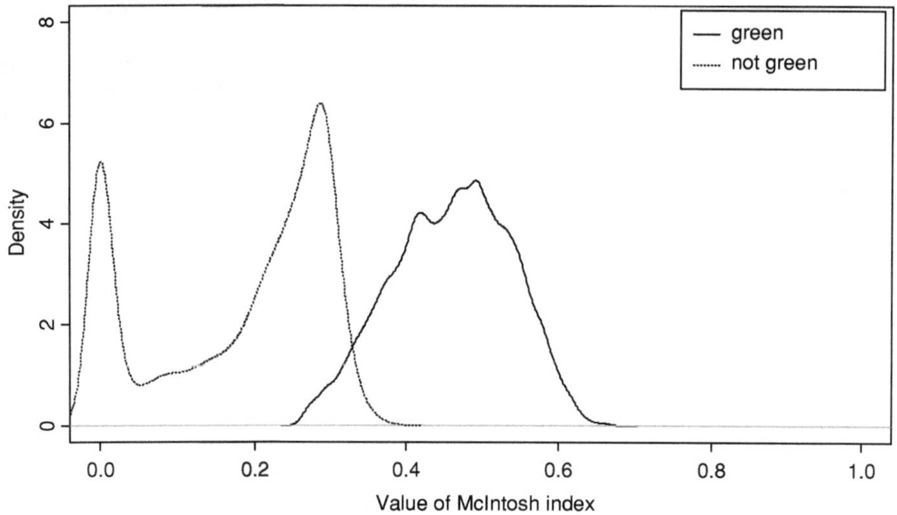

Fig. 21.3 Kernel density estimate for distributions of the McIntosh (MI) index for the sample of Polish FADN farms based on crop diversification criteria. *Source* own calculations based on FADN data

where

n.min number of crops with a share above required minimum

n.max number of crops with a share above maximum

Results of the proposed formula (Fig. 21.4) strongly depend on a number of crops, so the modified index should not be considered as a general biodiversity measure and its use should be limited to the EC proposal-related considerations.

Requirements of the EC proposal allow for the minimum 5 % share of a crop and its maximum share cannot exceed 70 %. The theoretical maximum of the index is ln n.min. In this case, among farms from the FADN sample, the maximum value of an indicator was 2.5 while the minimum was −0.35.

The critical value of the index for delimitation of farms in line with the greening proposal is 0.5. All farms with an index below this value could be considered as "not-green."

In addition to graphical illustrations, basic descriptive statistics for the indices considered have also been presented (Table 21.1). To compare the predictive quality of the analyzed biodiversity indices for assessing whether farm cropping structure fulfills "greening" requirements, the authors decided to use the logit model. After fitting the logit model for each biodiversity index, McFadenn R^2 was calculated.

$$R^2_{MF} = 1 - \frac{\log \mathrm{L}_M}{\log \mathrm{L}_0}$$

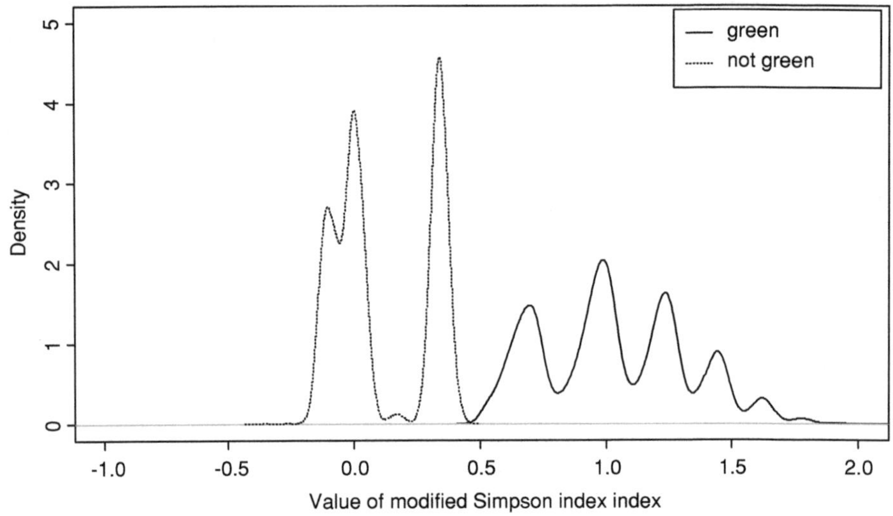

Fig. 21.4 Kernel density estimate for distributions of a modified Simpson index for the sample of Polish FADN farms based on crop diversification criteria. *Source* own calculations based on FADN data

where L_M is the value of the likelihood function for the model being estimated, and L_0 is the likelihood for a model with no predictors.

Results in Table 21.1 confirm observations from Figs. 21.1 to 21.3. Over 22 % of farms in the FADN sample could not be delimited as "green" or "not-green" using the Shannon, Simpson, or McIntosh indices. Despite the index used, the number of overlapping farms in the McFadden R^2 index is similar. This might suggest that none of the indices is precise enough to be used for verification of farms according to the diversification criteria of the CAP greening proposal. Proposed modification of the Simpson index allows for proper delimiting of single farms in this case.

Indices calculated for single-farm records were aggregated and compared with a share of "not-green" area for each of the 16 NUTS 2 regions in Poland. Results are shown in Table 21.2.

A relatively high correlation between values of all indices and "not-green" areas is observed. Using both Pearson and Spearman's correlation coefficients give similar results. It shows that biodiversity indices based on the single-farm records correspond to a large extent to the area of the farms. In this comparison, the Simpson index shows the best performance.

The same set of indices has been calculated based on statistical data on cropping structure in all NUTS 2 regions (Table 21.3). Along with the procedure used above, the correlation of indices based on average cropping structure with a share of area in not-green farms in the FADN population has been calculated.

Table 21.1 Predictive quality of biodiversity indices in relation to greening criteria

Index	Green farms		Not-green farms		Number of overlapping farms	McFadenn R^2
	Minimum	Maximum	Minimum	Maximum		
Shannon	0.75	2.37	0	1.12	2 481	0.8887
Simpson b	0.45	0.9	0	0.61	2 237	0.8344
McIntosh	0.26	0.68	0	0.37	2 237	0.8349
Simpson b modified	0.49	2.15	−0.35	0.42	0	1.0

Source own calculations based on FADN data

Table 21.2 Biodiversity indicators calculated for NUTS 2 based on single FADN data (area weighted average)

Region	Shannon–Weiner index	Simpson b index	McIntosh index	Simpson b modified	% of area in not-green farms in the FADN population
NUTS 2	H	1-SI	MI	SI_mod	NG_j
dolnośląskie	1.19	0.63	0.40	0.81	17.73
kujawsko-pomorskie	1.39	0.70	0.46	1.04	8.00
lubelskie	1.23	0.64	0.41	0.85	13.16
lubuskie	1.30	0.65	0.43	0.93	15.03
łódzkie	1.35	0.68	0.45	0.99	9.59
małopolskie	1.26	0.65	0.42	0.90	14.04
mazowieckie	1.27	0.65	0.42	0.90	13.33
opolskie	1.30	0.66	0.43	0.95	9.60
podkarpackie	1.34	0.67	0.43	0.96	10.43
podlaskie	1.19	0.63	0.40	0.82	16.20
pomorskie	1.20	0.62	0.40	0.84	18.93
śląskie	1.29	0.65	0.43	0.91	19.47
świętokrzyskie	1.40	0.69	0.46	1.02	8.63
warmińsko-mazurskie	1.14	0.61	0.39	0.78	23.22
wielkopolskie	1.32	0.67	0.44	0.97	11.67
zachodniopomorskie	1.28	0.66	0.43	0.93	11.94
Poland	1.28	0.66	0.43	0.92	13.18
Pearson's correlation with NG_j	−0.85	−0.89	−0.86	−0.86	X
Spearman's rank correlation with NG_j	−0.85	−0.91	−0.88	−0.87	X

Source own calculations based on FADN data

The results show that the correlation between biodiversity indices calculated based on aggregated data on cropping structure and the share of "not-green" farms is very low. Spearman's rank correlation shows slightly higher relations between indices and share of "not-green" areas. Although the hypothesis of a lack of

Table 21.3 Biodiversity indicators calculated, based on statistical (aggregated) data for NUTS 2

Region	Shannon–Weiner index	Simpson b index	McIntosh index	Simpson b modified index	% of area in "not-green" farms in FADN population
NUTS 2	H	1-SI	MI	SI_mod	NG_j
dolnośląskie	2.07	0.83	0.59	1.49	17.73
kujawsko-pomorskie	2.34	0.89	0.67	1.74	8.00
lubelskie	2.28	0.89	0.67	1.60	13.16
lubuskie	2.27	0.89	0.67	1.74	15.03
łódzkie	2.28	0.89	0.67	1.73	9.59
małopolskie	2.27	0.88	0.66	1.84	14.04
mazowieckie	2.31	0.89	0.67	1.59	13.33
opolskie	1.99	0.84	0.60	1.50	9.60
podkarpackie	2.24	0.88	0.66	1.72	10.43
podlaskie	2.05	0.84	0.60	1.50	16.20
pomorskie	2.28	0.89	0.67	1.73	18.93
śląskie	2.25	0.90	0.69	1.76	19.47
świętokrzyskie	2.33	0.90	0.69	1.76	8.63
warmińsko-mazurskie	2.23	0.89	0.66	1.84	23.22
wielkopolskie	2.32	0.90	0.69	1.76	11.67
zachodniopomorskie	2.22	0.89	0.66	1.72	11.94
Poland	2.36	0.90	0.69	1.76	13.18
Pearson's correlation with NG_j	−0.18	−0.12	−0.12	0.09	X
Spearman's rank correlation with NG_j	−0.41	−0.19	−0.19	0.10	X

Source own calculations based on FADN data

correlation has not been tested, it should be pointed out that the critical value sample for the 16 elements is 0.497 for Pearson's correlation and 0.503 for Spearman's. Thus, the value of the correlation coefficient, −0.41, in the case of Shannon's index does not prove a relation between indices calculated based on aggregated cropping structure and share of "not-green" farm area in the FADN population.

Using this method for verification of CAP greening requirements could lead to incorrect conclusions. Biodiversity indices calculated based on aggregated cropping structure for the region with the highest "not-green" area, which is Warmińsko-Mazurskie NUTS 2, are very similar to those calculated for the Łódzkie NUTS 2 region, which is one of the regions with the lowest share of "not-green" land.

21.4 Conclusions

Overlapping of density functions illustrates the shortcomings of the common biodiversity indices for this specific application. However, it might also be concluded that fulfilling diversification criteria proposed by the EC does not result in increasing crop diversity measured by the most popular biodiversity indices.

Modification of the Simpson index seems to provide a good basis for verifying fulfillment of the greening criteria based on single-farm records.

Aggregating biodiversity indices for regions by using single-farm records results in a decreased accuracy, but differences in regional indices still explain to a large extent the share of the area in "not-green" farms.

However, there is a very small correlation between biodiversity indices calculated based on aggregated NUTS 2 crop structure and share of not-green land. This suggests that biodiversity indices used to assess crop diversity calculated for large aggregates based on aggregated crop structure are strongly biased and should not be used for verification of diversification criteria defined in the CAP reform proposal.

References

Britz, W., & Witzke, P. (Ed.). (2012). CAPRI model documentation 2012. Bonn University.

European Commission. (2011). Legal proposals for the CAP after 2013—Agriculture and rural development. European Commission Proposal of 12 October 2011, Brussels.

EUROSTAT (2013). Database Agriculture/Regional Agricultural Statistics [table:agr_r_crops]. http://epp.eurostat.ec.europa.eu/portal/page/portal/statistics/search_database.

FADN. (2013). Farm Accountancy Data Network. http://ec.europa.eu/agriculture/rica/.

GUS—Central Statistical Office Poland (2011). Uprawy rolne i wybrane element metod produkcji roślinnej. PowszechnySpisRolny2010. Warsaw, pp 1–152.

Hiep, C., & Engels, P. (1974). Comparing species diversity and evenness indices. *Journal of the Marine Biological Association of the United Kingdom, 54*, 559–563.

Jones and Stokes Associates. (1987). *Sliding toward extinction: the state of California's natural heritage*. CA: Sacramento.

Khan, S. A. (2006). *Methodology for Assessing Biodiversity.*, Centre of Advanced Study in Marine Biology Parangipettai: Annamalai University.

Majewski, E., Czekaj, S., & Was, A. (2013). *Impacts of Greening of the Common Agricultural Policy on Financial Results of Cereal Farms in Poland After the Year 2013.*, Sustainable Development and Agribussiness Warsaw: Warsaw University of Life Science.

McFadden D. (1973). Conditional logit analysis of qualitative choice behaviour. In P. Zarembka (Ed.), *Frontiers in Econometrics*, (pp. 105–142). New York: Academic Press.

McIntosh, Robert P. (1967). An index of diversity and the relation of certain concepts to diversity. *Ecology, 48*, 392–404.

Shannon, C. E. (1948). A mathematical theory of communication. *The Bell System Technical Journal, 27*, 379–423, 623–656.

Simpson, E. H. (1949). Measurement of diversity. *Nature, 163*, 688.

Turkmen G., & Kazanci N. (2010). Applications of various diversity indices to benthic macroinvertebrate assemblages in streams of a natural park in Turkey. Ankara: Hacettepe University. http://balwois.com/balwois/administration/full_paper/ffp-1765.pdf.

Chapter 22
Economic Efficiency of Production Systems in the Gharb Irrigated Area (Morocco) Affected by Access to Water Resources

R. Harbouze, Ph. Le Grusse, A. Bouaziz, J. C. Mailhol, P. Ruelle and M. Raki

Abstract The aim of this work is to calculate and compare the economic efficiency indices of irrigated farms and the level of optimisation of irrigation water used for the main crops of the Gharb area. To this end, a Data Envelopment Analysis (DEA) model was used to calculate efficiency indices. The survey covered 49 farms with different crop systems (vegetables, citrus crops, cereals, forage, sugar beet and sugar cane) and different irrigation systems (drip, sprinkler and gravity-fed). The results show that the most efficient farms are both those affected by water stress and those with "unlimited" access to water resources (private pumping). On the other hand, 73 % of the farms are inefficient, indicating that the majority of farmers do not have a good grasp of the available technology.

22.1 Introduction

The major work undertaken by Farrell (1957) for estimating production efficiency was initiated by the evaluation of technical efficiency suggested by Debreu (1951) and Koopmans' definition of efficiency: '*A feasible input-output vector is said to be technically efficient if it is technologically impossible to increase any output and/or reduce any input without simultaneously reducing another output and/or*

R. Harbouze (✉)
Montpellier 1 University, Montpellier, France
e-mail: harbouze@iamm.fr

R. Harbouze · Ph. Le Grusse
CIHEAM-IAM, Mediterranean Agronomic Institute of Montpellier, Montpellier, France

R. Harbouze · A. Bouaziz · M. Raki
Hassan II Agronomy and Veterinary Medicine Institute, Rabat, Morocco

J. C. Mailhol · P. Ruelle
Joint Research Unit G-Eau, Cemagref, Montpellier, France

C. Zopounidis et al. (eds.), *Agricultural Cooperative Management and Policy*, Cooperative Management, DOI: 10.1007/978-3-319-06635-6_22,

one other input (Koopmans 1951)'. Koopmans enabled the first significant step towards border econometrics. Farrell's innovation lies in the application of the efficiency calculated by Debreu '*with the resource load factor which calculates the maximum equiproportional reduction of all the factors of production, making it possible to maintain the existing level of production* within each manufacturing unit of a sector'.

The concept of efficiency is often used to characterise resource use; one can say that efficiency is a ratio representing the performance of a process which transforms a set of inputs into a set of outputs. It corresponds to the difference between the maximum possible production, taking into account the inputs consumed and the actual production (Boussemart 1994).

In terms of comparative analysis, the limit of production refers to best practices. The variation of each observation compared to this limit represents its degree of inefficiency. The variation may be due to a lack of competition, which means that farms may operate below their capacity if they are protected on the market (Bachta and Chebil 2002). Other explanations point to the effects of non-physical inputs (information, creative knowledge, etc.) on the efficiency of the farms (Muller 1974).

Today, in the new context of water scarcity and climate change, governments limit the over-exploitation of water resources and encourage the use of alternative resources (reuse of sewage, desalination, etc.). Groundwater merits peculiar attention due to its relative importance, especially in the coastal zone of the Gharb. Indeed, due to the many different uses to which they are put, groundwater resources and aquifers in different parts of the world are increasingly over-exploited.

It is consequently important to consider another system of resource management and to move away from supply management to demand management. In this way it will be possible to reduce total water demand by diversifying crops and sources of income and by introducing crops and activities with a higher added value, lower water requirements, higher income potential and more significant financial capacities (Bouaziz and Belabbes 2002). Farmers thus need to be encouraged to make more efficient use of irrigation water in growing crops. Demand management, particularly of large irrigation areas, represents considerable potential for water saving and conservation in the context of limited water resources and increasing mobilisation costs.

In Morocco, irrigated areas (1.454.000 ha) account for about 11 % of the total agricultural area. In the past, these areas profited as they had been declared a priority for investment. They thus contribute to approximately 45 % of the agricultural-added value, accounting for 75 % of agricultural exports and generating more than a third of the employment in rural areas (Belghiti 2008). The irrigated sector consumes nearly 92 % of the water used. Water therefore needs to be used wisely to ensure technical, economic and social optimisation and especially to preserve it for future generations.

This study was conducted in the Gharb irrigated area and was based on a survey of a sample of 49 farms, with the aim of:

- assessing the economic efficiency indices of the farms;
- identifying farms that could be used as management models for economically inefficient farms;
- understanding how farming systems affect the level of efficiency of these farms.

22.2 Materials and Methods

22.2.1 The Gharb Area

The Gharb area is located in north-western Morocco. It is bounded to the west by the Atlantic Ocean and the dunes of the Sahel, to the north by the hills of pre-Rif, to the east by the Sais plateau and to the South by the Maamora Forest. Its elevation ranges from 4 to 25 m. It is a coastal area with sandy soils, sharing borders with the alluvial plains, and the central Sebou region (the main river) with clay and loamy soils.[1] It covers a total area of 616,000 ha. The uncultivated area (infrastructure, rangeland and arid uncultivated land) comprises about 228,000 ha, of which 122,000 ha consists of forest. The 388,000 ha of agricultural land are divided into:

- 250,000 ha of irrigable land, of which 107,000 ha are equipped with large hydraulic systems and 12,000 ha with small and medium hydraulic systems, and
- 138,000 ha of rain-fed cultivated land.

The area includes three irrigation sectors:

- *The first section (PTI)* a net area of 35,858 ha was fully equipped between 1972 and 1978. Most of the area is located on the left bank of the Sebou river and part of it is on the right bank of Oued Beht (sectors P7 and P8). The areas along the Sebou river are supplied by lift stations that serve networks of canals. Sectors P7 and P8 are supplied by the pumping station and the Boumaiz canal (between Sebou and Beht). Only sector P7 is irrigated by sprinklers (2,558 ha), while the remaining 33,300 ha are irrigated by gravity.
- *The second section (STI)* covers an area of 37,000 ha. It includes sectors C1 to C4, N1-N5 and N9. These sectors are irrigated by gravity, while other sectors are equipped with sprinklers.
- *The third section (TTI)* is irrigable land but most of it has not yet been equipped. The net area to be equipped by the end of the project will exceed 110,000 ha. It includes areas Z1 to Z6, E1 to E5 and N10. Areas Z1, Z2 and N10 are part of the coastal area called **Mnasra**.

---

[1] Gharb Regional Agricultural Development Agency.

This region is characterised by a range of different crops: sugar, cereals, vegetables, forage crops, grain legumes, oil crops and fruits (citrus, etc.).

22.2.1.1 The Survey and Farm Sample

The survey covered 49 farms in three areas of Gharb (Fig. 22.1):

- 18 farms in the central area, mainly in sectors S9, N4, N5, P7 and P8. The main crops in this area are cereals, sugar (sugar beet and sugar cane) and forage crops. The two main irrigation systems are surface and sprinkler systems.
- 18 farms in the Beht zone. The main crops in this area are citrus fruits, cereals and forage crops. Surface irrigation systems are mainly used in the area.
- 13 farms[2] in the coastal zone, where the main crops are bananas, strawberries and vegetables under drip irrigation.

The farms were chosen to reflect the diversity of the three areas, particularly in terms of farm size (<5 ha, 5–10 ha and >10 ha), land tenure systems (private property, collective, estates, agrarian reform and *Guich*) and irrigation systems (drip, surface and sprinkler or pump[3]).

In terms of size, 38 % of the farms sampled in the coastal zone had less than 5 ha, and 38 % had more than 10 ha. Farms of between 5 and 10 ha account for only 23 % of the sample. In the central and Beht areas, more than half of the farms surveyed had less than 5 ha (Fig. 22.2).

Direct land tenure was observed in 94 % of our survey in the central area, 67 % in the Beht area and 61 % in the coastal zone. The remaining land was under indirect tenure: land rent or run by associations for part of the production.

22.2.1.2 Production Systems and Access to Water

In the coastal zone, the main crops are vegetables and fruit (bananas and strawberries), followed by cereals and oil crops. The central area is more oriented to the production of sugar beet and sugar cane, cereals and fodder (corn fodder and Berseem clover). The **Beht** area is mostly used for citrus cultivation in addition to the same main crops as in the central area. Livestock is the main source of cash income for most farmers. Meat production plays a very important role in the production systems of the farms we surveyed in the central zone and the Beht zone. Cattle farming occupies 72–61 % in these two areas, respectively.

---

[2] Six of the 13 farms were surveyed by Mamounata SEMDE, Rural Engineer, Hassan II Agronomy and the Veterinary Medicine Institute.

[3] In this irrigation method, a worker brings water to the plot with a plastic hose and tries to cover the whole acreage. This practice is very common in groundnut cultivation in the sandy coastal area.

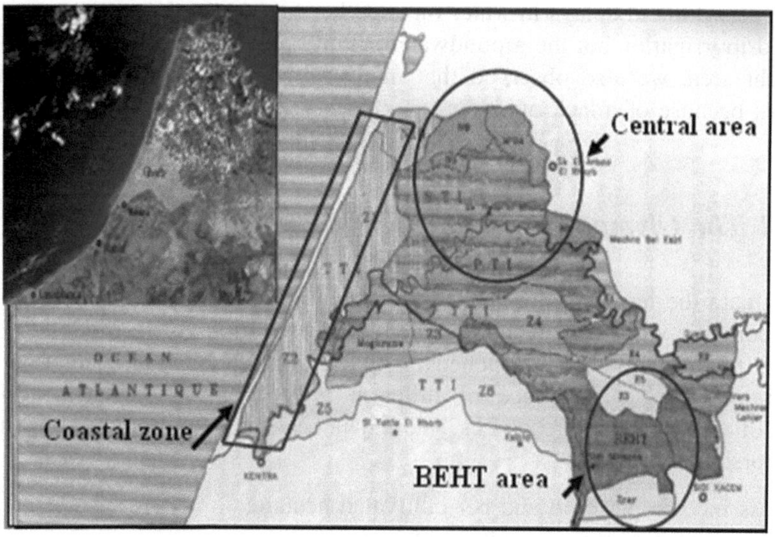

Fig. 22.1 Gharb area (*source* GRAAD 2007)

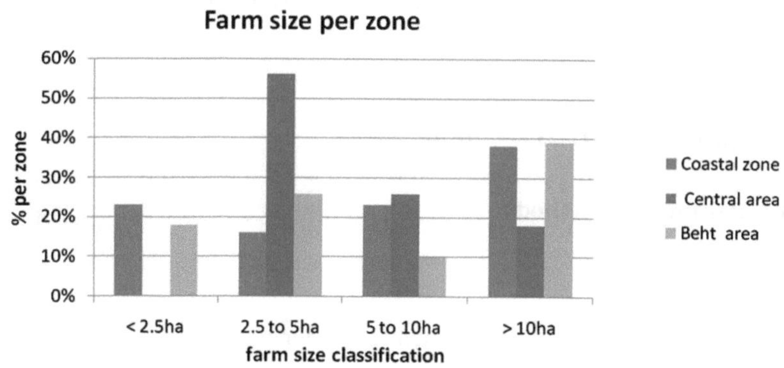

Fig. 22.2 Farm size in the 3 zones

In terms of water access, the farmers only use groundwater in the coastal zone, where private pumping is very widespread. Most farmers use drip irrigation. Less fortunate farmers use the 'pumpman' technique. In the central area, they use the state water supply for irrigation: 56 % by gravity feed; 39 % by sprinklers and 5 % use both gravity and drip irrigation through a pool supplied by the public network.

In the Beht area, which is facing a serious water shortage, all the farms use the state water supply, 83 % irrigate by gravity feed, 11 % use both gravity feed and sprinklers, and one farm uses gravity and drip irrigation. The latter is generally supplied from a well (which in theory is not allowed in an equipped area) in order to cope with water shortage problems, the irregular supply from the state water supply network and delays in water turns. Other farmers in this region would like to use drip irrigation but the groundwater salinity prevents them from doing so. In the Beht area, we also observed that many farmers irrigate from the drainage network because of a lack of water.

22.2.2 The Olympe Platform

To evaluate the levels of access to water resources for the 49 farms in different agro-economic situations, techno-economic investigations were modelled on the Olympe platform[4] (Penot and Deheuvels 2007). This enabled an evaluation of the techno-economic performance of different production systems with different means of access to water resources. These data enabled us to produce conventional indicators:

- Gross income per crop and per cultivated hectare;
- Net margin per crop per cultivated hectare;
- Water efficiency of the main crops grown and volume of water provided (in m^3);
- Production expenditure, structural expenditure per crop per hectare and the proportion of the water cost in relation to the overall expenditure.

22.2.3 Efficiency Measurement

22.2.3.1 Parametric Methods

Pragmatic Approach

The simplest way of measuring economic efficiency is by means of the partial productivity index, usually that of the labour force. This approach ignores the presence of other factors which affect the average and marginal productivity (Lau and Yotopoulos 1971).

Econometric Approach

Farrell also put forward a measure related to the outputs, implemented by Timmer (1971). This measurement corresponds to the ratio of the observed output to the optimal output for a given level of inputs. It coincides with input measurement when technology enables a constant yield of scale (Afriat 1972).

[4] http://www.olympe-project.net/

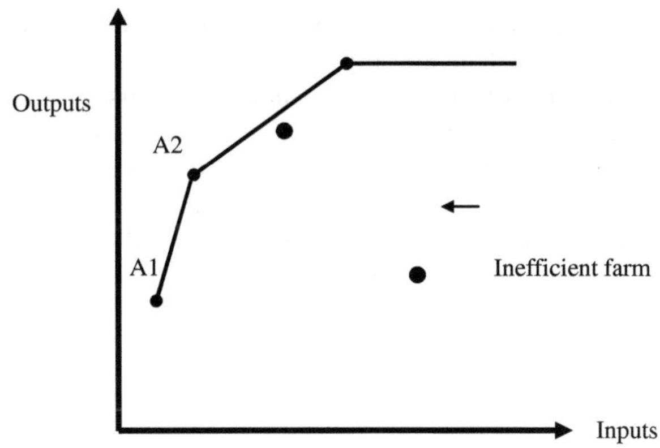

Fig. 22.3 Linear envelope per item

22.2.3.2 Non-parametric Methods: Data Envelopment Analysis as a Means of Measuring Efficiency

The approach developed from an original proposal by Farrell (1957) is described as non-parametric because it was built by mathematically programming an envelope of observations, but with no vector of the parameters being estimated. Charnes et al. (1978, 1981) generalised Farrell's proposal and implemented it as an operational tool while allowing for estimation of the production function by an envelope curve formed of the segments of the right-hand side joining the effective entities (A1, A2..) (Fig. 22.3).

The DEA method is based on linear programming to identify empirical production functions. It determines the limit of efficiency from the point of view of best practices and compares all similar units in a given population. Each unit is regarded as a decision-making unit (DMU) that transforms inputs into outputs.

The efficiency formula for the DEA method for a given decision-making unit k is as follows:

$E_k = \underline{\text{summon}}$ balanced outputs/$\underline{\text{summon}}$ balanced inputs

$$E_k = W_1 * \text{Out1} + W_2 * \text{Out2} + \ldots\ldots/V_1 * \text{Inp1} + V_2 * \text{Inp2} + \ldots\ldots$$

With: W (weights of outputs) and V (weights of inputs).

The DEA method calculates the separated weighting for each unit, assuming that the weightings give the best result for the unit concerned. The basic ideas of this method for each DMU k are as follows:

- To maximise E_K, under the constraint $E_k \leq 1$ for all the DMU of the population concerned.
- All weightings are positive.

In this work, we used the form ratio of the Charnes, Cooper and Rhodes (CCR) model (Liu et al. 2013). This model makes it possible to measure the unused factors available to the farm and the potential production not achieved by the farm during the production process.

Specifically, DMU 1 consumes amounts $Xj = (\text{xij})$ of inputs ($i = 1$, m) and produces amounts $Y\ J = (\text{y rj})$ of outputs ($r = 1,\ldots.,$ s). For these constants, which generally take the form of observations, we assume that $X\ ij > 0$ and $y\ rj > 0$. The matrix (sxn) of output measurements is indicated by Y, and the matrix (mxn) of input measurements is indicated by X (Badillo and Paradi 1999).

The form ratio of the model is as follows:

For DMU_0

$$\text{Max}_{u,v} h_0(u,v) = \sum_{r=1}^{s} u_r y_{r0}$$

Under the constraint

$$\sum_{i=1}^{m} v_i x_{i0} = 1$$

$$\sum_{r=1}^{s} u_r y_{rj} - \sum_{i=1}^{m} v_i x_{ij} \leq 0$$

$$u_r, v_i \geq \varepsilon$$

where:

h_O represents the technical efficiency of the farm according to the definition of Farrell (1957)

X, Y are matrices of the observed outputs and inputs, respectively;

y_0 is the vector of the observed outputs of the farm used for efficiency evaluation;

x_0 is the vector of the observed inputs of the farm used for efficiency evaluation

v_i and u_i are the weights determined by the solution of the problem (Vermersch et al. 1995), i.e. by the data on all the DMU used, e.g. unit of reference, x_{i0} and y_{r0} observed values of DMU_0

By convention, it is accepted that the number of DMU must be equal to or higher than three times the number of inputs and outputs (Raab and Lichty 2002).

For evaluating the efficiency scores, we used one output (farm income) and eight aggregates of inputs: To calculate the efficiency indices, the prices of the inputs must be known. Consequently, aggregation is generally based on monetary value.

Inputs:

- *Land.* Because of the presence of several types of soil and of land tenure, it was not possible to compare the farms without taking these factors into account. To solve this problem, we used the cost of renting land as the basis of the weighting factor.

- *Labour.* For each farm and crop management operation we determined the number of working days per category (family or external labour). This input is expressed as a total number of days. We assumed that the cost of one working day was 35 DH/day for family members and 50 DH/day for external labour).
- *Mechanisation.* This aggregate corresponded to a number of cultivation operations (tillage, seedbed preparation and harvesting including transport of the yield). For each operation, the expenditure was collected at the farm level.
- *Water irrigation.* The expenditure included the cost of water delivered by the irrigation office and the cost of using water from private wells. In terms of rainfall, we assumed that the farms received equal amounts of rain and consequently we did not take it into consideration. The water input was expressed in m^3/ha.
- *Seeds.* We introduced this input by allocating the market price to each seed type.
- *Pesticides.* Expenditure took into account all pesticides used during crop management (in DH).
- *Fertilisers.* In this aggregate, we included all the expenditure on the various fertilisers used during the crop cycle (in DH).

22.3 Results and Discussion

In this section, we first analyse the results of the DEA method for the various parts of the Gharb area (data are for the year 2007). First, the farms were grouped into efficiency classes to show the behaviour of the farmers and how they used their inputs. Second the farms were grouped by types of crop and we then identified the causes of the variations observed.

22.3.1 Farmers' Efficiency Scores by Zone

In the coastal zone, the farmers scored an average economic efficiency of around 73 %, with an average deviation of 23 %. A total of 54 % of the farms were not efficient. Each efficiency score measures the proportional reduction in inputs with no corresponding reduction in outputs. For example, farmers in the coastal zone were able to reduce their inputs by an average of 27 % while maintaining constant outputs.

In the Beht area, the average economic efficiency score of the 19 farmers was around 62 %, with an average deviation of 18 %. A total of 78 % of the farms were inefficient. This result (efficiency score) was a little lower than in the coastal zone (54 %), with a larger number of dots representing farmers located away from the efficiency limit on the scatter graph.

The central area had the lowest mean economic efficiency score at 58 % with an average deviation of 16 %. A total of 83 % of the farms were inefficient.

22.3.1.1 Efficiency Classes and Average Inputs/ha per Zone

The most efficient class (80–100 %) for the coastal zone has the highest income/ha (Table 22.1). The farmers manage to produce an average of 14 DH/m^3 of water but they also use the most inputs/ha, as well as an average of almost 7,600 m^3 of water per hectare. Their income/inputs ratio was also the highest. It would have been interesting to compare these results with the water requirement at the ETM[5] during the year under examination, which incidentally has not been specified (for example in 2006 it was useless to irrigate sugar beet in the Gharb).

The least efficient farmers (20–40 %) were those with the lowest income/ha, and those who used the least inputs/ha. These farmers did not use much irrigation water (less than half that used by the most efficient class), and the water used had a total value of only 5 DH/m^3.

The efficiency levels of these farmers and the ratio of income to inputs were coherent, i.e. the most efficient farmers were those who had the highest income/input ratio and also the farmers who produced more outputs with proportionally fewer inputs.

In the Beht area, the most efficient farmers only obtained an income value of 27283 DH/ha (five times less than in the coastal zone) (Table 22.2). These farmers only used 6 DH/m^3 of water.

In terms of water productivity, the farmers in the 40–60 % efficiency class displayed better water productivity with 7 DH/m^3; this is higher than 3 out of the 4 classes in the coastal zone.

In the central area, the least efficient farmers are those who use the most inputs/ha (Table 22.3), which is not justified by the level of production they achieve. Conversely, the most efficient farmers are those who use the least inputs/ha (6–10 % less). This situation contrasted with that observed in the coastal zone and in the Beht area, where the most efficient farms were those who used the most inputs/ha.

In terms of water productivity, the farmers in the central area did not achieve the threshold value of 3 DH/m^3. This is very poor compared with the other areas.

22.3.1.2 Analysis of Crop Efficiency Scores

Our first analysis examined the production systems of the classes corresponding to the extreme scores. At the lowest level of efficiency (20–40 %), the farmers in this class used a cropping system based on either a cereal or sugar and earned more than 50 % of their income with these crops. In the 80–100 % efficiency class, 80 % of the most efficient farms were those which produced vegetables. We can thus conclude that it would be useful to continue the analysis based on the type of crops.

---

[5] *ETM* Evapotranspiration maximum.

Table 22.1 Efficiency classes and average inputs/ha (coastal zone)

Efficiency class (%)	Fertilisers (DH/ha)	Seed (DH/ha)	Plant protection products (DH/ha)	Labour (WD/ha)	Mechanisation (DH/ha)	Water (m^3/ha)	Land (DH/ha)	Income (DH/ha)	Income/m^3 (DH/m^3)	Inputs (DH/ha)	Income/ inputs
80–100	11417	7252	6026	126	6271	7583	4000	102930	14	43170	2,3
60–80	5670	3691	1512	112	1463	8645	4000	31500	4	24575	1,4
40–60	2133	2388	1148	54	2059	3819	3403	22988	6	14923	1,5
20–40	1288	899	835	31	2220	2286	3250	11953	5	10723	1,1

Table 22.2 Efficiency classes and average inputs/ha (Beht)

Efficiency class (%)	Fertilisers (DH/ha)	Seed (DH/ha)	Plant protection products (DH/ha)	Labour (JT/ha)	Mechanisation (DH/ha)	Water (m^3/ha)	Land (DH/ha)	Income (DH/ha)	Income/m^3 (DH/m^3)	Inputs (DH/ha)	Income/inputs
80–100	3051	268	540	67	2607	4940	2100	27283	6	13365	2,1
60–80	1694	634	719	41	2102	4285	2500	20683	5	11235	1,8
40–60	1641	865	1051	31	1909	2191	2700	15263	7	10354	1,4
20–40	1461	832	495	22	2388	2064	2750	9037	6	9693	0,9

Table 22.3 Efficiency classes and average inputs/ha (Central area)

Efficiency class (%)	Fertilisers (DH/ha)	Seed (DH/ha)	Plant protection (DH/ha)	Labour (JT/ha)	Mechanisation (DH/ha)	Water (m^3/ha)	Land (DH/ha)	Income (DH/ha)	Income/m^3 (DH/m^3)	Inputs (DH/ha)	Income/ inputs
80–100	522	159	88	11	1235	9261	3000	26170	3	10009	2,6
60–80											
40–60	1312	1450	342	15	1635	4227	3214	13340	3	10603	1,2
20–40	1122	1456	570	33	1174	3153	4000	6696	2	11057	0,6

Sugar beet turned out to be the least efficient crop (Table 22.4), producing the lowest income/ha and using the most inputs/ha. In the 20–40 % efficiency class, sugar beet used irrigation water the most efficiently, which can be explained by the fact that these farms used less irrigation water than farms in the other efficiency classes. On the other hand, apart from mechanisation (and irrigation), other inputs are higher than those of the other classes.

In the most efficient class, sugar beet generated the highest income/ha and used less inputs/ha than in the other efficiency classes.

In the most efficient class, sugar cane procured the highest income/ha and used the most inputs/ha, i.e. just the opposite of sugar beet, where the most efficient class used the least inputs/ha) (Table 22.5).

The best water efficiency was obtained by sugar cane in the most efficient class: 2.8 DH/m^3 of irrigation water.

The 80–100 % efficiency class included only 25 % of the farmers (who thus had efficient crops). This result shows that the crops were very badly managed in terms of inputs, as the majority of the sugar crops in our sample were inefficient. This raises the question of the way in which these crops were managed.

The observations made for sugar beet are also valid for cereals and citrus fruits (Tables 22.6 and 22.7). For example, the most efficient class was that which procured the highest income/ha and used the least inputs. The most efficient citrus fruits made the best use of irrigation water: 6 DH/m^3.

Vegetable crops were the most efficient (Table 22.8) and procured the highest income/ha of all the other crops (2–7 times higher). Vegetable crops also made the best use of irrigation water, which reached 12.9 DH/ha in the 80–100 % efficiency class.

We observed that in order to be efficient, vegetable crops had to be grown intensively and well-managed (the more the inputs, the higher the income). The same was true for sugar cane.

22.3.2 Example of Decision-Making Aid for the Farmers

Table 22.9 summarises the results from three farms which illustrate how they can be used for decision making. Farm 1 does not manage its sugar crops very well: efficiency is low for sugar cane and very low for sugar beet, whereas cereals are very efficient. The general inefficiency of this farm (efficiency = 0.3) is due to the fact that the sugar crops occupied more than 60 % of the total surface area of the farm. To become more efficient, we need to see how this farmer could manage his sugar crops like farm 3, where the sugar cane management is highly efficient.

Farm 2 is efficient (efficiency = 1) and its indices are good for two crops, but there is nevertheless room for improving watermelon cultivation.

At this stage, we can say that a farm is inefficient because the farmer has managed some crops badly, but we cannot answer the question of why some crops are inefficient.

Table 22.4 Income and inputs/ha per efficiency class for sugar beet

Crop	Efficiency class (%)	Income (DH/ha)	Fertilisers (DH/ha)	Seed (DH/ha)	Plant protection products (DH/ha)	Labour (WD/ha)	Mechanisation (DH/ha)	Water (m^3/ha)	Income/m^3 (DH/m^3)	Inputs (DH/ha)	Income/inputs
Sugar beet	80–100	23140	930	900	1624	35	1575	3800	6,1	8154	2,8
	60–80	21505	2050	1150	950	69	1564	3015	7,1	9636	2,2
	40–60	20045	4056	1100	1281	83	930	2706	7,4	11625	1,7
	20–40	17400	5193	1575	1927	94	800	2242	7,8	13906	1,3

Table 22.5 Income and inputs/ha per efficiency class for sugar cane

Crop	Efficiency class (%)	Income (DH/ha)	Fertilisers (DH/ha)	Seed (DH/ha)	Plant protection products (DH/ha)	Labour (JT/ha)	Mechanisation (DH/ha)	Water (m^3/ha)	Income/m^3 (DH/m^3)	Inputs (DH/ha)	Income/inputs
Sugar cane	80–100	28917	2520	4733	800	50	1198	10389	2,8	16196	1,8
	60–80	15000	3125	1555	620	47	1000	11340	1,3	13615	1,1
	40–60	20167	1177	3200	452	50	2438	9647	2,1	13841	1,5
	20–40	18375	2395	1623	785	54	1225	9070	2,0	12453	1,5

Table 22.6 Income and inputs/ha per efficiency class for cereals

Crop	Efficiency class (%)	Income (DH/ha)	Fertilisers (DH/ha)	Seed (DH/ha)	Plant protection products (DH/ha)	Labour (JT/ha)	Mechanisation (DH/ha)	Water (m^3/ha)	Income/m^3 (DH/m^3)	Inputs (DH/ha)	Income/ inputs
Cereals	80–100	9301	990	631	401	15	1644			4191	2,2
	60–80	7370	869	711	536	20	1398			4214	1,7
	40–60	7969	1259	722	554	16	2240			5336	1,5

Table 22.7 Income and inputs/ha per efficiency class for citrus crops

Crop	Efficiency class (%)	Income (DH/ha)	Fertilisers (DH/ha)	Seed (DH/ha)	Plant protection products (DH/ha)	Labour (JT/ha)	Mechanisation (DH/ha)	Water (m^3/ha)	Income/m^3 (DH/m^3)	Inputs (DH/ha)	Income/inputs
Citrus fruits	80–100	42500	1710	1550	2600	17	1975	7000	6,1	14030	3,0
	60–80	45000	6030	3500	750	132	450	8100	5,6	19400	2,3
	40–60	28000	3141	1850	1205	97	600	7237	3,9	13809	2,0
	20–40										

Table 22.8 Income and inputs/ha per efficiency class for vegetable crops

Crop	Efficiency class (%)	Income (DH/ha)	Fertilisers (DH/ha)	Seed (DH/ha)	Plant protection products (DH/ha)	Labour (JT/ha)	Mechanisation (DH/ha)	Water (m^3/ha)	Income/m^3 (DH/m^3)	Inputs (DH/ha)	Income/inputs
Vegetable crops	80–100	71462	6609	4536	3504	65	1382	5534	12,9	21072	3,4
	60–80	37007	4219	3624	2431	80	1257	5200	7,1	16931	2,2
	40–60	35277	4078	5254	2502	116	1864	5400	6,5	20358	1,7
	20–40	15750	3431	2893	1079	60	1050	4400	3,6	12752	1,2

Table 22.9 Crop and farm efficiency (DEA) for three farms

Farm	Crop	Area (ha)	DEA crop	DEA farms
1	Soft wheat	1	1.0	0,30
	Durum wheat	3.2	0.8	
	Sugar beet	4.7	0.35	
	Sugar cane	1.5	0.5	
2	Watermelon	3	0.5	1
	Strawberry	10.5	1.0	
	Melon	4	0.9	
3	Berseem clover	1	0.7	1
	Sugar cane	4	1.0	

To summarize, the results of the efficiency index analysis first showed that 73 % of the farms in the entire sample had very low efficiency scores (54 % for the coastal zone, 78 % for Beht area and 83 % for the central area) suggesting that:

- the majority of farmers do not master the available technology;
- the technical level of the farmers in one and the same zone is very similar; and
- using the available technology, the farmers in one and the same zone allocate their resources in similar ways. This implies that the farmers have the same technical information.

The efficiency of the farms in the coastal zone does not enable them to outclass farms in the other two areas. Admittedly 46 % of these farms are efficient, but the most efficient ranked only fifth out of the 49 farms in the sample, whereas four farms in the water-deficient Beht area took the first four places.

Indeed, although farmers in the Beht area lacked water for irrigation, this did not prevent them from being as efficient as farmers in the coastal zone, where there is no lack of water; this can be explained by the fact that the farmers in Beht have already acquired experience in optimally managing the available resources.

In the central area, the farmers were unable to farm as intensively as in the coastal zone and in Beht. Their income/ha and water optimisation are very low. This situation has encouraged the farmers to save on inputs, while in the coastal zone and in Beht farmers tended to intensify their modes of production.

22.4 Conclusion

We calculated and compared the economic efficiency of farming systems in different situations with respect to access to water resources in the Gharb plain. Our sample comprised differences in farming systems and in modes of access to water resources. The farmers in the coastal zone grew more vegetables and fruit using drip irrigation and had the highest total income and income/ha. In this zone, farmers diversified because water was abundant and accessible thanks to private pumping directly from the water table. The crops cultivated in this zone made the most efficient use of irrigation water, for example strawberries with a value of 16.3 Dirhams/m^3.

In the central and Beht areas, access to water is through the public irrigation network. Farmers in the central area (North sectors) used sprinklers, while farmers in Beht faced a water shortage and most farms were irrigated by gravity. Farms in the two areas presented similarities in their performance and in their use of irrigation water.

In the coastal zone, the most efficient farmers were those who made intensive use of production factors, because they had free access to water and were thus able to diversify. In the central area, where farmers had little opportunity to diversify, they managed inputs (efficiency is related to input control, i.e. load management).

However, examination of the individual results of farmers revealed that the four most efficient farmers of the sample were farmers in the Beht area, where water resources are limited.

The next step will be to confirm these results and to obtain partial efficiencies for each input. These partial efficiencies will enable us to explain why a particular crop is inefficient. It will then be possible to formulate more appropriate recommendations for farmers.

This approach should enable us to start by evaluating negative externalities such as environmental aspects (leaching of nitrates on a farm scale), and then to evaluate the efficiency of environmental outputs and to compare them with economic outputs.

References

Afriat, S. N. (1972). Efficiency estimation of production functions. *International Economic Review, 13*(3), 568–598.

Bachta, M. S., & Chebil, A. (2002). Efficacité technique des exploitations céréalières de la plaine du Sers-Tunisie. *New Medit, 1*(2), 41–45.

Badillo, P. Y., & Paradi J. C. (1999). *La méthode DEA: Analyse des performances*, Paris: Hermès science-publications.

Belghiti, M. (2008). L'agriculture irriguée au MAROC: Enjeux et marges de progrès. *Hommes, Terre et Eaux, 139*, 35–39.

Bouaziz, A., & Belabbes, K. (2002). Efficience productive de l'eau en irrigué au Maroc. *Hommes, Terre & Eaux, 32*(124), 57–72.

Boussemart, J. P. (1994). Diagnostique de l'efficacité productive par la méthode DEA : Application à des élevages porcins. *Cahiers d'Economie et Sociologie Rurales, 31*(4), 44–58.

Charnes, A., & Cooper, W. W., et al. (1978). Measuring the efficiency of decision making units. *European Journal of Operational Research, 2*(6), 429–444.

Charnes, A., Cooper, W. W., et al. (1981). Evaluating program and managerial efficiency : An application of data envelopment analysis to program follow through. *Management Science, 27*(6), 668–697.

Debreu, G. (1951). The coefficient of resource utilization. *Econometrica, 19*(3), 273–292.

Farrell, M. J. (1957). The Measurement of Productive Efficiency. *Journal of the Royal Statistical Society. Series A (General), 120*(3), 253–290.

Koopmans, T. C. (1951). Efficient allocation of resources. *Econometrica, 19*(4), 455–465.
Lau, L. J., & Yotopoulos, P. A. (1971). A test for relative efficiency and application to Indian agriculture. *The American Economic Review, 61*(1), 94–109.
Liu, J. S., & Lu, L. Y. Y., et al. (2013). Data envelopment analysis 1978–2010: A citation-based literature survey. *Omega, 41*(1), 3–15.
Muller, J. (1974). On sources of measured technical efficiency: The impact of information. *American Journal of Agricultural Economics, 56*(4), 730–738.
Penot, E., & Deheuvels O. (2007). Modélisation économique des exploitations agricoles, Modélisation, simulation et aide à la décision avec le logiciel Olympe. L'Harmattan, 9–21.
Raab, R. L., & Lichty, R. W. (2002). Identifying subareas that comprise a greater metropolitan area: The criterion of county relative efficiency. *Journal of Regional Science, 42*(3), 579–594.
Timmer, C. P. (1971). Using a probabilistic frontier production function to measure technical efficiency. *Journal of Political Economy, 79*(4), 776–794.
Vermersch, D., & Dervaux, B., et al. (1995). Efficacité technique et gains potentiels de productivité des exploitations céréalières françaises. Économie & prévision: 117–127.

MIX
Papier aus verantwortungsvollen Quellen
Paper from responsible sources
FSC® C105338

FSC
www.fsc.org

If you have any concerns about our products,
you can contact us on
ProductSafety@springernature.com

In case Publisher is established outside the EU,
the EU authorized representative is:
Springer Nature Customer Service Center GmbH
Europaplatz 3, 69115 Heidelberg, Germany

Printed by Libri Plureos GmbH
in Hamburg, Germany